T0269580

CAMBRIDGE LIBRARY COLLECTION

Books of enduring scholarly value

Botany and Horticulture

Until the nineteenth century, the investigation of natural phenomena, plants and animals was considered either the preserve of elite scholars or a pastime for the leisured upper classes. As increasing academic rigour and systematisation was brought to the study of 'natural history', its subdisciplines were adopted into university curricula, and learned societies (such as the Royal Horticultural Society, founded in 1804) were established to support research in these areas. A related development was strong enthusiasm for exotic garden plants, which resulted in plant collecting expeditions to every corner of the globe, sometimes with tragic consequences. This series includes accounts of some of those expeditions, detailed reference works on the flora of different regions, and practical advice for amateur and professional gardeners.

A Synopsis of the British Flora

The botanist and horticulturalist John Lindley (1799–1865) worked for Sir Joseph Banks, and was later instrumental in saving the Royal Horticultural Society from financial disaster. He was a prolific author of works for gardening practitioners but also for a non-specialist readership, and many of his books have been reissued in this series. This 1829 work is a classification of British plants, using the 'natural' system of the French botanist Antoine Laurent de Jussieu, which Lindley firmly supported, believing that the Linnaean system was both inaccurate and had 'almost disappeared from every country but our own'. Lindley describes genera and species in English, but using a uniform, standard vocabulary, and gives the alternative Latin names proposed by taxonomists including Smith, Curtis, Linnaeus, and the *Hortus Kewensis*. He also offers tables showing the components of each genus, and substantial indexes giving both Latin and English common names of the plants discussed.

Cambridge University Press has long been a pioneer in the reissuing of out-of-print titles from its own backlist, producing digital reprints of books that are still sought after by scholars and students but could not be reprinted economically using traditional technology. The Cambridge Library Collection extends this activity to a wider range of books which are still of importance to researchers and professionals, either for the source material they contain, or as landmarks in the history of their academic discipline.

Drawing from the world-renowned collections in the Cambridge University Library and other partner libraries, and guided by the advice of experts in each subject area, Cambridge University Press is using state-of-the-art scanning machines in its own Printing House to capture the content of each book selected for inclusion. The files are processed to give a consistently clear, crisp image, and the books finished to the high quality standard for which the Press is recognised around the world. The latest print-on-demand technology ensures that the books will remain available indefinitely, and that orders for single or multiple copies can quickly be supplied.

The Cambridge Library Collection brings back to life books of enduring scholarly value (including out-of-copyright works originally issued by other publishers) across a wide range of disciplines in the humanities and social sciences and in science and technology.

A Synopsis of the British Flora

Arranged According to the Natural Orders

John Lindley

CAMBRIDGE
UNIVERSITY PRESS

University Printing House, Cambridge, CB2 8BS, United Kingdom

Cambridge University Press is part of the University of Cambridge.
It furthers the University's mission by disseminating knowledge in the pursuit of education, learning and research at the highest international levels of excellence.

www.cambridge.org
Information on this title: www.cambridge.org/9781108076715

This edition first published 1829
This digitally printed version 2015

ISBN 978-1-108-07671-5 Paperback

A

SYNOPSIS

OF

THE BRITISH FLORA;

ARRANGED ACCORDING TO

The Natural Orders.

LONDON:
Printed by A. & R. Spottiswoode,
New-Street-Square.

A

SYNOPSIS

OF

THE BRITISH FLORA;

ARRANGED ACCORDING TO

The Natural Orders:

CONTAINING

VASCULARES, OR FLOWERING PLANTS.

BY

JOHN LINDLEY, F.R.S. L.S. AND G.S.

MEMBER OF THE IMPERIAL ACADEMY NATURÆ CURIOSORUM OF BONN ;
OF THE BOTANICAL SOCIETY OF RATISBON ;
AND OF THE PHYSIOGRAPHICAL SOCIETY OF LUND ;
CORRESPONDING MEMBER OF THE LINNÆAN SOCIETY OF PARIS ;
ASSISTANT SECRETARY OF THE HORTICULTURAL SOCIETY ;
AND
PROFESSOR OF BOTANY IN THE UNIVERSITY OF LONDON.

LONDON:

PRINTED FOR

LONGMAN, REES, ORME, BROWN, AND GREEN,

PATERNOSTER-ROW.

1829.

TO

SIR JOHN SAUNDERS SEBRIGHT, BART. M.P.

OF BEECHWOOD, IN HERTFORDSHIRE;

THIS WORK

IS VERY RESPECTFULLY INSCRIBED,

BY

THE AUTHOR.

PREFACE.

In submitting this little book to the public, it is right that something should be said in explanation of the reasons that have led to its preparation; especially as it may to some appear unnecessary, after the many useful, and, in several respects, very excellent, works, that have been already published, in illustration of the Flora of Great Britain.

These have all, with the exception of the Flora Scotica of Dr. Hooker, been arranged upon the principles of a system, which, whatever popularity it may, from particular circumstances, have acquired, and however useful it may have been found in communicating a knowledge of the names of things, does certainly not now tend to the advancement of science, or to an accurate knowledge of things themselves. Of course I allude to the system of Linnæus; a system which has almost disappeared from every country but our own, and which ought now to find no other place in science, than among the records of things whose fame has passed away. Hence all our British Floras are, in this view of the case, essentially defective, with the exception already made. I might therefore, without impropriety, stop at this point of my explanation; but, in addition to the fundamental error now adverted to, there is another of even more importance. The technical language in which these works are written is far from accurate; terms are applied in them vaguely and erroneously, and they so abound with mistakes, most of which are at variance with all correct notions of the structure of plants, that they are totally unfit to be placed in the hands of students. To these observations the Flora Scotica of Dr. Hooker, and the Flora Edinensis of Dr. Greville, are honourable exceptions.

When, therefore, I came to consider what work it would be most proper for me to recommend to the Botanical Class in the University of London, I could not but be forcibly struck with these reflections; if I selected any of our general Floras, they would not only not answer my purpose in explaining that system which alone ought to be taught in detail, in the present state of botanical science, but they would also be totally at variance with the statements I should find it my duty to make to the class. If I made choice of the Floras of either my friends Hooker or Greville, those works would be continually, from their confined range, inapplicable to the circumstances of a Class of Botany in London.

For these reasons I determined upon preparing a work upon those principles, of which the greater part of Europe has now become the advocate, and which might, if destitute of all other merit, at least possess the recommendation of being commensurate with the present state of botanical knowledge. In effecting this, I have proposed to myself the following objects:—

Firstly, to reduce the language in which plants are described to a uniform standard, in correspondence with the purest principles of the science, but divested of unnecessary technicalities. By this it is hoped that some impediments will be removed from the path of the student, and that his mind will be kept continually alive to those beautiful theories of structure, which not only constitute one of the most interesting departments of Botany, but give its chief claim to the name of a science.

In the second place, I have attempted to render the

nomenclature of genera and species conformable to that of Continental writers of the highest authority. Prejudices in favour of ancient errors have, in some cases, been allowed among us to take the place of modern discoveries; and, in other instances, we have continued to employ names, in the use of which we have perhaps been, strictly speaking, in the right, but which it has become inconvenient to retain, in opposition to the rest of Europe. In this, however, I hope such discretion has been exercised as has protected the rights of English Botanists, wherever they have deserved protection.

Many genera appear now, for the first time, in an English Flora, and to some it may seem that this is the reverse of an improvement. Undoubtedly, if the British Flora is considered without reference to other countries, a less minute subdivision of some orders might suffice; but when it is viewed, not as an isolated Flora, but as a small fragment of the vegetation of the world, such an opinion will, probably, be changed. Besides, our daily experience shows us that excessive analysis is far preferable to excessive synthesis, especially for the purposes of students; the former leads to no other inconvenience, than that of increasing the degree of investigation which plants must receive to be understood: the latter has a constant tendency to render investigation superficial, and characters confused. Upon the niceties by which the genera of many orders, such as Gramineæ and Orchideæ, are distinguished, some of the most curious parts of Botany will be found to depend.

In species I have chiefly followed the English Flora of Sir James Smith. That work is, undoubtedly, the best which we yet have seen upon species, and must always remain a standard book for British Botanists. In proof of the esteem in which I hold it, it has been made the basis of the present work, and I have generally adopted its specific characters, the phraseology in which they were expressed having been carefully revised. I trust, however, that several things will be found altogether new to the Botanist; and that the following pages will show that, even in so well-beaten a field as that of Great Britain, many interesting discoveries, even in the species of Flowering Plants, will yet reward the active and accurate observer. The most remarkable additions are, *Erica ciliaris*, found in Cornwall, and *Molinia depauperata*, from the Clova Mountains. Of the former, I have very lately received specimens from near Penryn, in Cornwall, through the kindness of Sir Charles Lemon, of Carclew, so that that very interesting plant is now confirmed to our Flora. To the genus *Rubus* I have made large additions; but it is probable that still larger yet remain to be made. The species have been hitherto overlooked by our British Botanists; and I do not doubt, that they would have escaped my notice also, if I had not enjoyed the opportunity of studying them in a living state, in the garden of the Horticultural Society. In this noble collection they are all growing: I have observed them for several years, and I am satisfied that their distinctions are permanent and important.

But, above all things, I have laboured to remove the difficulties that at present attend the study of the natural affinities of plants, both in this country and elsewhere. For this purpose I have prefixed to each class and order analytical tables of their contents; in these tables the most important or obvious characters are employed to distinguish one thing from another, and the less important peculiarities are kept out of sight. By which means I hope that the subject will be found simplified, and that as great a facility will have been given to acquiring an incipient knowledge of Botany, as can be offered even by the Linnæan system. It must, however, be borne in mind, that these tables are entirely artificial, and have not been constructed with reference to any thing beyond what is to be

A

SYNOPSIS

OF

THE BRITISH FLORA.

A

SYNOPSIS

OF

THE BRITISH FLORA.

CLASSES.

1. VASCULARES.

Syn. Phanerogamia, or Phænogamia *of authors*—Cotyledoneæ *Juss.* — Embryonatæ *Rich.*—Vasculares *Dec.*

Formed of cellular tissue, woody fibre, and spiral vessels. Embryo furnished with cotyledons. Epidermis with cuticular pores. Sexual organs developed.

2. CELLULARES.

Syn. Cryptogamia *Linn.* — Acotyledoneæ *Juss. Dec.*—Exembryonatæ or Arhizæ *Rich.* — Cellulares *Dec.*—Nemea, Cellularia *Fries.* — Acotyledoneæ and Pseudo-cotyledoneæ *Agardh.*

Formed of cellular tissue only, rarely with woody fibre. Embryo destitute of cotyledons. Epidermis without cuticular pores. Sexual organs none.

Class I. VASCULARES.

Substance of the plant composed of cellular tissue, woody fibre, and spiral vessels. Leaves formed with parenchyma, and veins consisting of woody fibre and spiral vessels. Epidermis with cuticular pores. Flowers consisting of floral envelopes, stamens, and pistilla. Seeds distinctly attached to a placenta covered with a testa, and containing an embryo with one or more cotyledons; germinating at two fixed points, viz. the plumule and radicle.

SUBCLASSES.

1. Dicotyledones.

Syn. Dicotyledones *Juss.* — Dicotyledoneæ or Exogenæ *Dec.* — Exorhizeæ and Synorhizeæ *Rich.* — Phanerocotyledoneæ or Seminiferæ *Agardh, aph.* 74.

Trunk consisting of bark, wood, and pith, in concentric layers. Leaves with reticulated veins. Cotyledons two or more, opposite; radicle naked.

2. Monocotyledones.

Syn. Monocotyledones *Juss.* — Monocotyledoneæ or Endogenæ *Dec.* — Endorhizeæ *Rich.* — Cryptocotyledoneæ or Graniferæ *Agardh, aph.* 73.

Trunk consisting of a homogeneous substance, with no distinction of bark, wood, and pith. Leaves with parallel veins. Cotyledon one, or, if two, alternate; radicle enclosed in a sheath.

Subclass I. Dicotyledones.

Trunk more or less conical, formed of three parts one within the other, viz. the bark, the wood, and the pith, of which the wood is enclosed between the two others; increasing by an annual deposition of new wood and cortical matter between the wood and bark. Leaves always articulated with the stem, with branching reticulated veins, often opposite and divided. Flowers generally with a distinct calyx and a quinary division of the floral envelopes. Embryo with two or more opposite cotyledons, which often become green and leaf-like after germination; radicle naked, i. e. elongating into a root without penetrating any external case.

DIVISIONS.

1. Dichlamydeæ. Calyx and corolla both present; occasionally imbricated and confounded with each other.
2. Monochlamydeæ. Calyx only present. Corolla none.
3. Achlamydeæ. Flowers destitute of calyx and corolla.

Division I. Dichlamydeæ.

ANALYSIS OF THE ORDERS.

§ 1. Polypetalous.

* Stamens *hypogynous*; or inserted round the sides of the ovarium, if that organ is inferior. (*Thalamifloræ.*)

† *Ovaria in more than one row; or, if solitary, then the stamens opposite the petals.*

Stamens indefinite in number - - - - - - 1. Ranunculaceæ.
Stamens definite in number - - - - - - 2. Berberedeæ.

†† *Ovarium solitary. Fruit one-celled; or placentæ parietal.*

Disk large, adhering to the stamens and the ovarium - 3. Nymphæaceæ.
Disk small, or none
Sepals 2, deciduous
Corolla regular - - - - - - - - 4. Papaveraceæ.
Corolla irregular - - - - - - - 5. Fumariaceæ.
Sepals several
Stamens tetradynamous - - - - - 6. Cruciferæ.
Stamens not tetradynamous
irregular - - - - - - - 7. Violaceæ.
regular
indefinite, (embryo inverted) - - 8. Cistineæ.
definite, (embryo erect)
Sepals distinct; vernation circinate 9. Droseraceæ.
Sepals united in a tube; vernation straight - - - - - } 10. Frankeniaceæ.

††† *Ovarium solitary; or several in one row. In the former case, fruit many-celled, or, if one-celled, with a free central placenta.*

Anthers one-celled
Flowers regular - - - - - - - - 12. Malvaceæ.
Flowers irregular - - - - - - - - 11. Polygaleæ.

Anthers two-celled
Flowers ecalcarate regular
Seeds indefinite in number
Stamens indefinite (polyadelphous) - - 13. Hypericineæ.
Stamens definite - - - - - - 14. Caryophylleæ.
Seeds definite in number
Fruit dehiscent
with valves - - - - - - - 15. Lineæ.
with elasticity
Styles deciduous; albumen abundant - - - - - - } 19. Oxalideæ.
Styles persistent, rigid; albumen none - - - - - - } 18. Geraniaceæ.
Fruit indehiscent
Ovarium stipitate - - - - - 16. Tiliaceæ.
Ovarium sessile - - - - - - 17. Acerineæ.
Flowers calcarate irregular - - - - - - 20. Balsamineæ.

** Stamens *perigynous.* (*Calyciflora.*)

† *Ovarium superior.*

Embryo curved round albumen
Leaves with scarious stipulæ - - - - - - 21. Paronychiæ.
Leaves destitute of stipulæ
Seeds comose - - - - - - - - 22. Tamariscineæ.
Seeds naked - - - - - - - - 23. Portulaceæ.
Embryo not curved round albumen, generally straight
Seeds indefinite in number
Divisions of the calyx 4-5 - - - - - 25. Saxifrageæ.
Divisions of the calyx 6 - - - - - - - 26. Salicariæ.
Seeds definite in number
Hypogynous scales at the base of the carpella - 24. Crassulaceæ.
Hypogynous scales none
Flowers papilionaceous - - - - - 31. Leguminosæ.
Flowers regular
Carpella concrete
Leaves pinnated - - - 30. Staphyleaceæ.
Leaves simple
Ovula pendulous - - - 28. Ilicineæ.
Ovula erect or ascending
Stamens opposite the petals 27. Rhamneæ.
Stamens alternate with the petals - - - - } 29. Celastrineæ.
Carpella distinct - - - - - - 32. Rosaceæ.

†† *Ovarium inferior.*

Seeds definite in number
Ovula erect
Arborescent. Leaves stipulate - - - - 33. Pomaceæ.
Herbaceous. Leaves exstipulate - - - 36. Circæaceæ.
Ovula pendulous
Flowers axillary - - - - - - - - 37. Halorageæ.
Flowers umbellate - - - - - - - - 38. Umbelliferæ.
Seeds indefinite in number
Divisions of calyx 5 - - - - - - - - 34. Grossulaceæ.
Divisions of calyx 4 - - - - - - - - 35. Onagrariæ.

§ 2. Monopetalous.

* *Ovarium inferior.*

Flowers regular
Fruit two or more celled
few-seeded
Leaves whorled, stem square, albumen corneous - - - - - - - - } 39. Stellatæ.
Leaves opposite, stem round, albumen none 45. Valerianeæ.
many-seeded
berried - - - - - - - - - 42. Vaccinieæ.
capsular - - - - - - - - - 43. Campanulaceæ.
Fruit one-celled
Anthers connate - - - - - - - 47. Compositæ.
Anthers distinct
Stamens alternate with the petals - - 46. Dipsaceæ.
Stamens opposite the petals - - - 41. Loranthеæ.

Flowers irregular
Fruit many-seeded. Anthers connate - - - - 44. LOBELIACEÆ.
Fruit few-seeded. Anthers distinct - - - - 40. CAPRIFOLIACEÆ.

** *Ovarium superior.*

Flowers regular
Fruit few-seeded
four-lobed - - - - - - - - - - 48. BORAGINEÆ.
undivided
Calyx plaited - - - - - - - - 52. PLUMBAGINEÆ.
Calyx imbricated
Pericarp membranous, dehiscing transversely - - - - - - - 51. PLANTAGINEÆ.
Pericarp dehiscing by valves
Æstivation plicate. Stem twining 49. CONVOLVULACEÆ.
Æstivation imbricate. Stem erect 50. POLEMONIACEÆ.
Pericarp indehiscent - - - - - 53. OLEINÆ.
Fruit many-seeded
Stamens opposite the petals - - - - - 59. PRIMULACEÆ.
Stamens alternate with the petals
Anthers dehiscing by pores
Arborescent - - - - - - 54. ERICEÆ.
Herbaceous - - - - - - 55. PYROLEÆ.
Anthers dehiscing lengthwise by valves
Leaves alternate - - - - - - 58. SOLANEÆ.
Leaves opposite
Ovarium single - - - - - 57. GENTIANEÆ.
Ovarium double - - - - - 56. APOCYNEÆ.
Flowers irregular
Ovarium four-lobed. Stem square - - - - 65. LABIATÆ.
Ovarium undivided
many-seeded
one-celled with a free central placenta - 60. LENTIBULARIÆ.
two-celled
Stem leafy green - - - - - 61. SCROPHULARINEÆ.
Stem scaly brown - - - - - 62. OROBANCHEÆ.
few-seeded
Flowers ringent. Seeds albuminous - - 63. MELAMPYRACEÆ.
Flowers labiate or regular. Seeds exalbuminous - - - - - - - 64. VERBENACEÆ.

Order 1. RANUNCULACEÆ *Juss.*

Sepals 3-6, hypogynous, generally imbricate in æstivation, occasionally valvate or duplicate.

Petals 5-15, hypogynous, in one or more rows, distinct, occasionally deformed in consequence of metamorphosis in the stamens.

Stamens indefinite in number, hypogynous; *anthers* adnate, in the true genera turned outwards.

Pistilla numerous, seated on a torus, one-celled or united into a single many-celled pistillum; *ovarium* one or more seeded, the *ovula* adhering to the inner edge; *style* one to each ovarium, short, simple.

Fruit either consisting of dry nuts or caryopsides; or baccate with one or more seeds; or follicular with one or two valves.

Seeds albuminous. *Embryo* minute. *Albumen* corneous.

Herbs or very rarely *shrubs*. *Leaves* alternate or opposite, generally divided, with the petiole dilated and forming a sheath half clasping the stem. *Hairs*, if any, simple. *Inflorescence* variable.

ANALYSIS OF THE GENERA.

§ *Genuine Ranunculaceæ.* Anthers bursting outwardly.

Æstivation of calyx valvate or induplicate - - -	1. CLEMATIS.
Æstivation of calyx imbricate	
Fruit one-seeded	
Seed pendulous	
Involucrum none	
Sepals and petals undistinguishable -	2. THALICTRUM.
Sepals and petals distinct - - -	3. ADONIS.
Involucrum three-leaved - - - -	4. ANEMONE.
Seed erect	
Sepals lengthened at the base - - -	5. MYOSURUS.
Sepals not lengthened at the base - -	6. RANUNCULUS.
Fruit many-seeded	
Flowers regular	
Calyx deciduous	
Petals and sepals distinct - - -	7. TROLLIUS.
Petals and sepals not distinguishable -	8. CALTHA.
Calyx persistent - - - - -	9. HELLEBORUS.
Flowers irregular	
Petals all calcarate - - - - -	10. AQUILEGIA.
Only two petals calcarate - - - -	11. DELPHINIUM.
Petals ecalcarate - - - - -	12. ACONITUM.

§ *Spurious Ranunculaceæ.* Anthers bursting inwardly.

Calyx deciduous - - - - - - - -	13. ACTÆA.
Calyx persistent - - - - - - - -	14. PÆONIA.

---

§ *Genuine Ranunculaceæ.*

1. CLEMATIS *Linn.* TRAVELLER'S JOY.

Involucrum none, or shaped like a calyx immediately beneath the flower. *Sepals* 4-8, coloured. *Petals* none or shorter than the sepals. *Cariopsides* numerous, terminating in a bearded tail. — Roots *perennial.* Leaves *exactly opposite.* Dec.

1. C. *Vitalba* Linn. E. B. 9. 612.
Leaves pinnate; leaflets heart-shaped, partly cut. Petioles twining, permanent. Panicles forked, not longer than the leaves. *Smith.*
In hedges. — Shrub. *July.*

2. THALICTRUM *Linn.* MEADOW-RUE.

Sepals and *petals* undistinguishable, petaloid, and very deciduous. *Stamens* numerous. *Ovaries* 4-15. *Styles* short. *Cariopsides* stalked, either having elevated ribs, or being triangular with winged angles, or inflated. *Embryo* very minute, with converging cotyledons. — *Caulescent* herbaceous plants, *with yellow fasciculate* roots, *a fistular* stem, *compound* leaves, and *panicled* flowers.

1. T. *alpinum* Linn. E. B. 4. 262.
Stem perfectly simple and almost naked, with a simple terminal cluster. *Smith.*
In elevated moist alpine pastures; on most of the highest mountains in Wales and Scotland. *Smith.* — Perennial. *June.*

2 T. *minus* Linn. E. B. 1. 11.
Leaves doubly pinnate; leaflets ternate, three-cleft, glaucous on both sides. Flowers panicled, pendulous. Stem zigzag. Stipulas rounded. *Smith.*

In chalky pastures, especially such as are rather mountainous; or in shell sand on the sea coast. *Smith.* — Perennial. *June, July.*

3. T. *majus* Crantz. E. B. 9. 611.
Leaves triply pinnate; leaflets ternate, lobed, glaucous beneath. Branches of the panicle aggregate, somewhat umbellate. Flowers drooping. Stipulas crescent-shaped, notched. *Smith.*

On bushy hills in the North of England. — Perennial. *June, July.*

4. T. *flavum* Linn. E. B. 6. 367.
Stem erect, furrowed, leafy. Leaves doubly pinnate; partly three-lobed. Panicle compound, close, corymbose. Flowers and stamens erect. *Smith.*

In wet meadows, and about the banks of rivers and ditches, common. — Perennial. *June, July.*

3. ADONIS *Linn.* PHEASANT'S-EYE.

Calyx of 5 adpressed sepals. *Petals* 5-15 with a naked claw. *Stamens* numerous, inserted at the base of a torus. *Cariopsides* numerous, arranged in a spike, tipped by the hardened style. — Caulescent herbaceous plants *with finely multifid* leaves, *and solitary scarlet or yellow* flowers. Involucrum *none.*

1. A. *autumnalis* Linn. E. B. 5. 308.
Petals about eight, inversely heart-shaped. Fruit ovate. Stem branched. *Smith.*
A. æstivalis *Withering.*

In corn fields, but not common. — Annual. *May—October.*

4. ANEMONE *Linn.*

Involucrum of 3 cut leaves distant from the flower. *Sepals* and *petals* 5-15 in number, coloured, passing gradually into each other so that they cannot be distinguished.

1. A. *Pulsatilla* Linn. Pasque-flower. E. B. 1. 51.
Flower solitary, nearly upright. Involucrum in deep linear segments. Petals six, erect. Fruit with feathery tails. Leaves doubly pinnate, cut, with linear lobes. *Smith.*
A. pratensis *Sibth.*

In high open chalky pastures. — Perennial. *April, May.*

2. A. *nemorosa* Linn. E. B. 5. 355.
Flowers solitary. Petals six, elliptical. Fruit pointed, without tails. Involucrum of three ternate or quinate, stalked, lobed, and cut leaves. *Smith.*

n groves, thickets, and heathy ground, abundantly. — Perennial. *April.*

3. A. *apennina* Linn. E. B. 15. 1062.
Flower solitary. Petals numerous, lanceolate. Fruit pointed, without tails. Involucrum of three ternate, stalked, deeply cut leaves. *Smith.*

In groves in the central part of England, but rare. — Perennial. *April.*

4. A. *ranunculoides* Linn. E. B. 21. 1484.
Flowers solitary, or in pairs. Petals five, elliptical. Fruit pointed, without tails. Involucrum of three, somewhat stalked, deeply cut leaves.

In groves, very rare. Near King's Langley, Herts; and Wrotham, Kent; and near Abbot's Langley. — Perennial. *April.*

5. MYOSURUS *Linn.* MOUSE-TAIL.

Calyx of 5 sepals, distinct at the base, or elongated downwards considerably. *Petals* 5 with a filiform tubular claw. *Stamens* 5-20. *Ovaries*, and afterwards *fruits*, several, triquetrous, very much crowded, inserted in a spike up the elongated receptacle, and terminated by the straight style. *Dec.*

1. M. *minimus* Linn. E. B. 7. 435.
Stems the length of the leaves or longer. Appendages of the calyx somewhat leafy. *Dec.*

In corn fields, on a gravelly soil. — Annual. *May.*

6. RANUNCULUS *Linn.* CROWFOOT, or BUTTERCUP.

Calyx of 5 sepals, which are not elongated at the base. *Petals* 5-10, with a nectariferous scale at the base. *Stamens* numerous. *Cariopsides* ovate, somewhat compressed, ending in a short horn or mucro, arranged in a globose or cylindrical head. — Roots *fascicled.*

* *Leaves simple.*

1. R. *Flammula* Linn. E. B. 6. 387.
Leaves ovate-lanceolate, bluntish, stalked. Stem reclining. Roots fibrous. Fruit smooth. *Smith.*

In watery places, common. — Perennial. *June—September.*

2. R. *Lingua* Linn. E. B. 2. 100.
Leaves lanceolate, pointed, nearly sessile, somewhat serrated. Stem erect, many-flowered. Root fibrous. Fruit smooth. *Smith.*

In marshes and ditches, not common. — Perennial. *July.*

3. R. *gramineus* Linn. E. B. 33. 2306.
Leaves linear-lanceolate, many-ribbed, entire. Stem erect, very smooth, with few flowers. Roots tuberous. *Smith.*

In dry alpine pastures in Wales. — Perennial. *May, June.*

4. R. *Ficaria* Linn. Pilewort. E. B. 9. 584.
Leaves heart-shaped, angular, stalked, smooth. Petals numerous, elliptic-oblong. *Smith.*
Ficaria ranunculoides *Dec.*

In meadows, bushy places, and about hedge banks, every where. — Perennial. *April.*

** *Leaves lobed, or cut.*

5. R. *auricomus* Linn. Goldilocks. E. B. 9. 624.
Radical leaves kidney-shaped, deeply three-cleft, notched; stem-leaves divided to the base into linear segments. Stem many-flowered. Calyx coloured. *Smith.*

In dry groves, bushy and shady places, not uncommon. — Perennial. *April, May.*

6. R. *sceleratus* Linn. E. B. 10. 681.
Stem erect, hollow, much branched. Leaves smooth; lower ones palmate; upper fingered. Fruit oblong, very numerous, minute. *Smith.*

Common in watery places. — Annual. *June—August.*

7. R. *alpestris* Linn. E. B. 34. 2390.
Leaves very smooth; radical ones somewhat heart-shaped, obtuse, in three deep lobed segments; those of the stem lanceolate, entire. Flower mostly solitary. Calyx smooth. *Smith.*

In moist places, about two or three rocks, on the Clova mountains of Angusshire, rarely flowering. *Sm.* — Perennial. *May.*

8. R. *bulbosus* Linn. E. B. 8. 515.
Calyx reflexed. Flower-stalks furrowed. Stem upright, many-flowered. Leaves compound. Root bulbous. Fruit smooth. *Smith.*
In pastures and meadows, common. — Perennial. *May.*

9. R. *Philonotis* Ehrh. E. B. 21. 1504.
Calyx reflexed, pointed. Stem upright, many-flowered, hairy. Leaves ternate. Root fibrous. Fruit tuberculated. *Smith.*
R. hirsutus *Curtis.*
β smaller in all its parts.
R. parvulus *Linn.*

In moist meadows, and waste or cultivated ground that is liable to be overflowed, frequent. β on wet commons. — Annual. *June—October.*

10. R. *repens* Linn. E. B. 8. 516.
Calyx spreading. Flower-stalks furrowed. Shoots creeping. Leaves compound, cut; the uppermost entire.

In meadows, pastures, and waste places, very common. — Perennial. *June—Aug.*

11. R. *acris* Linn. E. B. 10. 652.
Calyx spreading. Flower-stalks round and even. Leaves in three deep lobed and cut segments; those of the uppermost linear and entire. Stem erect, covered with close hairs. *Smith.*

In meadows and pastures very common. — Perennial. *June, July.*

12. R. *arvensis* Linn. E. B. 2. 135.
Fruit very prickly at the sides. Leaves once or twice deeply three cleft, with linear-lanceolate segments. Stem erect, much-branched, many-flowered. *Smith.*

In corn fields, not uncommon. — Annual. *June.*

13. R. *parviflorus* Linn. E. B. 2. 120.
Fruit armed at the sides with hooked prickles. Leaves simple, hairy, sharply cut; upper ones three-lobed. Stem prostrate. *Smith.*
In gravelly fields and under hedges. — Annual. *May, June.*

14. R. *hederaceus* Linn. E. B. 28. 2003.
Fruit wrinkled. Leaves roundish kidney-shaped, with three or five lobes, entire, smooth. Stem creeping. *Smith.*
In shallow rivulets, ditches, and pools. — Perennial. *May—August.*

15. R. *aquatilis* Linn. E. B. 2. 101.
Stem floating. Leaves in capillary segments under water, above water three-parted, with cuneiform lobes toothed at the end. Petals obovate, larger than the calyx. Fruit hispid with rigid bristles. *Dec.*

In ditches and ponds, common. -- Perennial. *May, June.*

16. R. *pantothrix* Dec.
Stem floating. Leaves wholly in capillary segments. Petals obovate, larger than the calyx. Fruit smooth. *Dec.*
R. circinatus *Sibth.*
R. fluviatilis *Wiggers.*

In ditches and ponds, common. — Perennial. *May, June.*

7. TROLLIUS *Linn.* Globe-flower.

Calyx coloured of 5-10-15 sepals which are deciduous and petaloid. *Petals* 5-10, small, tubular at the base, one-lipped. *Stamens* and *ovaries* numerous. *Follicles* numerous, sessile, subcylindrical, many-seeded. — *Upright* herbaceous plants, *with palmate multifid* leaves *and fascicled* roots. *Dec.*

1. T. *europæus* Linn. E. B. 1. 28.
Sepals 15, converging into a globe. Petals 5-10, the length of the stamens. *Dec.*

In shady, mountainous, rather moist situations. — Perennial. *May, June.*

8. CALTHA *Linn.* Marsh-marigold.

Calyx and *petals* 5 in number, undistinguishable from each other, coloured. *Stamens* indefinite in number. *Ovaries* 5-10. *Follicles* 5-10, compressed, spreading, many-seeded. — *Perennial very smooth* herbaceous *plants.*

1. C. *palustris* Linn. E. B. 8. 506.
Stem erect. Leaves heart-shaped, rounded. *Smith.*
β *minor;* stem usually one-flowered, ascending. *Dec.*

In marshy meadows, and about the margins of ponds, rivers, and brooks, everywhere. β In similar situations, but much less frequent. — Perennial. *March, April.* β *May.*

2. C. *radicans* Forst. E. B. 31. 2175.
Stem reclining, creeping. Leaves triangular, somewhat heart-shaped, sharply crenate. *Smith.*

By the sides of lakes and rivulets in Scotland. — Perennial. *May, June.*

9. HELLEBORUS *Linn.* Hellebore.

Calyx persistent, of 5 sepals, which are roundish, obtuse, large, and often green. *Petals* 8-10, very short, tubular, narrow, and nectariferous at the base. *Stamens* 30-60. *Ovaries* 3-10. *Stigmas* terminal, orbicular. *Follicles* coriaceous; *seeds* arranged in a double row, elliptical, umbilicated. *Dec.*

1. H. *viridis* Linn. E. B. 3. 200.
Stem many-flowered, leafy. Leaves digitate. Petals spreading. *Smith.*

In woods and thickets. In Oxfordshire, Cambridgeshire, Sussex, and other chalk countries. — Perennial. *April, May.*

2. H. *fœtidus* Linn. Bear's-foot, or Setter-wort. E. B. 9. 613.
Stem many-flowered, leafy. Leaves pedate. Petals converging. *Smith.*

In thickets and waste ground, on a chalky soil. — Perennial. *March, April.*

10. AQUILEGIA *Linn.* COLUMBINE.

Calyx of 5 sepals, deciduous, petaloid. *Petals* 5, gaping upwards, their upper lip large and flat, their lower very small; each elongated downwards into a hollow spur, callous at the apex, and projecting between the sepals. *Ovaries* 5. *Follicles* the same number, erect, many-seeded, pointed by the styles.

1. A. *vulgaris* Linn. E. B. 5. 297.
Spurs incurved. Follicles villous. Stem leafy, many-flowered, smooth, as well as the leaves. Styles not longer than the stamens.
β. Stems one-flowered; spurs less curved.
A. alpina *Huds.*

In meadows, pastures, and thickets. β. In more mountainous situations. — Perennial. *June, July.*

11. DELPHINIUM *Linn.* LARKSPUR.

Calyx deciduous, petaloid, irregular; the upper sepal elongated at the base into a spur. *Petals* 4; the two upper ones elongated at the base into appendages contained within the spur. *Dec.*

1. D. *Consolida* Linn. E. B. 26. 1839.
Stem erect, nearly smooth, branching in a straggling manner. Flowers few, in long racemes. Pedicels longer than the bracteæ. Follicles smooth. *Dec.*

In sandy or chalky corn-fields. — Annual. *June, July.*

12. ACONITUM *Linn.* WOLF'S-BANE, or MONKSHOOD.

Calyx petaloid, irregular, deciduous, or withering; the upper sepal concave and helmet-shaped. The two upper *petals* with long claws, expanding into a sac at the apex and concealed beneath the helmet. — Leaves *palmate*. Dec.

1. A. *vulgare* Dec.
Upper petal arched at the back; lateral ones hairy at the inner side. Ovaries three, smooth. Leaves deeply five-cleft, cut, with linear segments, furrowed above. *Smith.*
A. Napellus *Smith.*

By the side of the river Teme, Herefordshire; and still more abundantly on the banks of a brook, running into that river. *Smith.* — Perennial. *June, July.*

---

§§ *Spurious Ranunculaceæ.*

13. ACTÆA *Linn.* BANE-BERRIES, or HERB CHRISTOPHER.

Calyx deciduous, of 4 sepals. *Petals* 4. *Fruit* many-seeded. — *Perennial* herbs. *Dec.*

1. A. *spicata* Linn. E. B. 13. 918.
Cluster dense, ovate. Petals the length of the stamens. *Smith.*

In the north-west corner of Yorkshire, as about Malham Cove, Clapham, Askrigg, and the base of Ingleborough hill. *Sm.* — Perennial. *May, June.*

14. PÆONIA *Linn.* Pæony.

Calyx of 5 sepals, foliaceous, unequal. *Petals* 5-10, nearly orbicular. *Stamens* indefinite in number. *Disk* fleshy, surrounding the ovaries. *Carpella* 2-5, with fleshy stigmas formed by two plates; changing into coriaceous *follicles*. *Seeds* roundish, shining. — Roots *fascicled.* Cauline leaves *twice ternate.* Flowers *large, white, or purple.*

1. P. *corallina* Retz. E. B. 22. 1513.
Leaves twice ternate; leaflets ovate, undivided, smooth. Follicles downy, recurved. *Smith.*

Abundant in the rocky clefts of the Steep Holmes, in the Severn. — Perennial. *May, June.*

Order 2. Berberideæ *Vent.*

Sepals 3-4-6, deciduous, in a double row, surrounded externally by petaloid scales.

Petals hypogynous, either equal to the sepals in number, and opposite to them, or twice as many, generally with an appendage at the base in the inside.

Stamens equal in number to the petals, and opposite to them; *anthers* generally with two separate cells, opening elastically with a valve from the bottom to the top.

Ovarium solitary, one-celled; *style* rather lateral; *stigma* orbicular.

Fruit berried or capsular.

Seeds attached to the bottom of the cell on one side, 1-2 or 3; *albumen* between fleshy and corneous; *embryo* straight in the axis.

Shrubs or *herbaceous perennial* plants, for the most part smooth.

1. BERBERIS *Linn.* Berberry.

Sepals 6, in a double row, externally scaly. *Petals* 6, with two glands at the base of each. *Fruit* fleshy, one-celled, 2-3-seeded. — Shrubs, *with spiny* stems *and* leaves, *and yellow racemose* flowers.

1. B. *vulgaris* Linn. E. B. 1. 49.
Thorns three-cleft. Clusters pendulous. Leaves obovate-oblong, with bristly serratures. Petals entire. *Smith.*

In hedges, and on bushy calcareous hills. — Shrub. *May, June.*

2. EPIMEDIUM *Linn.* Barrenwort.

Sepals 4, with two bracteolæ. *Petals* 4, with a scale at the base. *Pod* oblong, two-valved, 1-celled, many seeded. — Herbaceous *plants, with compound* leaves.

1. E. *alpinum* Linn. E. B. 7. 438.
Radical leaves none; stem-leaf twice ternate. *Smith.*
In mountain thickets, rare. — Perennial. *May.*

Order 3. Nymphæaceæ *Dec.*

Sepals and *petals* numerous, imbricated, passing gradually into each other, the former persistent, the latter inserted upon the disk which surrounds the pistillum.

Stamens numerous, inserted above the petals into the disk; *filaments* petaloid; *anthers* adnate, bursting inwards by a double longitudinal cleft.

Disk large, fleshy, surrounding the pistilla either wholly or in part.

Ovarium polyspermous, many-celled, with the stigmata radiating from a common centre upon a sort of flat urceolate cup.

Fruit a many-celled capsule.

Seeds very numerous, attached to spongy dissepiments, and enveloped in a gelatinous arillus. *Albumen* farinaceous. *Embryo* small on the outside of the base of the albumen, enclosed in a membranous bag; *cotyledons* foliaceous.

Herbs, with peltate or cordate fleshy leaves, growing in quiet waters.

1. NYMPHÆA *Linn.*

Sepals 4. *Petals* and *stamens* numerous, inserted into a disk which surrounds the sides of the ovarium, and adheres to it. *Stigmata* radiating.

1. N. *alba* Linn. White Water-lily. E. B. 3. 160.
Leaves heart-shaped, entire; even beneath. Petals elliptic-oblong. Rays of the stigma sixteen, recurved. Root horizontal. *Smith.*
In clear pools and slow rivers. — Perennial. *July.*

2. NUPHAR *Smith.*

Sepals 5-6. *Petals* 10-18, inserted along with the numerous stamens into a disk which surrounds the base of the ovarium. *Stigmata* radiating.

1. N. *lutea* Smith. Yellow Water-lily. E. B. 3. 159.
Sepals 5. Border of the stigma entire. Footstalks two-edged. Lobes of the leaves meeting each other. *Smith.*
In rivers and pools frequent. — Perennial. *July.*

2. N. *pumila* Hoffm. E. B. 32. 2292.
Sepals five. Border of the stigma toothed. Footstalks two-edged. Lobes of the leaves rather distant. *Smith.*
N. minima *E. Bot.*
N. Kalmiana *Hook. Fl. Scot.* not of others.
In the highland lakes of Scotland — Perennial. *July.*

Order 4. Papaveraceæ *Juss.*

Sepals 2, deciduous.

Petals hypogynous, either 4, or some multiple of that number, inserted in a cruciate manner.

Stamens hypogynous, either 8, or some multiple of four, generally very numerous, inserted in four parcels, one of which adheres to the base of each petal; anthers 2-locular, innate.

Ovarium solitary; *style* short or none; *stigmata* alternate with the placentæ, 2 or many; in the latter case stellate upon the flat apex of the ovarium.

Fruit one-celled, either siliquiform with 2 parietal placentæ, or capsular with several placentæ.

Seeds numerous. *Albumen* between fleshy and oily. *Embryo* minute, straight at the base of the albumen, with plano-convex *cotyledons*.

Herbaceous plants or *shrubs* with a milky juice. *Leaves* alternate, more or less divided. *Peduncles* long, one-flowered. *Flowers* never blue.

ANALYSIS OF THE GENERA.

lacentæ several	
Stigmata radiating connected - - - - - -	1. Papaver.
Stigmata radiating distinct - - - - - -	2. Meconopsis.
Placentæ two	
connate and dividing the pod into 2 cells - - -	3. Glaucium.
distinct	
Seeds not crested. Pod with 2-4 valves - -	4. Römeria.
Seeds crested. Pod with 2 valves - - -	5. Chelidonium.

---

1. PAPAVER *Linn.* Poppy.

Sepals 2, convex. *Petals* 4. *Stamens* numerous. *Style* none. *Stigmata* 4-10, radiating, sessile on the top of the ovarium. *Capsule* one-celled, dehiscing by minute valves concealed beneath the projecting rim of the top. *Placentæ* projecting into the cavity, and forming incomplete dissepiments. — Herbaceous plants, *with divided* leaves and *white milky juice; the* peduncles *inflexed before flowering.*

* *Capsules bristly.*

1. P. *hybridum* Linn. E. B. 1. 43.
   Capsule nearly globular, furrowed, bristly. Calyx hairy. Stem leafy, many-flowered. Leaves doubly pinnatifid. *Smith.*
   In sandy or chalky fields, but rare. — Annual. *July.*

2. P. *Argemone* Linn. E. B. 9. 643.
   Capsule club-shaped, ribbed, bristly. Calyx slightly hairy. Stem leafy, many-flowered. Leaves doubly pinnatifid. *Smith.*
   β. maritimum
   Stem one-flowered.

P. maritimum *Withering.*

In corn fields and their borders, on gravelly or sandy ground. β. In sandy ground near the sea. — Annual. *June, July.*

3. P. *nudicaule* Linn.

Capsule hispid, obovate-oblong. Sepals bristly. Peduncles very long, arising from the root. Leaves pinnatifid, with toothed or cut lobes. *Dec.*

On the north-west coast of Ireland. — Perennial. *June.*

** *Capsules smooth.*

4. P. *dubium* Linn. E. B. 9. 644.

Capsule smooth, oblong, angular. Stem many-flowered, hairy Bristles on the flower-stalks close-pressed. Leaves doubly pinnatifid. *Smith.*

In cultivated fields, especially on a light soil. — Annual. *June, July.*

5. P. *Rhœas* Linn. E. B. 9. 645.

Capsule smooth, nearly globular. Stigma many-rayed. Stem many-flowered, rough, like the flower-stalks, with spreading bristles. Leaves pinnatifid, cut. *Smith.*

In corn-fields, a troublesome weed. — Annual. *June, July.*

6. P. *somniferum* Linn. E. B. 30. 2145.

Capsule nearly globular, smooth as well as the calyx and stem. Leaves notched, clasping the stem, glaucous. *Smith.*

On sandy ground in fens. Annual. *July.*

2. MECONOPSIS *Dec.*

Sepals 2. *Petals* 4. *Stamens* numerous. *Style* short. *Stigmas* 4-6, radiating, convex, distinct. *Capsule* one-celled, dehiscing by 4-6 valves at the top. *Placentæ* narrow, scarcely projecting. — Perennials, *with yellow* juice. *Dec.*

1. M. *cambrica* Dec. E. B. 1. 66.

Capsule smooth, with 5-6 valves. Leaves numerous, stalked, pinnate, cut.

Papaver cambricum *Linn.*

In rocky humid places in Wales and Westmoreland. — Perennial. *June.*

3. GLAUCIUM *Juss.* HORNED-POPPY.

Petals 4. *Stamens* numerous. *Pod* long, 2-valved, with the placentæ meeting in the middle, and forming a spongy dissepiment, which divides the cavity of the pod into two cells. *Seeds* destitute of a crest. — Biennials, *with glaucous, scabrous, pinnatifid* leaves, *and yellow milky* juice.

1. G. *luteum* Scopoli. E. B. 1. 8.

Stem smooth. Stem-leaves wavy. Pod roughish, with minute tubercles. *Smith.*

G. flavum *Crantz.*

Chelidonium Glaucium *Linn.*

On the sandy sea coast. — Biennial. *July, August.*

2. G. *corniculatum* Curtis. E. B. 20. 1433.
Stem hairy. Stem-leaves pinnatifid, cut. Pod rough, with upright bristles. *Smith.*
G. phœniceum *Smith.*
In sandy fields, or on the sea coast, a very rare, or perhaps doubtful, native. *Smith.* — Annual. *June, July.*

4. RÖMERIA *Medicus.*

Petals 4. *Stamens* numerous. *Pod* long, 2-3-4-valved; the valves opening from the top to the bottom. *Placentæ* distinct. *Seeds* pitted, destitute of a crest. — Annuals, *with yellow milky* juice, *and purple* flowers. *Dec.*

1. R. *hybrida* Dec. E. B. 3. 201.
Pods 3-4-valved, erect, with rigid bristles at the apex. Leaves doubly pinnatifid, linear, smooth.
Chelidonium hybridum *Linn.*
Glaucium violaceum *Smith.*
In corn-fields. — Annual. *May, June.*

5. CHELIDONIUM *Linn.* CELANDINE.

Sepals 2, smooth. *Petals* 4. *Stamens* numerous. *Pod* long, two-valved; the placentæ not connate; the valves opening from bottom to top. *Seeds* crested. — Herbaceous plants, *with smooth, brittle, tender* leaves, *and an acrid yellow juice.*

1. Ch. *majus* Linn. E. B. 22. 1581.
Peduncles umbellate. Leaves pinnatifid, with rounded segments, the lobes of which are toothed. Petals elliptical, entire. *Dec.*
In waste ground and thickets, especially on a chalky soil. — Perennial. *May, June.*

2. Ch. *laciniatum* Miller.
Peduncles umbellate. Leaves pinnatifid; their segments cut into linear, jagged, acute lobes. Petals serrate or cut. *Dec.*
Ch. majus β. *Smith.*
Found at Wimbledon, in Surrey, according to Dillenius. — Perennial. *May, June.*

Order 5. FUMARIACEÆ *Dec.*

Sepals 2, deciduous.
Petals 4, cruciate, parallel; the two outer, either one or both, saccate at the base; the two inner callous and coloured at the apex, where they cohere and enclose the anthers and stigma.
Stamens 6, in two parcels, opposite the outer petals; *anthers* membranous, all two-celled, except by abortion.
Ovarium superior, one-celled; *ovula* horizontal; *style* filiform; *stigma* with two or more points.
Fruit various; either an indehiscent one or two-seeded nut, or valved polyspermous pod.

Seeds horizontal, shining, with an arillus. *Albumen* fleshy. *Embryo* minute, out of the axis; in the indehiscent fruit straight; in those which dehisce, somewhat arcuate.

Herbaceous plants, with brittle stems and a watery juice. *Leaves* multifid.

1. CORYDALIS *Dec.*

Petals 4, of which one is calcarate at the base. *Pod* 2-valved, compressed, many-seeded. *Dec.*

1. C. *bulbosa* Dec. E. B. 21. 1471.
Stem mostly simple, erect. Leaves twice ternate. Bracteas palmate, longer than each flower-stalk. *Smith.*
Fumaria solida *Smith.*
About Kendal, and in other parts of Westmoreland; also at Perry Hall, near Birmingham. At Wickham, Hampshire. — Perennial. *April, May.*

2. C. *lutea* Dec. E. B. 9. 588.
Pods nearly cylindrical, shorter than their stalks. Stem angular erect. Bracteas minute. Spur short, rounded. *Smith.*
Fumaria lutea *Smith.*
C. capnoides β *Dec.*
On old walls; very rare. — Perennial. *May.*

3. C. *claviculata* Dec. E. B. 2. 103.
Pods lanceolate, undulated. Stem climbing. Foot-stalks ending in branched tendrils. *Smith.*
Fumaria claviculata *Smith.*
In bushy, shady situations. — Annual. *June, July.*

2. FUMARIA *Tourn. Dec.* FUMITORY.

Petals 4, the three upper connate at the base, the intermediate one being saccate. *Fruit* indehiscent, monospermous, not pointed by the style. *Dec.*

1. F. *officinalis* Linn. E. B. 9. 589.
Cluster rather lax. Pods single-seeded, globose, abrupt, on upright stalks, twice as long as the bracteas. Stem spreading. Segments of the leaflets lanceolate. *Smith.*
In cultivated ground, and about hedges; common. — Annual. *May—August.*

2. F. *parviflora* Lam. E. B. 9. 590.
Cluster lax. Pods single-seeded, globose, pointed. Stem spreading. Segments of the leaflets linear, channelled. *Smith.*
In fields in the south of England. — Annual. *August, September.*

3. F. *capreolata* Linn. E. B. 14. 943.
Cluster rather lax. Pods single-seeded, globose. Stem climbing by means of the twisting footstalks. Leaflets wedge-shaped, lobed. *Smith.*
Fumaria media *Loisel.*
In cultivated fields; common about Liverpool. — Annual. *June—September.*

Order 6. Cruciferæ *Juss.*

Sepals 4, deciduous, cruciate.

Petals 4, cruciate, alternate with the sepals.

Stamens 6, of which two are shorter, solitary, and opposite the lateral sepals, occasionally toothed; and four longer, in pairs, opposite the anterior and posterior sepals; generally distinct, sometimes connate, or furnished with a tooth on the inside.

Disk with various green glands between the petals and the stamens and ovarium.

Ovarium superior, unilocular, with parietal placentæ often meeting in the middle, and forming a spurious dissepiment. *Stigmata* two, opposite the placentæ.

Fruit a siliqua or silicula, one-celled, or spuriously two-celled; one, or many-seeded; dehiscing by two valves separating from the septum; or indehiscent.

Seeds attached in a single row by a funiculus to each side of the placentæ, generally pendulous. *Albumen* none. *Embryo* with the radicle folded upon the cotyledons.

Herbaceous plants, annual, biennial, or perennial, very seldom suffruticose. *Flowers* usually yellow or white: seldom purple.

ANALYSIS OF THE SUBORDERS, TRIBES, AND GENERA.

SUBORDERS.

Radicle applied to the edges of the cotyledons (O = *accumbent*) - - PLEURORHIZEÆ.
Radicle applied to the back of the cotyledons (O || *incumbent*)
  Cotyledons straight O || - - - - - - NOTORHIZEÆ.
  Cotyledons folded lengthwise O>> - - - ORTHOPLOCEÆ.
  Cotyledons doubled twice transversely O || || || - DIPLECOLOBEÆ.

Suborder PLEURORHIZEÆ.

TRIBES.

Fruit dehiscent
  a siliqua or pod - - - - - - - - i. Arabideæ.
  a silicula
    Septum broad; valves flat or concave - - ii. Alyssineæ.
    Septum narrow; valves navicular - - - iii. Thlaspideæ.
Fruit indehiscent - - - - - - - - - iv. Cakilineæ.

---

i. Arabideæ
  Calyx with two sacs at the base
    Stigmas converging; thickened or cornute at the back 1. Mathiola.
    Stigmas two-lobed or capitate - - - - - 2. Cheiranthus.

Calyx equal at the base
Pod taper short - - - - - - - - - - - 3. NASTURTIUM.
Pod 4-cornered - - - - - - - - - 4. BARBAREA.
Pod linear, with flat valves
Seeds in two rows - - - - - - - 5. TURRITIS.
Seeds in one row
Valves with one rib - - - - - - 6. ARABIS.
Valves without a rib - - - - - - 7. CARDAMINE.
Pod lanceolate - - - - - - - - - - 8. DENTARIA.

ii. ALYSSINEÆ.
Seeds solitary - - - - - - - - - - - 9. GLYCE.
Seeds several
Valves flat or nearly so
Petals entire - - - - - - - - 10. DRABA.
Petals two-parted - - - - - - - 11. EROPHILA.
Valves ventricose - - - - - - - - 12. COCHLEARIA.

iii. THLASPIDEÆ
Seeds two or more in each cell
Valves of the silicula winged - - - - - - 13. THLASPI.
Valves of the silicula not winged
Silicula elliptical entire - - - - - - 14. HUTCHINSIA.
Silicula emarginate - - - - - - - 15. TEESDALIA.
Seed one in each cell - - - - - - - - 16. IBERIS.

iv. CAKILINEÆ - - - - - - - - - - - 17. CAKILE.

Suborder NOTORHIZEÆ.

TRIBES.

Fruit dehiscent
a siliqua - - - - - - - - - - - v. SISYMBRIEÆ.
a silicula
Septum broad; valves flat or convex - - vi. CAMELINEÆ.
Septum narrow; valves cruciate - - - vii. LEPIDINEÆ.
Fruit indehiscent - - - - - - - - - viii. ISATIDEÆ.

---

v. SISYMBRIEÆ
Calyx with two sacs at the base - - - - - - 18. HESPERIS.
Calyx equal
Pod taper - - - - - - - - - - - - - 19. SISYMBRIUM.
Pod 4-cornered
Calyx lax - - - - - - - - - - 20. ALLIARIA.
Calyx closed - - - - - - - - - 21. ERYSIMUM.

vi. CAMELINEÆ - - - - - - - - - - - - 22. CAMELINA.

vii. LEPIDINEÆ
Silicula ventricose, one-seeded - - - - - - - 23. CORONOPUS
Silicula emarginate, winged, many-seeded - - - 24. CAPSELLA
Silicula ovate, one-seeded - - - - - - - - - 25. LEPIDIU

viii. ISATIDEÆ - - - - - - - - - - - 26. ISATIS.

Suborder ORTHOPLOCEÆ.

TRIBES.

Fruit dehiscent
a siliqua - - - - - - - - - - - ix. BRASSICEÆ.
a silicula - - - - - - - - - x. VELLEÆ.
Fruit indehiscent - - - - - - - - xi. RAPHANEÆ.

ix. Brassiceæ
Pod taper
Calyx closed - - - - - - - - - - 27. Brassica.
Calyx spreading - - - - - - - - - - 28. Sinapis.
Pod compressed linear - - - - - - - - - 29. Diplotaxis.

x. Velleæ - - - - - - - - - - - - - 30. Carrichtera.

xi. Raphaneæ
Fruit a silicula - - - - - - - - - - - - 31. Crambe.
Fruit a siliqua - - - - - - - - - - - - 32. Raphanus.

Suborder DIPLECOLOBEÆ.

xii. Subularieæ - - - - - - - - - - - - 33. Subularia.

Suborder Pleurorhizeæ *Dec.* O=

Cotyledons flat, accumbent. Radicle lateral. Seeds compressed.

Tribe i. *Arabideæ* Dec.

Siliqua dehiscent; septum linear, something broader than the seeds. Seeds oval, compressed, often bordered. Cotyledons flat, accumbent, parallel with the septum. *Dec.*

1. MATTHIOLA *R. Brown.* Stock.

Siliqua taper. *Stigmata* connivent, thickened, or connate at the back. *Calyx* with two sacs at the base. *Seeds* compressed, in one row, generally bordered. *Cotyledons* flat. *Dec.*

1. M. *incana* R. Brown. E. B. 27. 1935.
Stem shrubby, upright, branched. Leaves lanceolate, obtuse, entire, hoary. Pods without glands. *Smith.*
Cheiranthus incanus *Linn.*
On maritime cliffs in the south of England. — Shrub. *May, June.*

2. M. *sinuata* R. Brown. E. B. 7. 462.
Stem herbaceous, spreading. Leaves downy, glandular, obtuse, sinuated; those of the branches undivided. Pods rough with prominent glands. *Smith.*
Cheiranthus sinuatus *Linn.*
Ch. tricuspidatus *Hudson.*
On the sandy sea-coasts of Wales and Cornwall. — Biennial. *August.*

2. CHEIRANTHUS *Linn.*

Siliqua taper or compressed. *Stigma* 2-lobed or capitate. *Calyx* with two sacs at the base. *Seeds* in one row, ovate, compressed *Dec.*

1. Ch. *Cheiri* Linn. Wall-flower. E. B. 27. 1934.
Leaves lanceolate, acute; most hoary beneath, with simple close

hairs. Stem shrubby. Branches angular. Style prominent. *Smith.*

Ch. fruticulosus *Linn. Smith.*

On old walls. — Shrub. *April, May.*

3. NASTURTIUM *R. Brown.*

Siliqua nearly taper, shortened, or declinate. *Stigma* almost two-lobed. *Calyx* equal at the base, spreading. *Seeds* small, irregularly attached in two rows, not bordered. *Dec.*

1. N. *officinale* R. Brown. WATER-CRESS. E. B. 12. 855.
Leaves pinnate; leaflets roundish-heart-shaped, wavy. *Smith.*
Sisymbrium Nasturtium *Linn.*
n clear springs, rivulets, and ponds, very common and abundant. — Perennial. *June, July.*

2. N. *sylvestre* R. Brown. E. B. 33. 2324.
Leaves pinnate; leaflets lanceolate, deeply serrated or cut. Root creeping. *Smith.*
Sisymbrium sylvestre *Linn.*
In gravelly wet meadows, about the margins of rivers and ditches. — Perennial. *June—September.*

3. N. *terrestre* R. Brown. E. B. 25. 1747
Leaves pinnatifid, unequally toothed. Root tapering. Petals scarcely so long as the calyx. Pod curved. *Smith.*
Sisymbrium terrestre *Smith.*
N. palustre *Dec.*
About the banks of ditches, and in damp meadows. — Annual. *June—September.*

4. N. *amphibium* R. Brown. E. B. 26. 1840.
Leaves oblong, pinnatifid, or serrated. Roots fibrous. Petals longer than the calyx. Pod elliptical. *Smith.*
Sisymbrium amphibium *Linn.*
Common in meadows and by the side of rivers and ditches. — Perennial. *June—August.*

4. BARBAREA *R. Brown.*

Siliqua with four angles, slightly compressed: the valves not pointed at the apex, concave, keeled. *Calyx* equal at the base. *Dec.*

1. B. *vulgaris* R. Brown. BELLEISLE CRESS. E. B. 7. 443.
Lower leaves lyrate, the terminal lobe roundish; upper obovate, toothed. *Smith.*
Erysimum Barbarea *Linn.*
Common in waste ground. — Perennial. *May—August.*

2. B. *præcox* R. Brown. E. B. 16. 1129.
Lower leaves lyrate; upper deeply pinnatifid, with linear-oblong entire segments. *Smith.*
Erysimum præcox *Smith.*
In watery grassy places, or on the banks of ditches, rare. — Biennial. *April—October.*

5. TURRITIS *Linn.* TOWER-MUSTARD.

Siliqua linear; the valves flat. *Seeds* in two rows in each cell. — Flowers *white or whitish.* Dec.

1. T. *glabra* Linn. E. B. 11. 777.
Radical leaves toothed, rough; the rest entire, clasping the stem, smooth. *Smith.*

On banks and by road sides. — Annual. *May, June.*

6. ARABIS *Linn.*

Siliqua linear; valves flat, with a single rib in the middle. *Seeds* in one row in each cell, oval or orbicular, compressed. *Cotyledons* flat. — Flowers *white, unfrequently pink.* Dec.

1. A. *thaliana* Linn. E. B. 13. 901.
Leaves hairy, more or less toothed; radical ones stalked, oblong. Stamens not much shorter than the petals. Stem branched. Pods pointing upwards. *Smith.*

On walls, dry banks, cottage roofs, and dry sandy ground, every where. — Annual. *April.*

2. A. *stricta* Hudson. E. B. 9. 614.
Leaves toothed, obtuse, bristly; radical ones somewhat lyrate. Stems hairy. Petals nearly erect. Calyx smooth. *Smith.*

On St. Vincent's rocks near Bristol, and elsewhere in that neighbourhood. — Perennial. *May.*

3. A. *hispida* Linn. E. B. 7. 469.
Radical leaves lyrate or hastate, smooth or bristly, tufted; stem-leaves lanceolate, entire, scattered, mostly smooth. Petals spreading. Root branched at the crown. *Smith.*
Cardamine petræa *Hudson.*
C. hastulata *E. Botany.*
Arabis Crantziana *Willd.*
Arabis petræa *Lam.*

On lofty alpine rocks of Wales and Scotland, in moist places. — Perennial. *July.*

4. A. *ciliata* R. Brown. E. B. 25. 1746.
Leaves somewhat toothed, smooth on both sides, distinctly fringed and bearded; radical ones obovate. Stem simple. *Smith.*
Turritis alpina *Linn.*
Turritis ciliata *Willd.*

On cliffs near the sea, in Ireland. — Biennial. *July, August.*

5. A. *hirsuta* R. Brown. E. B. 9. 587.
Leaves toothed and bristly. Stem rough, with simple spreading hairs. Pods quite erect, with slightly keeled valves. *Smith.*
Turritis hirsuta *Linn.*

On old walls, stony banks, or rocks, uncommon. — Perennial. *May.*

6. A. *Turrita* Linn. E. B. 3. 178.
Leaves toothed, clasping the stem. Flower-stalks the length of the calyx, each with a leafy bractea. Pods linear, flat, thick-edged, recurved in one direction. *Smith.*
A. umbrosa *Crantz.*

On the walls of Trinity and St. John's colleges, Cambridge, of Magdalen college, Oxford, and of the castle of Cliesh, Kinross-shire. — Biennial. *May.*

7. CARDAMINE *Linn.*

Siliqua linear; valves flat, nerveless, usually dehiscing with elasticity. *Seeds* ovate, not bordered; umbilical cords slender. *Dec.*

* *Leaves simple.*

1. C. *bellidifolia* Linn. E. B. 33. 2355.
Leaves simple, ovate, slightly wavy, entire; the radical ones much shorter than their foot-stalks. Style short, conical. *Smith.*

Gathered wild in Scotland by Mr. Milne. *Withering.* — Perennial. *August.*

** *Leaves pinnate.*

2. C. *impatiens* Linn. E. B. 2. 80.
Leaves pinnate; leaflets lanceolate, mostly cut. Stipulas fringed. *Smith.*

In shady, rather moist, rocky situations, in the north of England; rare in Scotland. — Annual. *May, June.*

3. C. *hirsuta* Linn. E. B. 7. 492.
Leaves pinnate, without stipulas; leaflets stalked, roundish-oblong, notched. *Smith.*
C. flexuosa *Withering.*
C. parviflora *Lightfoot.*

In waste or cultivated ground, especially in moist shady places, very frequent. — Annual. *March—June.*

4. C. *pratensis* Linn. Ladies'-smock. E. B. 11. 776.
Leaves pinnate, without stipulas; leaflets of the radical ones roundish and toothed; those of the stem-leaves lanceolate, entire. Petals with a tooth upon the claw. *Smith.*

Common in meadows and moist pastures. — Perennial. *April, May.*

5. C. *amara* Linn. E. B. 14. 1000.
Leaves pinnate, without stipulas; leaflets of the lowermost roundish; of the rest toothed or angular. Stem creeping at the base. Style obliquely elongated. *Smith.*

In watery places, by the sides of rivers and brooks, but not common. — Perennial. *April, May.*

8. DENTARIA *Linn.* CORALWORT.

Siliqua lanceolate; valves flat, without ribs, often dehiscing with elasticity; placentæ not winged. *Umbilical* cords broad. *Seeds* ovate, not bordered, in one row. *Dec.*

1. D. *bulbifera* Linn. E. B. 5. 309.
Lower leaves pinnated; upper simple, with axillary bulbs.

Near Mayfield, Sussex. In the Old Park Wood, near Harefield, Middlesex, abundantly. In woods between Beconsfield and Wickham, plentifully. On the north sides of the High Rocks, Tonbridge Wells, and elsewhere in that neighbourhood. — Perennial. *April, May.*

Tribe ii. *Alyssineæ* Dec.

Silicula dehiscing longitudinally; septum broad, oval, membranous; valves flat or concave. Seeds compressed, often bordered. Cotyledons flat, accumbent, parallel with the septum

9. GLYCE.

Silicula nearly ovate; valves flattish; cells one-seeded; the umbilical cords adhering by the base to the septum. *Seeds* usually bordered. *Calyx* spreading. *Petals* entire. Hypogynous *glands* 8. *Filaments* not toothed. *R. Brown.*

1. G. *maritima* Sweet Alyssum. E. B. 25. 1729.
Alyssum maritimum *Willd.*
A. minimum *Linn.*
A. halimifolium *Bot. Mag.*
Koniga maritima *R. Brown.*
On cliffs and other places near the sea. — Annual. *July, August.*
I am unwillingly obliged to reject Mr. Brown's name of Koniga, on account of its too close resemblance in sound and meaning to Königia. The meaning of the word now proposed is obvious.

10. DRABA *Linn.* WHITLOW-GRASS.

Silicula sessile, oval or oblong; valves flat or convex. *Seeds* numerous, not margined. *Calyx* equal. *Petals* entire. All the *stamens* without teeth. *Dec.*

1. D. *aizoides* Linn. E. B. 18. 1271.
Stalks solitary, naked. Petals slightly notched, twice the length of the calyx. Leaves lanceolate, rigid, keeled, fringed. *Smith.*
On walls and rocks in South Wales. — Perennial. *March, April.*

2. D. *rupestris* R. Br. E. B. 19. 1338.
Stalk nearly leafless. Petals undivided. Silicula elliptic-oblong. Leaves lanceolate, slightly toothed, fringed with simple hairs. *Smith.*
D. hirta *Linn.*
On rocks in the Highlands of Scotland. — Perennial. *May, June.*

3. D. *incana* Linn. E. B. 6. 388.
Stem-leaves numerous, hoary, like the stem, with close starry pubescence. Silicula elliptic-oblong, oblique or twisted, longer than the hairy partial stalks. *Smith.*
D. contorta *Dec.*
On alpine limestone rocks in the north. — Biennial. *May, June.*

4. D. *muralis* Linn. E. B. 13. 912.
Stem branched. Leaves heart-shaped, toothed, hairy. Silicula elliptical, obtuse, flat, shorter than the partial stalks. *Smith.*
On the shady sides of limestone mountains, or on walls, rare. — Annual. *April, May.*

11. EROPHILA *Dec.*

Silicula oval or oblong; valves flat. *Seeds* numerous, not bordered. *Calyx* equal. *Petals* divided in two. *Stamens* not toothed. — Annuals, *with naked scapes and white* flowers. *Dec.*

1. E. *vulgaris* Dec. E. B. 9. 586.
Silicula elliptical, shorter than the pedicel. Scapes with 5-10 flowers.
Draba verna *Linn.*
On walls, everywhere. — Annual. *March, April.*

12. COCHLEARIA *Linn.* SCURVY-GRASS.

Silicula sessile, ovate, globose, or oblong; valves ventricose. *Seeds* numerous, not bordered. *Calyx* equal, spreading. *Petals* entire. *Stamens* not toothed. — Flowers *white*. Leaves *usually fleshy*. Dec.

1. C. *officinalis* Linn. E. B. 8. 551.
Radical leaves roundish; those on the stem oblong and somewhat sinuated. Silicula globose. *Smith.*

On the sea coast, in stony or muddy situations, abundantly. — Annual. *May.*

2. C. *grœnlandica* Linn. E. B. 34. 2403.
Leaves kidney-shaped, fleshy, entire; uppermost oblong. Silicula globose. *Smith.*

On the mountains of Clova, Angus-shire, and at Loch-ne-gare. *Mr. G. Don.* — Annual. *August.*

3. C. *anglica* Linn. E. B. 8. 552.
Radical leaves ovate, entire; those on the stem sessile, lanceolate, and toothed. Silicula elliptical, strongly reticulated with veins. *Smith*

On the muddy sea shore, and about the mouths of large rivers, in many places. — Annual. *May.*

4. C. *danica* Linn. E. B. 10. 696.
Leaves all triangular and stalked. Silicula elliptical, reticulated with veins. *Smith.*
β. *integrifolia;* cauline leaves very few; radical reniform: all entire. *Dec.*

On the sea coast, in a muddy soil, but rare. β. At the Giants' Causeway in Ireland. — Annual. *May, June.*

5. C. *Armoracia* Linn. Horse-radish. E. B. 33. 2323.
Radical leaves oblong, crenate; those of the stem lanceolate, either cut or entire. *Smith.*

In waste places; generally an outcast from gardens. — Perennial. *May.*

Tribe iii. *Thlaspideæ* Dec.

Silicula dehiscing; septum very narrow; valves keeled, navicular. Seeds oval, sometimes bordered. Cotyledons flat, accumbent, at right angles with the septum.

13. THLASPI *Linn.*

Silicula emarginate; valves navicular, winged at the back; cells two or many-seeded. *Petals* equal. *Calyx* equal at base. *Dec.*

1. T. *arvense* Linn. Mithridate Mustard, or Penny Cress. E. B. 24. 1659.
Silicula orbicular, nearly flat, shorter than its stalk. Leaves smooth, oblong, toothed. Stem erect. *Smith.*

In cultivated or waste ground, but not common. — Annual. *June, July.*

2. T. *perfoliatum* Linn. E. B. 33. 2354.
Silicula inversely heart-shaped. Stem-leaves heart-shaped, rather sharp at the base, clasping the branched stem. Style very short. *Smith.*

Among the stone-pits about Burford in Oxfordshire. — Annual. *April, May.*

3. T. *alpestre* Linn. E. B. 2. 81.
Stem-leaves arrow-shaped. Stems simple. Style prominent beyond the margin of the obovate abrupt silicula. *Smith.*

In mountainous pastures in the north of England, among limestone rocks and lead mines *Smith.* — Perennial. *June, July.*

14. HUTCHINSIA *R. Brown.*

Silicula elliptical; valves navicular, not winged; cells 2-seeded, seldom many-seeded. *Calyx* equal. *Petals* equal. *Dec.*

1. H. *petræa* R. Brown. E. B. 2. 111.
Leaves pinnate, entire. Petals scarcely equal to the calyx. Stigma sessile. Seeds two in each cell. *Smith.*
Lepidium petræum *Linn.*

On limestone rocks and walls, especially in the south of Britain. — Annual. *March, April.*

15. TEESDALIA *R. Brown.*

Silicula oval, emarginate; valves navicular; cells two-seeded. *Stamens* with a scale at the base.

1. T. *Iberis* Dec. E. B. 5. 327.
Petals unequal.
Iberis nudicaulis *Linn.*
T. nudicaulis *R. Br.*

In dry barren gravelly fields. — Annual. *May.*

16. IBERIS *Linn.* CANDY-TUFT.

Two exterior *petals* larger than the others. *Silicula* much compressed, truncate-emarginate. *Seeds* ovate, pendulous. *Dec.*

1. I. *amara* Linn. E. B. 1. 52.
Stem herbaceous. Leaves lanceolate, acute, partly notched. Flowers in oblong clusters. *Smith.*

In chalky fields, but rare. — Annual. *July.*

Tribe iv. *Cakilineæ* Dec.

Siliqua or silicula separating transversely into joints, with 1 or 2 cells, and 1 or 2 seeds. Seeds not bordered. Cotyledons flat, accumbent, parallel with the septum when there is one.

17. CAKILE *Tournefort.*

Silicula with two joints, compressed; the upper joint ensiform or ovate. *Seeds* solitary in each joint; that of the upper erect, of the lower pendulous. *Dec.*

1. C. *maritima* Willd. Sea Rocket. E. B. 4. 231.
Joints of the silicula two-edged; the upper one arrow-shaped. Leaves fleshy, pinnatifid, obtuse. *Smith.*
Bunias Cakile *Linn.*

On the sandy sea coast frequent. — Annual. *June—September.*

Suborder Notorhizeæ *Dec.* O ||

Cotyledons flat, incumbent. Radicle lying upon the back of the cotyledons. Seeds ovate, not bordered.

Tribe v. *Sisymbrieæ* Dec.

Siliqua 2-celled, dehiscing lengthwise; valves concave or keeled. Seeds ovate or oblong, not bordered. Cotyledons flat, incumbent, at right angles with the septum.

18. HESPERIS *Linn.*

Siliqua nearly taper, or somewhat 4-cornered. *Stigmas* 2, erect, converging. *Calyx* with 2 sacs at the base. *Seeds* oblong, somewha 3-cornered. *Stamens* without teeth. *Dec.*

1. H. *matronalis* Linn. Dame's Violet. E. B. 11. 731.
Partial flower-stalks the length of the calyx. Leaves ovate-lanceolate, toothed. Stem upright, slightly branched. Pods smooth, irregularly tumid, equilateral, nearly erect. *Smith.*
H. inodora *Linn.*
In hilly pastures, especially near rivulets, but rare. — Perennial. *May, June.*

19. SISYMBRIUM *Linn.*

Siliqua nearly taper, sessile on a torus. *Stigmas* 2, nearly distinct, or combined in a head. *Calyx* equal at base. *Seeds* ovate or oblong. *Cotyledons* flat, incumbent, sometimes oblique. *Stamens* without teeth. *Dec.*

1. S. *officinale* Scopoli. Hedge-mustard. E. B. 11. 735.
Pods pressed close to the main stalk, awl-shaped, downy. Leaves runcinate, hairy. Stem rough, with reflexed bristles. *Smith.*
In waste ground, by road sides, and on banks, common. — Annual. *June, July.*

2. S. *Irio* Linn. London Rocket. E. B. 23. 1631.
Leaves runcinate, toothed, smooth as well as the stem. Pods erect. *Smith.*
In waste ground, or on banks and heaps of rubbish, chiefly about London. — Annual. *July, August.*

3. S. *Sophia* Linn. Flixweed. E. B. 14. 963.
Leaves doubly pinnatifid, a little hairy. Petals smaller than the calyx. *Smith.*
About waste ground, frequent. — Annual. *July—September.*

20. ALLIARIA *Adanson.*

Siliqua nearly taper, somewhat 4-cornered, in consequence of its projecting ribs. *Calyx* lax. *Seeds* rather cylindrical. *Cotyledons* linear-oblong, flat. *Dec.*

1. A. *officinalis* Dec. Jack by the Hedge, or Sauce Alone.
E. B. 12. 796.
Leaves cordate. Pods prismatical, much longer than the pedicels. *Dec.*
Erysimum Alliaria *Linn.*
Common in hedges. — Annual. *May.*

21. ERYSIMUM *Linn.*

Siliqua 4-cornered. *Calyx* closed. *Cotyledons* flat, oblong. *Dec.*

1. E. *cheiranthoides* Linn. E. B. 14. 942.
Leaves lanceolate, obscurely toothed, roughish with close forked bristles. Pods erect, on horizontal stalks. Stigma almost sessile. *Smith.*
Cheiranthus erysimoides *Huds.*
In turnip-fields, gardens, osier-holts, and hedges, not uncommon. — Annual. *July.*

2. E. *orientale* R. Brown. E. B. 26. 1804.
Leaves elliptic-heart-shaped, obtuse, clasping the stem; radical ones obovate; all smooth, glaucous, undivided, entire. *Smith.*
Brassica orientalis *Linn.*
Erysimum perfoliatum *Dec.*
In fields and on cliffs near the sea. — Annual. *June.*

Tribe vi. *Camelineæ* Dec.

Silicula with concave valves; septum elliptical in its chief diameter. Seeds ovate. Cotyledons flat, incumbent, at right angles with the septum.

22. CAMELINA *Crantz.*

Silicula obovate or roundish; valves ventricose, dehiscing along with part of the style; cells many-seeded. *Style* filiform. *Seeds* oblong, not bordered *Dec.*

1. C. *sativa* Crantz. Gold of Pleasure. E. B. 18. 1254.
Silicula obovate, bordered, twice as long as the style. Leaves lanceolate-arrow-shaped. *Smith.*
Myagrum sativum *Linn.*
Alyssum sativum *Smith.*
In cultivated fields, chiefly among flax, with whose seeds it is often introduced from abroad, but does not long propagate itself with us spontaneously. *Smith.* — Annual. *June.*

Tribe vii. *Lepidineæ* Dec.

Silicula with a very narrow septum; valves keeled, or very concave. Seeds solitary or very few, ovate, not bordered. Cotyledons flat, incumbent, parallel with the septum.

23. CORONOPUS *Gærtn.*

Silicula double; valves ventricose or slightly carinate, scarcely dehiscing, one-seeded. *Seeds* roundish, three-cornered. *Cotyledons* incumbent, linear. — Racemes *opposite the leaves.* Flowers *white.* Dec.

1. C. *Ruellii* Gærtn. E. B. 24. 1660.
Silicula undivided, crested with little sharp points. Style prominent Leaves pinnatifid, subdivided. *Smith.*
Senebiera coronopus *Dec.*
Cochlearia coronopus *Linn.*
Common in waste ground, and by waysides. — Annual. *June—September.*

2. C. *didyma* Smith. E. B. 4. 248.
Silicula cloven, of two round wrinkled lobes. Style scarcely discernible. Leaves pinnatifid, partly notched. *Smith.*
Senebiera didyma *Smith.*
S. pinnatifida *Dec.*
Lepidium didymum *Linn.*
About Exeter, Milford Haven, Truro, and Penryn. — Annual. *July.*

24. CAPSELLA *Dec.*

Silicula triangular, wedge-shaped at the base; valves navicular, apterous; cells many-seeded. — Racemes *terminal.* Flowers *white.*

1. C. *Bursa Pastoris* Dec. Shepherd's Purse. E. B. 21. 1485.
Thlaspi Bursa Pastoris *Linn.*
Common every where in waste places. — Annual. *March—November.*

25. LEPIDIUM *Linn.*

Silicula ovate, or somewhat cordate; valves keeled, or occasionally ventricose, dehiscing; cells 1-seeded. *Seeds* somewhat triquetrous, or compressed. — Racemes *terminal.* Flowers *white.* Dec.

1. L. *latifolium* Linn. E. B. 3. 182.
Leaves ovate-lanceolate, undivided, serrated. *Smith.*
In salt marshes, and wet sandy shady situations under cliffs, near the sea. — Perennial. *July.*

2. L. *ruderale* Linn. E. B. 23. 1595.
Stamens two. Petals none. Leaves smooth; lower ones pinnatifid, toothed; upper linear, entire. Silicula notched. *Smith.*
In waste ground, especially near the sea, in a muddy or calcareous soil. — Annual. *June.*

3. L. *campestre* R. Brown. E. B. 20. 1385.
Silicula scaly, notched; bordered at the summit. Style very short. Stem-leaves arrow-shaped, toothed. *Smith.*
Thlaspi campestre *Linn.*
In cultivated fields. — Annual. *July.*

4. L. *hirtum* Smith. E. B. 26. 1803.
Silicula often hairy, not scaly, bordered at the summit. Style prominent. Stem-leaves arrow-shaped, slightly toothed. *Smith.*
Thlaspi hirtum *Linn.*
In fields on hilly ground. — Perennial. *June.*

Tribe viii. *Isatideæ* Dec.

Silicula with indistinct or indehiscent keeled valves, 1-celled, 1-seeded, with an imperfect septum. Seeds ovate, oblong. Cotyledons flat, incumbent, parallel with what should be the septum.

26. ISATIS *Linn.*

Silicula elliptical, flat, 1 celled, 1-seeded; valves keeled, navicular, scarcely dehiscing. *Seed* pendulous, oblong. — Flowers *small, yellow.* Dec.

1. I. *tinctoria* Linn. Dyer's Woad. E. B. 2. 97.
Radical leaves copiously crenate; those of the stem entire. Silicula abrupt, smooth, thrice as long as broad. *Smith.*

In cultivated fields, and about their borders, but rare. — Biennial. *July.*

Suborder ORTHOPLOCEÆ *Dec.* O > >

Cotyledons incumbent, folded lengthwise, so as to receive the radicle in the folds. Seeds generally round, never bordered.

Tribe ix. *Brassiceæ* Dec.

Siliqua dehiscing lengthwise; septum linear. Seeds globose. Cotyledons folded together.

27. BRASSICA *Linn.*

Siliqua with valves dehiscing lengthwise; dissepiment linear. *Seeds* globose. *Cotyledons* doubled together. *Dec.*

1. B. *Napus* Linn. Rape, or Cole-seed. E. B. 30. 2146.
Root spindle-shaped. Leaves smooth; upper ones lanceolate, heart-shaped at their base, clasping the stem; lower ones lyrate, toothed. *Smith.*

In corn fields, waste ground, and on ditch banks. — Biennial. *May.*

2. B. *Rapa* Linn. Common Turnip. E. B. 31. 2176.
Root fleshy, orbicular, depressed. Radical leaves lyrate, rough; those of the stem smooth; the uppermost entire. *Smith.*

In cultivated fields and their borders, more or less completely naturalised. — Biennial. *April.*

3. B. *campestris* Linn. Wild Navew. E. B. 22. 2234.
Root tapering. Radical leaves lyrate, rough; stem-leaves smooth, clasping, oblong, partly pinnatifid; all somewhat glaucous. *Smith.*

Abundant by the sides of rivers, marsh ditches, &c. — Annual. *June, July.*

4. B. *oleracea* Linn. Cabbage. E. B. 9. 637.
Root cylindrical, fleshy. Leaves glaucous, waved, lobed, partly lyrate, all perfectly smooth. Pod without a beak. *Smith.*

On cliffs near the sea. — Biennial. *May, June.*

5. B. *monensis* Hudson. E. B. 14. 962.
Leaves glaucous, deeply pinnatifid, nearly smooth; lobes oblong, unequally toothed. Stem simple, smooth. Pods quadrangular; beak lodging two or three seeds. *Smith.*

On the sandy sea-coast, but not frequent. — Perennial. *June, July.*

28. SINAPIS *Linn.*

Siliqua rather taper; valves ribbed. *Style* small, short, acute. *Seeds* in one row, roundish. *Calyx* spreading. — Flowers *yellow*. Dec.

1. S. *arvensis* Linn. Charlock. E. B. 25. 1748.
Pods with many angles, rugged, longer than their own awl-shaped beak. Leaves toothed; partly lyrate, or hastate. *Smith.*
In corn-fields, a very troublesome weed; abundant in waste ground newly disturbed. — Annual. *May.*

2. S. *alba* Linn. White Mustard. E. B. 24. 1677.
Pods bristly, rugged, spreading, shorter than their own flat two-edged beak. Leaves lyrate *Smith.*
In cultivated as well as waste ground, by road sides, &c. — Annual. *June.*

3. S. *nigra* Linn Brown Mustard. E. B. 14. 969.
Pods quadrangular, smooth, slightly beaked, close-pressed to the stalk. Lower leaves lyrate; upper linear-lanceolate, entire, smooth. *Smith.*
In fields, waste ground, and on banks by road sides. — Annual. *June, July.*

29. DIPLOTAXIS *Dec.*

Siliqua compressed, linear. *Seeds* in two rows, ovate. *Calyx* equal at the base. — Flowers *yellow or white*. Calyxes *usually covered with soft down*. Dec.

1. D. *tenuifolia* Dec. E. B. 8. 525.
Pods stalked, erect. Style filiform, short, very rough. Upper leaves entire; lower pinnatifid; lobes linear, entire, or pinnatifid. *Dec.*
Sisymbrium tenuifolium *Linn.*
Sinapis tenuifolia *R. Br.*
On old walls and heaps of rubbish. — Perennial. *June—October.*

2. D. *muralis* Dec. E. B. 16. 1090.
Pods sessile, erect. Style short, somewhat filiform. Radical leaves toothed or lyrate, smooth. Stems almost leafless, ascending. *Dec.*
Sisymbrium murale *Linn.*
Sinapis muralis *R. Brown.*
In barren ground near the sea. — Annual. *August, September.*

Tribe x. *Velleæ* Dec.

Silicula with concave valves dehiscing lengthwise; septum elliptical. Seeds globose. Cotyledons folded together.

30. CARRICHTERA *Dec.*

Stamens all distinct. *Style* ovate, flat, foliaceous. — Flowers *small, pale-yellow*. Siliculæ *pendulous, with inflexed stalks*. Dec.

1. C. *Vellæ* Dec. E. B. 21. 1442.
Vella annua *Linn.*
Found in Ray's time, by a Mr. Lawson, on Salisbury Plain, not far from Stonehenge. A doubtful native. — Annual. *June.*

Tribe xi. *Raphaneæ* Dec.

Silicula or siliqua separating transversaly into one or few-seeded joints or cells. Seeds globose. Cotyledons folded together.

31. CRAMBE *Linn.*

Silicula with two joints, the lower abortive, the upper globose, one-seeded. *Cotyledons* thick, somewhat foliaceous, deeply emarginate. — Flowers *white*. Dec.

1. C. *maritima* Linn. Sea Kale. E. B. 13. 924.
Longer filaments toothed. Leaves roundish, sinuated, wavy, toothed, glaucous, very smooth as well as the stem. *Smith.*
On the sandy sea coast. — Perennial. *May, June.*

32. RAPHANUS *Linn.*

Siliqua divided across into many cells, or separating into several pieces. *Seeds* in one row, globose, pendulous. *Cotyledons* rather thick, doubled together. *Dec.*

1. R. *Raphanistrum* Linn. Jointed Charlock. E. B. 12. 856.
Pods jointed, striated, of one cell. Leaves lyrate.
In corn fields, a troublesome weed. — Annual. *June, July.*

2. R. *maritimus* Smith. Sea Radish. E. B. 23. 1643.
Pods jointed, deeply furrowed, of one cell. Radical leaves interruptedly lyrate, serrated.
By the sea-side. — Biennial. *May, June.*

Suborder Diplecolobeæ *Dec.* O || || ||

Cotyledons incumbent, linear, folded twice in a transverse direction (*bicrures*). Seeds depressed.

Tribe xii. *Subularieæ* Dec.

Silicula oval; septum elliptical; valves convex; cells many-seeded; stigma sessile; Cotyledons twice folded.

33. SUBULARIA *Linn.*

Character the same as of the tribe.

1. S. *aquatica* Linn. Awl-wort. E. B. 11. 732.
On the sandy or gravelly bottoms of alpine lakes, under water. — Annual. *July.*

Order 7. VIOLACEÆ *Juss.*

Sepals 5, persistent, with an imbricate æstivation, usually elongated at the base.

Petals 5, hypogynous, equal or unequal, usually with an obliquely convolute æstivation.

Stamens 5, alternate with the petals, inserted on an hypogynous disk, often unequal; *anthers* bilocular, bursting inwards, either separate or cohering; *filaments* dilated, elongated beyond the anthers; two, in the irregular flowers, generally furnished with an appendage or gland at their base.

Ovarium one-celled, many-seeded or one-seeded, with three parietal placentæ opposite the three outer sepals; *style* single, usually declinate, with an oblique hooded *stigma*.

Capsule of three valves, bearing the placentæ in their axis.

Embryo straight, erect, in the axis of fleshy *albumen*.

Herbaceous plants or low *shrubs*. *Leaves* simple, usually alternate, stipulate, entire, with an involute *vernation*.

1. VIOLA *Linn.*

Sepals unequal, auricled. *Petals* unequal, the lower spurred. *Stamens* on the apex of a 5-toothed torus; two lower *anthers* with processes at their back. *Capsule* 3-valved, opening with elasticity.

1. V. *hirta* Linn. E. B. 13. 894.
Stem none. Leaves heart-shaped, rough with hairs, as well as their foot-stalks. Sepals obtuse. Lateral petals with a hairy central line. *Smith.*
In groves and thickets, on a chalky or limestone soil. — Perennial. *April.*

2. V. *odorata* Linn. Violet. E. B. 9. 619.
Stem none, producing runners. Leaves heart-shaped, nearly smooth, as well as their foot-stalks. Sepals obtuse. Lateral petals with a hairy central line.
In woods, hedges, and pastures, frequent. — Perennial. *March, April.*

3. V. *palustris* Linn. E. B. 7. 444.
Stem none. Leaves kidney-shaped, smooth. Root creeping. Lateral petals with a hairy central line. *Smith.*
In mossy bogs, or on sandy turfy heaths, chiefly, though not exclusively, in the northern and mountainous counties. *Smith.* — Perennial. *April.*

4. V. *canina* Linn. Dog Violet. E. B. 9. 620.
Stem at length ascending, channelled. Leaves oblong-heart-shaped. Calyx acute. Stipules serrated. Bracteas awl-shaped, entire. *Smith.*
In groves, thickets, hedges, and heathy ground, common. — Perennial. *April, August.*

5 V. *lactea* Smith. E. B. 7. 445.
Stem ascending, round. Leaves ovate-lanceolate. Stipulas jagged. Bracteas lanceolate, somewhat serrated.
On mountainous boggy heaths. — Perennial. *May.*

6. V. *flavicornis* Smith.
Stem ascending, woody, somewhat angular, much branched. Leaves heart-shaped, coriaceous, smooth and even. Stipules and bracteas fringed. Sepals lanceolate. *Smith.*

In pastures and on banks, in a gravelly soil. — Perennial. *May, June.*

7. V. *tricolor* Linn. Pansy. Heart's-ease. E. B. 18. 1287.
Stem angular, diffuse, divided. Leaves oblong, deeply crenate. Stipules lyrate, pinnatifid. Bracteas obsolete. *Smith.*
β. Petals shorter than calyx.
Viola arvensis *Sibthorp.*

In cultivated fields. — Annual. *May, September.*

8. V. *lutea* Hudson. E. B. 11. 721.
Stem triangular, unbranched. Leaves ovate-oblong, crenate, fringed. Stipules lobed, palmate. Bracteas minute, scarcely toothed. Spur the length of the calyx. *Smith.*
V. grandiflora *Huds. ed.* 2.

In moist mountainous pastures. — Perennial. *May, September.*

Order 8. Cistineæ *Juss.*

Sepals 5, continuous with the pedicel, persistent, unequal; the three inner with a twisted æstivation.
Petals 5, hypogynous, very fugitive, twisted in æstivation in a direction contrary to that of the sepals.
Stamens indefinite in number, hypogynous, distinct; *anthers* innate.
Ovarium distinct, one or many-celled; *ovula* with a foramen at their apex; *style* single; *stigma* simple.
Fruit capsular, usually 3 or 5-valved, occasionally 10-valved, either one-celled with parietal placentæ in the axis of the valves, or imperfectly 5 or 10-celled with dissepiments proceeding from the middle of the valves, and approaching each other in the centre.
Seeds indefinite in number. *Embryo* inverted, either spiral or curved in the midst of mealy albumen.
Shrubs or *herbaceous* plants. *Branches* often viscid. *Leaves* usually entire, opposite or alternate, stipulate or exstipulate. *Racemes* usually unilateral.

1. HELIANTHEMUM *Tournefort.* Rock Rose.

Sepals 3, equal, with 2 occasional external ones. *Petals* 5. *Stigma* capitate. *Style* either wanting or present. *Capsule* of 3 valves. — *Trailing half* shrubby *plants.*

* *Dwarf shrubs, without stipules.*

1. H. *canum* Dunal. E. B. 6. 396.
Leaves opposite, ovate or oblong, petiolate, flat, hoary beneath. Racemes terminal, with bracteæ. Sepals 5, the inner with four

ribs. Style twisted at the base, reflexed; at the apex inflexed. Seeds blackish? *Bentham*

Cistus parvifolius *Linn.*

C. hirsutus *Huds.*

C. anglicus *Linn.*

On alpine rocks, rare. — Shrub. *May, June.*

** *Herbaceous, without stipulæ.*

2. H. *guttatum* Miller. E. B. 8. 544.

Annual, erect. Leaves oblong-lanceolate or linear, the lower opposite, the upper alternate. Racemes without bracteæ. Sepals 5. Style straight, very short. Stigma capitate. *Bentham.*

Cistus guttatus *Linn.*

In warm sandy pastures in the south. — Annual. *June, July.*

*** *Herbaceous, with stipulæ.*

3. H. *ledifolium* Willd. E. B. 34. 2414.

Herbaceous, downy, with stipulas. Leaves lanceolate. Flower-stalks solitary, erect, opposite to the leaves, shorter than the calyx. *Smith.*

Cistus ledifolius *Linn.*

C. salicifolius *Huds.*

Very rare in the south in sandy places. — Annual. *June, July.*

**** *Dwarf shrubs, with stipulæ.*

4. H. *surrejanum* Miller. E. B. 31. 2207.

Shrubby, procumbent, with stipulas. Leaves ovate-oblong, hairy, and dotted beneath. Petals lanceolate. *Smith.*

Cistus surrejanus *Linn.*

At Croydon, in Surrey. — Shrub. *July, August.*

5. H. *vulgare* Gærtner. E. B. 19. 1321.

Leaves opposite, ovate or oblong, nearly flat, green on the upper surface. Racemes terminal, with bracteæ. Sepals 5, the inner furrowed and scarious at the edge. Style bent at the base, somewhat clavate at the apex. Seeds black. *Bentham.*

Cistus Helianthemum *Linn.*

Cistus tomentosus *Smith.*

Common in hilly places. — Shrub. *July, August.*

6. H. *apenninum* Dec. E. B. 19. 1322.

Hoary in every part. Leaves opposite, ovate-oblong or oblong-linear, hoary on each side, more or less revolute at the edge. Racemes terminal, with bracteæ. Sepals 5, the inner furrowed and scarious at the edge. Style bent at the base, somewhat clavate at the apex. Seeds black. *Bentham.*

Cistus polifolius *Linn.*

On stony hills near the sea, very rare. *Smith.* — Shrub. *June, July.*

Order 9. Droseraceæ *Dec.*

Sepals 5, persistent, equal, with an imbricate æstivation.
Petals 5, hypogynous.
Stamens distinct, withering, either equal in number to the petals and alternate with them, or 2 3 or 4 times as many.
Ovarium single; *styles* 3-5, either wholly distinct, or slightly connected at the base, bifid or branched.
Capsule of one or three cells, and three or five valves, which bear the placentæ either in the middle or at their base.
Seeds either naked or furnished with arillus. *Embryo* straight, erect, in the axis of a fleshy or cartilaginous albumen.
Delicate *herbaceous plants*, often covered with glands. *Leaves* alternate with *stipulary* ciliæ *and* a circinate vernation. *eduncles,* when young, circinate.

1. DROSERA *Linn.* Sun-dew.

Sepals and *petals* 5, without appendages. *Stamens* 5. *Styles* 3-5, divided in two. — *Glandular* herbaceous *plants.* Dec.

1. D. *rotundifolia* Linn. E. B. 13. 867.
Leaves depressed, nearly orbicular, on hairy foot-stalks. Flower-stalks radical, racemose. *Smith.*
On mossy turfy bogs, frequent. — Perennial. *July, August.*

2. D. *longifolia* Linn. E. B. 13. 868.
Leaves obovate, erect, on naked foot-stalks. Flower-stalks radical, racemose. *Smith.*
On mossy turfy bogs, with the preceding. — Perennial. *July, August.*

3. D. *anglica* Hudson. E. B. 13. 869.
Leaves oblong, obtuse, erect, on naked foot-stalks. Flower-stalks radical, racemose. Styles 8. Capsules with four valves. *Smith.*
On bogs, but rare. — Perennial. *July, August.*

Note. — For Parnassia, a genus of doubtful station, many of the characters of which are those of Droseraceæ, see *Saxifrageæ.*

Order 10. Frankeniaceæ *St. Hilaire.*

Sepals 4-5, united in a furrowed tube, persistent, equal.
Petals alternate with the sepals, hypogynous, unguiculate, with appendages at the base of the limb.
Stamens hypogynous, either equal in number to the petals, and alternate with them, or having a tendency to double the number. *Anthers* roundish, versatile.
Ovarium superior. *Style* filiform, 2-fid or 3-fid.
Capsule one-celled, enclosed in the calyx, 2- 3- or 4-valved, many-seeded; *dehiscence* septicidal.

Seeds attached to the margins of the valves, very minute. *Embryo* straight, erect, in the midst of albumen (divided into two plates. *Gærtn. fil.*).

Herbaceous plants or *under-shrubs*. *Stems* very much branched. *Leaves* opposite, exstipulate, with a membranous sheathing base. *Flowers* sessile in the divisions of the branches, and terminal, embosomed in leaves.

1. FRANKENIA *Linn.* SEA-HEATH.

Style 3-fid, with oblong lobes, the inner surface of which is stigmatic. *Capsule* of 3 or 4 valves, many-seeded. *Dec.*

1. F. *lævis* Linn. E. B. 3. 205.
Flowers solitary. Leaves linear, revolute, crowded; fringed at the base. *Smith.*

In muddy salt marshes. — Perennial. *July.*

2. F. *pulverulenta* Linn. E. B. 31. 2222.
Leaves obovate, abrupt; downy and somewhat powdery beneath *Smith.*

On the sea coast of Sussex, very rare. — Annual. *July.*

Order 11. Polygaleæ *Juss.*

Sepals 5, persistent, unequal, the inner usually petaloid.

Petals 3-4, hypogynous, adhering to the tube of the stamens.

Stamens 8, monadelphous in two opposite equal parcels. *Anthers* one-celled, bursting by pores at the apex.

Ovarium single, generally 2-celled; *style* single, incurved; *stigma* funnel-shaped or two-lobed.

Fruit capsular or drupaceous, 1 or 2-celled; dehiscence loculicidal.

Seeds pendulous, solitary, with an arillus sometimes hairy or comose. *Embryo* straight; *albumen* usually copious and fleshy, occasionally absent; but, in that case, the inner coating of the testa is tumid.

Herbaceous plants or *shrubs*. *Leaves* mostly alternate, articulated with the stem. *Flowers* racemose. *Juice* of the root milky. *Bark* and *root* bitter.

1. POLYGALA *Linn.*

Sepals persistent, the two inner wing-shaped. *Petals* 3-5, adhering to the tube of the stamens; the lower carinate. *Capsule* compressed, elliptical, obovate, or obcordate. *Seeds* downy, without a coma, with a carunculate hilum. *Dec.*

1. P. *vulgaris* Linn. Milkwort. E. B. 2. 76.
Flowers crested. Bracteæ three, at the base of each flower-stalk,

deciduous. Wings about equal to the corolla. Stems ascending, simple, herbaceous. Leaves linear-lanceolate. *Smith.*

In gravelly and heathy pastures, very common. — Perennial. *June, July.* Many varieties of this species exist in Great Britain; they are well worth a careful examination.

Order 12. Malvaceæ *Juss.*

Sepals 5, very seldom 3 or 4, more or less united at the base, with a valvate æstivation, often provided with external bracteæ forming a kind of involucrum.

Petals of the same number as the sepals, hypogynous, with a twisted æstivation, either distinct or adhering to the tube of the stamens.

Stamens usually indefinite, sometimes of the same number as the petals, hypogynous; *filaments* monadelphous; *anthers* one-celled, reniform, bursting transversely.

Ovarium formed by the union of several carpella round a common axis, either distinct or coherent; *styles* the same number as the carpella, either united or distinct; *stigmata* variable.

Fruit either capsular or baccate; its carpella being either monospermous or polyspermous, sometimes united in one, sometimes separate or separable; dehiscence either loculicidal or septicidal.

Seeds sometimes hairy; *albumen* none; *embryo* with a straight radicle, and twisted and doubled cotyledons.

Herbaceous plants, *trees,* or *shrubs.* *Leaves* alternate, more or less divided, stipulate. *Hairs* stellate. *Peduncles* usually axillary.

1. MALVA *Linn.* Mallow.

Calyx surrounded by an involucrum, formed generally of 3 leaves, seldom of 5 or 6; bracteolæ oblong or setaceous. *Fruit* numerous, capsular, one-seeded, arranged in a circle. *Dec.*

1. M. *sylvestris* Linn. E. B. 10. 671.

Stem upright, herbaceous. Leaves with seven acute lobes. Footstalks and flower-stalks hairy. *Smith.*

About hedges, road sides, and in cultivated as well as waste ground, common. — Perennial. *May—August.*

2. M. *rotundifolia* Linn. E. B. 16. 1092.

Stems prostrate. Leaves roundish heart-shaped, bluntly five-lobed. Stalks when in fruit bent downwards. *Smith.*

β. *pusilla* Smith. E. B. 4. 241.

Petals not longer than the calyx.

M. parviflora *Hudson.*

In waste ground, and by way sides in towns or villages, frequent. β. Near Hithe, in Kent. — Annual. *June—September.*

3. M. *moschata* Linn. E. B. 11. 754.

Radical leaves kidney-shaped, cut; the rest in five deep, pinnatifid, jagged segments. Calyx hairy; its outer sepals linear-lanceolate. *Smith.*

In the grassy borders of fields, and by way sides, on a gravelly soil. — Perennial. *July, August.*

2. ALTHÆA *Linn.*

Calyx surrounded by an involucrum, having from 6 to 9 divisions. *Fruit* capsular, one-seeded, collected in a 5-lobed head. *Dec.*

1. A. *officinalis* Linn. Marsh-mallow. E. B. 3. 147.
Leaves simple, very soft and downy, slightly five-lobed. *Smith.*
In marshes, especially towards the sea, abundantly. — Perennial. *July—September.*

3. LAVATERA *Linn.*

Calyx surrounded by an involucrum, of from 3 to 6 divisions. *Fruit* capsular, one-seeded, collected in a circle round a common axis, which is dilated in various ways. *Dec.*

1. L. *arborea* Linn. E. B. 26. 1841.
Stem arboreous. Leaves downy, plaited, with seven angles. Stalks axillary, aggregate, single-flowered. *Smith.*
On maritime rocks, but rare. — Biennial. *July—October.*

Order 13. Hypericineæ *Juss.*

Sepals 4-5, either more or less cohering, or wholly distinct, persistent, unequal, with glandular dots.
Petals 4-5, hypogynous, with a twisted æstivation and oblique vernation, often having black dots.
Stamens indefinite, hypogynous, in three or more parcels; *anthers* versatile.
Ovary single, superior; *styles* several, rarely connate; *stigmata* simple.
Fruit a capsule or berry, of many valves and many cells; the inner edges of the former being curved inwards.
Seeds minute, indefinite, usually tapering; *embryo* straight, with an inferior *radicle* and no *albumen.*
Herbaceous plants, *shrubs,* or *trees,* with a resinous juice. *Leaves* opposite, dotted. *Flowers* generally yellow. *Inflorescence* variable.

1. HYPERICUM *Linn.* St. John's-wort.

Capsule membranous. *Styles* 3-5, sometimes variable in number. *Stamens* numerous, polyadelphous, occasionally reduced to almost a definite number. *Petals* 5. *Sepals* 5, more or less united at the base. — Herbaceous plants *or* shrubs. Leaves *opposite, often with pellucid dots,* or *black dots at the margin.* Dec.

1. H. *calycinum* Linn. E. B. 29. 2017.
Styles five. Flowers solitary. Stem shrubby, branched, quadrangular. Segments of the calyx obovate, obtuse, permanently spreading. Leaves oblong. *Smith.*
In bushy places in the west of Ireland and Scotland.—A doubtful native. Shrub. *July—September.*

2. H. *quadrangulum* Linn. St. Peter's-wort. E. B. 6. 370.
Styles three. Stem herbaceous, with four sharp angles. Leaves with copious pellucid dots. Segments of the calyx lanceolate. *Smith.*

Common in moist meadows and thickets, and about the banks of rivers. — Perennial. *July, August.*

3. H. *perforatum* Linn. E. B 5. 295.
Styles three. Stem two-edged. Leaves obtuse, with copious pellucid dots. Segments of the calyx lanceolate. *Smith.*

In groves, thickets, and hedges, abundantly. — Perennial. *July, August.*

4. H. *dubium* Leers E. B. 5. 296.
Styles three. Stem obscurely quadrangular. Leaves obtuse, nearly destitute of pellucid dots. Segments of the calyx elliptical. *Smith.*

H. delphinense *Villars.*
H. maculatum *Crantz.*

In rather mountainous groves and thickets. — Perennial. *July, August.*

5. H. *humifusum* Linn. E. B. 18. 1226.
Styles three. Flowers somewhat cymose. Stem compressed, prostrate. Leaves elliptical, smooth. Segments of the calyx ovate, leafy. *Smith.*

In sandy or gravelly, heathy, rather boggy, pastures, frequent. — Perennial. *July.*

6. H. *montanum* Linn. E. B. 6. 371.
Styles three. Calyx with dense, prominent, glandular serratures. Stem erect, round, smooth. Leaves ovate, naked, clasping the stem. *Smith.*

On wild bushy hills, on a gravelly or chalky soil. — Perennial. *July.*

7. H. *barbatum* Jacq. E. B. 28. 1986.
Styles three. Calyx and petals fringed and dotted. Stem erect, somewhat angular. Leaves ovate, naked, dotted, and glandular, clasping the stem. *Smith.*

By the side of a hedge, near the wood of Aberdalgy in Strathearn, Perthshire. — Perennial. *September, October.*

8. H. *hirsutum* Linn. E. B. 17. 1156.
Styles three. Calyx lanceolate, with glandular serratures. Stem erect, round. Leaves ovate, downy. *Smith.*

In thickets and hedges, chiefly on a dry chalky soil. — Perennial. *June, July.*

9. H. *pulchrum* Linn. E. B. 18. 1227.
Styles three. Calyx ovate, with glandular serratures. Stem erect, round. Leaves clasping the stem, heart-shaped, smooth. *Smith.*

In woods and bushy heathy places, on a clay soil, frequent. — Perennial. *July.*

10. H. *elodes* Linn. E. B. 2. 109.
Styles three. Calyx obtuse, glandular. Stem procumbent, creeping, round, shaggy, like the roundish obtuse leaves. Panicle of few flowers. *Smith.*

In spongy bogs. — Perennial. *July, August.*

2. ANDROSÆMUM *Allioni.*

Capsule berried, almost one-celled. *Calyx* divided into 5 pieces of unequal size. *Petals* 5. *Styles* 3. *Stamens* many, united at the base.—A shrub. Leaves *sessile*. Flowers *terminal*, stalked. *Dec.*

1. A. *officinale* Allioni. Tutsan. E. B. 18. 1225
Hypericum Androsæmum *Linn.*
In shady lanes and woods.—Shrub. *July, August.*

Order 14. CARYOPHYLLEÆ *Juss.*

Sepals 4-5, continuous with the peduncle; either distinct, or cohering in a tube, persistent.

Petals 4-5, hypogynous, unguiculate, inserted upon the pedicel of the ovarium; occasionally wanting.

Stamens twice as many as the petals, inserted upon the pedicel of the ovarium along with the petals; *filaments* subulate, sometimes monadelphous; *anthers* innate.

Ovarium stipitate on the apex of a pedicel (called the anthophorus); *stigmata* 2-5, sessile, filiform, papillose on the inner surface.

Capsule 2-5-valved, either 1-celled or 2-5-celled, in the latter case with a loculicidal dehiscence. *Placenta* central, in the 1-celled capsules distinct, in the 2-5-celled capsules adhering to the edge of the dissepiments.

Seeds indefinite in number, rarely definite; *albumen* mealy; *embryo* curved round the albumen; *radicle* pointing to the hilum.

Herbaceous plants, occasionally becoming *suffrutescent*. *Stems* tumid at the articulations. *Leaves* always opposite, and often connate at the base.

ANALYSIS OF THE GENERA.

Sepals united in a cylindrical tube (*Sileneæ*)
 Stigmata 2
  Calyx with bracteæ at the base - - - - - 1. DIANTHUS.
  Calyx naked at the base - - - - - - - - 2. SAPONARIA.
 Stigmata 3 - - - - - - - - - - - 3. SILENE.
 Stigmata 5
  Calyx-teeth simple - - - - - - - 4. LYCHNIS.
  Calyx-teeth foliaceous - - - - - - 5. AGROSTEMMA.
Sepals distinct, or cohering only at the base (*Alsineæ*)
 Capsule dehiscing with distinct valves
  Valves 2 - - - - - - - - - - - 6. BUFFONIA.
  Valves 3 - - - - - - - - - - - 7. CHERLERIA.
  Valves 6 - - - - - - - - - - - 8. SPERGULA.
  Valves 4 or 5
   Capsule with 4 cells - - - - - - - 9. ELATINE.
   Capsule with one cell - - - - - - 10. SAGINA.
 Capsule dehiscing at the apex with teeth
  Petals entire
   Sepals and petals 4 - - - - - - 11. MŒNCHIA.
   Sepals and petals 5 - - - - - - - 12. ARENARIA.

Petals toothed - - - - - - - - - 13. HOLOSTEUM
Petals bifid
Stigmata 5 - - - - - - - - - 14. CERASTIUM.
Stigmata 3
Stamens perigynous - - - - - - 15. LARBREA.
Stamens hypogynous - - - - - 16. STELLARIA.

Tribe 1. *Sileneæ* Dec.

Sepals united into a cylindrical 4 or 5-toothed tube.

1. DIANTHUS *Linn.*

Calyx tubular, 5-toothed, with from 2 to 4 opposite imbricated scales at the base. *Petals* 5, with long claws. *Stamens* 10. *Stigmata* 2. *Capsule* 1-celled. *Seeds* compressed, convex on one side, concave on the other, peltate. *Embryo* nearly straight. *Dec.*

* *Flowers aggregate.*

1. D. *Armeria* Linn. Deptford Pink. E. B. 5. 317.
Flowers aggregate, tufted. Bracteæ lanceolate, downy, as long as the calyx. Petals serrated.

In pastures, and about hedges, on a gravelly soil. — Annual. *July, August.*

2. D. *prolifer* Linn. E. B. 14. 956.
Flowers aggregate, capitate. Bracteæ ovate, obtuse, pointless, membranous, overtopping the calyx.
β. Flowers nearly solitary.
D. diminutus *W.*

In gravelly pastures, rare. — Annual. *July.*

** *Flowers solitary, several on the same stem.*

3. D. *Caryophyllus* Linn. Clove Pink, or Carnation. E. B. 3. 214.
Flowers solitary. Bracteæ almost rhomboid, very short. Petals notched, beardless.
β. Limb of petals slightly hairy near the throat.
D. arenarius *Hudson;* not of others.

On ruinous walls of old towns. — Perennial. *July.*

4. D. *deltoides* Linn. E. B. 1. 61.
Flowers solitary. Bracteæ ovate-lanceolate, acute, seldom more than two. Leaves bluntish, somewhat downy. Petals notched, smooth.
β. Leaves very glaucous; bracteæ generally 4; petals white, with a violet purple circle. *Smith.*
D. glaucus *Linn.*

In pastures, and the grassy borders of fields, on a gravelly or sandy soil. β. In the King's Park, Edinburgh, according to Lightfoot. — Perennial. *July—October.*

*** *Stems single-flowered, herbaceous.*

5. D. *cæsius* Smith. E. B. 1. 62.
Stems single-flowered. Bracteæ short, roundish. Leaves rough-edged. Petals unequally notched, hairy.
D. glaucus *Hudson.*

On dry limestone rocks, very rare. — Perennial. *June, July.*

2. SAPONARIA *Linn.* Soapwort.

Calyx tubular, 5-toothed, naked at the base. *Petals* with claws the length of the calyx. *Stamens* 10. *Stigmas* 2. *Capsule* 1-celled. *Dec.*

1. S *officinalis* Linn. E. B. 15. 1060.
   Calyx cylindrical. Leaves elliptic-lanceolate. *Smith.*
   Bootia vulgaris *Neck.*
   β. Upper leaves connate sheathing; corolla monopetalous.
   In meadows, by river sides, and under hedges. β. On sandy hills 7 miles to the north of Liverpool. — Perennial. *August, September.*

3. SILENE *Linn.*

Calyx tubular, 5-toothed, naked. *Petals* 5, unguiculate, generally having scales at the throat, with a bifid limb. *Stamens* 10. *Stigmas* 3. *Capsules* 3-celled at the base, dehiscing at the apex with 6 teeth. *Dec.*

* *Stem racemose, occasionally somewhat forked.*

1. S. *anglica* Linn. E. B. 17. 1178.
   Hairy and viscid. Petals slightly cloven. Flowers lateral, alternate, erect. Lower capsules spreading or reflexed. *Smith.*
   In cultivated fields, on a gravelly or sandy soil. — Annual. *June, July.*

2. S. *quinquevulnera* Linn. E. B. 2. 86.
   Hairy. Petals roundish, entire. Flowers lateral, alternate, erect, as well as the capsules. Calyx somewhat shaggy. *Smith.*
   Near Wrotham, Kent. *Hudson.* — Annual. *June, July.*

** *Stem forked; branches panicled.*

3. S. *inflata* Smith. E. B. 3. 164.
   Flowers copiously panicled, drooping. Petals cloven half way down, mostly without scales. Calyx smooth, inflated, reticulated. Stem erect. Leaves ovate, acute. *Smith.*
   Cucubulus Behen *Linn.*
   Lychnis Behen *Scop.*
   β. *maritima* Dec. Leaves ovate, lanceolate, pubescent, or hairy.
   In fields, pastures, and by way sides, common. β. Near Cromer, Norfolk. — Perennial. *July.*

4. S. *maritima* With. E. B. 14. 957.
   Flowers slightly panicled, or solitary, terminal. Petals cloven, each with a cloven acute scale. Calyx smooth, inflated, reticulated. Stem recumbent. Leaves lanceolate. *Smith.*
   S. amœna *Huds.*
   S. uniflora *Roth.*
   S. inflata β. *Hooker.*
   S. inflata uniflora *Otth.*
   On the sandy or stony sea-coast, as well as in the beds of alpine torrents. — Perennial. *August, September.*

*** *Stem, and branches if any, forked, leafy.*

5. S. *conica* Linn. E. B. 13. 922.
   Pubescent. Leaves linear, soft. Flowers solitary or panicled. Calyx short, conical, with 30 furrows. Petals cloven. Capsule ovate. *Dec.*
   In sandy fields, rare. — Annual. *July.*

6. S. *conoidea* Linn.
Stems pubescent. Leaves lanceolate linear, nearly smooth. Flowers solitary or panicled. Calyx long, conoid, with thirty furrows. Petals entire, obovate. Capsule lageniform. *Dec.*

In sandy fields. — Annual. *July.*

Introduced upon the authority of Hudson and Sir James Smith, in the Appendix to the English Flora, vol. iv. p. 267.

7. S. *noctiflora* Linn. E. B. 5. 291.
Stem forked. Petals cloven, each with a cloven abrupt scale. Calyx with ten hairy ribs; its teeth linear, almost as long as the tube. *Smith.*

In fields, on a sandy or gravelly soil. — Annual. *July.*

**** *Panicle forked, corymbose, leafless.*

8. S. *Armeria* Linn. E. B. 20. 1398.
Panicles forked, level-topped, many-flowered. Petals cloven, each with a double awl-shaped scale. Calvx and leaves smooth. Capsule not longer than its stalk. *Smith*

In fields or on old walls, a doubtful native. — Annual. *July, August*

***** *Stem panicled, imperfectly forked.*

9. S. *nutans* Linn. Nottingham Catchfly. E. B. 7. 465.
Panicle with drooping, unilateral, partly forked branches. Petals deeply cloven, with linear segments, and acute cloven scale. Leaves elliptic-lanceolate, downy. *Smith.*
β. Leaves broader.
S. paradoxa *Smith.*
Cucubalus viscosus *Huds.*

On limestone rocks, or chalky cliffs. *β.* On Dover cliffs. — Perennial. *June, July.*

10. S. *Otites* Smith. Spanish Catchfly. E. B. 2. 85.
Panicle with tufted, somewhat umbellate, upright branches. Flowers diœcious. Petals linear, undivided, naked. Leaves spatulate, roughish. *Smith.*
Cucubalus Otites *Linn.*

In dry sandy, or gravelly, open grassy fields, chiefly in Norfolk, Suffolk, or Cambridgeshire. — Perennial. *July, August.*

****** *Stems single-flowered.*

11. S. *acaulis* Linn. Moss Campion. E. B. 16. 1081.
Stems tufted, much branched. Leaves linear, acute, fringed at the base. Stalks terminal, solitary, single-flowered. Calyx smooth. *Smith.*
Cucubalus acaulis *Linn.*

On the summits of the loftiest mountains. — Perennial. *June, July.*

4. LYCHNIS *Linn.*

Calyx tubular, 5-toothed, naked. *Petals* 5, unguiculate, usually with scales at the throat. *Stamens* 10. *Stigmas* 5. *Capsule* 1-5-celled. *Dec.*

1. L. *Flos Cuculi* Linn. Ragged Robin. E. B. 8. 573.
Petals in four linear segments. Capsule roundish, of one cell. Stem rough with deflexed bristles. *Smith.*
In moist meadows, frequent. — Perennial. *June.*

2. L. *Viscaria* Linn. Red German Catchfly. E. B. 11. 788.
Viscid. Petals slightly cloven. Capsule stalked, of five cells. Leaves fringed at the base. *Smith.*
In dry fissures of rocks, but rare. — Perennial. *May, June.*

3. L. *alpina* Linn. E. B. 32. 2254.
Smooth. Petals cloven. Flowers densely corymbose. Capsule stalked, of five cells. Leaves linear-lanceolate, naked at the base. *Smith.*
Near the summits of the Clova mountains, Angusshire. — Perennial. *June, July.*

4. L. *sylvestris* Hoppe. E. B. 22. 1579.
Flowers red, usually diœcious, in dichotomous panicles. Petals half cloven, with narrow diverging lobes. Capsules roundish, with recurved valves. Leaves ovate or lanceolate. *Dec.*
L. diurna *Sibth.*
L. dioica rubra *Smith.*
Common in hedges. — Perennial. *May, June.*

5. L. *dioica* Linn. E. B. 22. 1580.
Flowers white, diœcious, in dichotomous panicles. Petals half cloven, with broad converging lobes. Capsules conical, with erect valves. Leaves ovate. *Dec.*
L. vespertina *Sibth.*
L. dioica alba *Smith.*
Common in hedges. — Perennial. *June—September.*

5. AGROSTEMMA *Linn.*

Calyx somewhat campanulate, coriaceous, with five foliaceous segments. *Stamens* 10. *Stigmas* 5. *Capsule* 1-celled.

1. A. *Githago* Linn. Corn Cockle. E. B. 11. 741.
Hairy. Calyx-teeth rising above the corolla. Petals undivided, without teeth. *Smith.*
Lychnis Githago *Dec.*
Githago segetum *Desf.*
In corn fields, a common weed. — Annual. *June, July.*

Tribe 2. *Alsineæ* Dec.

Sepals 4-5, distinct, or cohering only at the base.

6. BUFFONIA *Linn.*

Sepals 4. *Petals* 4, entire. *Stamens* 4. *Stigmas* 2. *Capsule* compressed, 1-celled, 2-valved, 2-seeded. *Dec.*

1. B. *annua* Dec. E. B. 19. 1313.
Stem loosely panicled at the base, with short straggling firm branches. Furrows of the calyx straight parallel. Capsule

scarcely so long as the calyx. Leaves subulate, dilated at the base. *Dec.*

B. tenuifolia *Linn.*

Found by Plukenet about Boston, Lincolnshire; and on Hounslow Heath, by Mr. Doody. — Annual. *June.*

7. CHERLERIA *Linn.*

Sepals 5. *Petals* 5, minute, emarginate. *Stamens* 10. *Stigmas* 3. *Capsule* of 3 cells (?) and 3 valves; each cell with two seeds.—*Smooth* herbaceous *plants, growing in tufts.* Leaves *small, clustered.* Flowers *on short stalks, pale, greenish white.* Dec.

1. C. *sedoides* Linn. Cyphel. E. B. 17. 1212.

On the loftiest mountains of Scotland, in moist spots near their summits not unfrequent. — Perennial. *July.*

8. SPERGULA *Linn.* Spurrey.

Calyx 5-parted. *Petals* 5, entire. *Stamens* 5-10. *Stigmas* 5. *Capsule* of one cell, 6 valves, and many seeds. *Dec.*

1. S. *arvensis* Linn. E. B. 22. 1535.

Leaves whorled. Stalks when in fruit reflexed. *Smith.*

β. Flowers usually pentandrous. Seeds bordered. E. B. 22. 1536.

S. pentandra *Linn.*

In sandy corn fields. — Annual. *June, July.*

2. S. *nodosa* Linn. E. B. 10. 694.

Leaves opposite, awl-shaped, smooth; upper ones clustered. Calyx without ribs. *Smith.*

In moist sandy or turfy ground. — Perennial. *July, August.*

3. S. *saginoides* Linn. E. B. 30. 2105.

Leaves opposite, awl-shaped, almost pointless, naked. Flower-stalks solitary, smooth, much longer than the leaves. *Smith.*

On the highland mountains of Scotland. — Perennial. *June.*

4. S. *subulata* Swartz. E. B. 16. 1082.

Leaves opposite, awl-shaped, bristle-pointed, fringed. Flower-stalks solitary, much longer than the leaves, slightly hairy. *Smith.*

S. laricina *Huds.*

S. saginoides *Curtis.*

On barren sandy heaths. — Perennial. *July, August.*

9. ELATINE *Linn.* Waterwort.

Calyx 3- or 4-parted. *Petals* 3 or 4, sessile. *Stamens* equal in number to the petals, or twice as many. *Stigmas* 4, capitate. *Capsule* with 4 cells, 4 valves, and many seeds. *Seeds* cylindrical. *Dec.*

1. E *tripetala* Smith. E. B. 14. 955.

Leaves opposite, rough with minute points. Flowers mostly three-cleft. *Smith.*

E. Hydropiper *E. B.* not of *Linn.*

? E. triandra *Hoffm.*

On the margins of ponds or ditches, in a sandy soil. About the eastern shore of Bomere pool, near Condover, Shropshire. Near Binfield, Berks. — Annual. *July, August.*

10. SAGINA *Linn.* PEARL-WORT.

Calyx 4-5-parted. *Petals* 4-5, or none. *Stamens* 4-5. *Capsule* with 4 or 5 valves, 1-celled, many-seeded. *Dec.*

1. S. *procumbens* Linn. E. B. 13. 880.
Stems procumbent, smooth. Leaves minutely pointed. Petals half as long as the calyx. *Smith.*
ery common. — Perennial. *May—August.*

2. S. *maritima* Don. E. B. 31. 2195.
Stems nearly upright, divaricated, smooth. Leaves obtuse, without bristles. Petals none. *Smith.*
On the sea-coast of Scotland, as well as of England and Ireland; also on the Highland mountains. *Smith.* — Annual. *May—August.*

3. S. *apetala* Linn. E. B. 13. 881.
Stems nearly upright, hairy. Leaves bristle-pointed, fringed. Petals obsolete, or wanting. *Smith.*
In dry, sandy, barren ground; on walls and waste places; very common. — Annual. *May, June.*

11. MŒNCHIA *Ehr.*

Sepals 4. *Petals* 4, entire. *Stamens* 4. *Stigmas* 4. *Capsule* cylindrical, 1-celled, many-seeded, with 8 or 10 teeth at the apex.

1. M. *glauca* Pers. E. B. 9. 609.
M. erecta *Smith.*
Sagina erecta *Linn.*
In pastures and heathy ground, on a barren gravelly soil. — Annual. *May.*

12. ARENARIA *Linn.*

Sepals 5. *Petals* 5, entire. *Stamens* 10, some of which are occasionally abortive. *Stigmas* 3. *Capsule* 1-celled, with 3 or 6 teeth at the apex, and many seeds. *Dec.*

* *Stipules none.*

1. A. *peploides* Linn. Sea Chickweed. E. B. 3. 189.
Leaves ovate, acute, fleshy. Calyx obtuse, without ribs. *Smith.*
On the sandy sea-coast frequent. — Perennial. *June, July.*

2. A. *trinervis* Linn. E. B. 21. 1483.
Leaves ovate, acute, stalked, ribbed. Calyx obscurely 3-ribbed, with a rough keel. *Smith.*
In shady bushy places, where the soil is rather moist. — Annual. *May, June.*

3. A. *serpyllifolia* Linn. E. B. 13. 923.
Leaves ovate, nearly sessile, rough. Sepals hairy; three outermost 5-ribbed.
On walls and dry sandy ground, common. — Annual. *July.*

4. A. *tenuifolia* Linn. E. B. 4. 219.
Leaves awl-shaped, pointed. Stem panicled. Capsules erect, of three valves. Petals lanceolate, shorter than the calyx. *Smith.*
In dry, barren sandy fields, and on walls, but not very frequent. — Annual. *June.*

5. A. *hirta* Wormskjold.
Leaves awl-shaped, bluntish, downy. Stems 1-3 flowered, hairy

Sepals acute, with 3 ribs, shorter than the capsule. Petals oblong, rather shorter than the calyx.

Alsine rubella *Wahl.*

Arenaria rubella *Smith.*

On the mountains of Breadalbane. — Perennial. *June.*

6. A. *verna* Linn. E. B. 8. 512.

Leaves awl-shaped, bluntish. Stem panicled. Sepals with three remote equal ribs; longer than the petals. *Smith.*

A. saxatilis *Huds.*

A. juniperina *and* laricifolia *With.*

A. cæspitosa *Ehr.*

In mountainous pastures in the north, among fragments of quartz and spar. — Perennial. *May—August.*

7. A. *fasciculata* Jacq. E. B. 25. 1744.

Leaves awl-shaped. Stem erect, straight, densely corymbose. Petals very short. Lateral ribs of the calyx dilated. *Smith.*

A. fastigiata *Smith.*

On rocks on the mountains of Angusshire and Fifeshire. — Annual. *June.*

8. A. *ciliata* Linn. E. B. 25. 1745.

Leaves spatulate, roughish; fringed at the base. Stems numerous, branched, procumbent, downy. Flowers terminal, solitary. Sepals with five or seven ribs.

A. multicaulis *Linn.*

Upon the limestone cliffs of a high mountain adjoining to Ben Bulben, in the county of Sligo. — Perennial. *August, September.*

** *Stipules membranous.*

9. A. *rubra* Linn. E. B. 12. 852.

Leaves linear, bristle-pointed. Stipules membranous, sheathing. Seeds compressed, angular, roughish. *Smith.*

A. campestris *L.*

In sandy fields abundantly. — Annual. *July, August.*

10. A. *media* Linn. E. B. 14. 958.

Leaves semicylindrical, fleshy, pointless. Stipules membranous, sheathing. Seeds compressed, bordered, smooth. *Smith.*

A. marina *Fl. Dan.*

On the sandy sea-coast, and in pastures adjacent. — Annual. *June, July.*

13. HOLOSTEUM *Linn.*

Sepals 5. *Petals* 5, toothed. *Stamens* 5, some of which are occasionally abortive. *Stigmas* 3. *Capsule* 1-celled, dehiscing at the apex in 6 teeth. *Embryo* doubled back in the albumen. *Dec.*

1. H. *umbellatum* Linn. E. B. 1. 27.

Flower stalks umbellate. Leaves ovate, acute. *Smith.*

On several walls and roofs about Norwich, especially in the northern part of the town; and about Bury. — Annual. *April.*

14. CERASTIUM *Linn.* MOUSE-EAR CHICKWEED.

Calyx 5-parted. *Petals* 5, bifid. *Stamens* 10. *Stigmas* 5. *Capsule* 1-celled, cylindrical or globose, dehiscing at the point; teeth 10, circinate or ascending. *Dec.*

1. C. *aquaticum* Linn. E. B. 8. 538.
Leaves cordate, the upper sessile. Flowers in loose dichotomous panicles. Petals bifid, scarcely longer than the calyx. Capsules deflexed ovate, longer than the calyx. *Bentham.*

In watery places. — Perennial. *July.*

2. C. *vulgatum* Linn. E. B. 11. 789.
Hairy, pale green. Leaves roundish-ovate, very blunt. Flowers in dense dichotomous panicles. Petals linear, with 2 teeth, scarcely longer than the calyx. Capsules ascending, oblong, about twice as long as the calyx: with subulate teeth. *Bentham.*

Very common in waste places. — Annual. *April, May.*

3. C. *viscosum* Linn. E. B. 11. 790.
Hairy, deep green. Leaves oblong lanceolate. Flowers in loose dichotomous panicles. Petals oblong, bifid, shorter or scarcely longer than the calyx. Capsules deflexed, incurved, about twice as long as the calyx: with lanceolate teeth. *Bentham.*

Very common in waste places -- Perennial. *May—September.*

4. C. *semidecandrum* Linn. E. B. 23. 1630.
Hoary and hairy. Stems erect, much branched. Flowers dichotomous, in corymbose panicles. Petals emarginate, shorter than the calyx. Capsules erect, oblong, straight, scarcely longer than the calyx. *Bentham.*

β. *tetrandrum.* E. B. 3. 166.
Stamens 4. Flowers 4-cleft. Petals inversely heart-shaped.

C. tetrandrum *Smith.*

Sagina cerastoides *Smith.*

γ. *pumilum.* Dwarf. Petals half divided.

C. pumilum *Curt.*

In waste places, and on old walls; γ at Croydon. — Annual. *March—June.*

5. C. *alpinum* Linn. E. B. 7. 472.
Pilose. Stems ascending. Leaves ovate or oval-oblong, obtuse. Flowers few, in loose dichotomous panicles. Peduncles smooth or pilose. Petals about twice as long as the calyx. Capsule oblong, cylindrical; when ripe about twice as long as the calyx. *Bentham.*

β. *piloso-pubescens* Bentham. E. B. 7. 473.
Hairs of the stem and leaves less woolly and thinner. Leaves rather longer.

C. latifolium *Smith,* not of *Linn.*, according to Mr. Bentham.

On the mountains of Scotland and Wales. — Perennial. *June, July.*

6. C. *arvense* Linn. E. B. 2. 93.
Stems ascending. Leaves lanceolate or linear, acute, or rather blunt. Flowers in dichotomous panicles. Peduncles pubescent with glands intermixed. Petals twice as long as the calyx. Capsule oblong, longer than the calyx. *Bentham.*

In fields and on banks — Perennial. *May—August.*

15. LARBREA *Aug. St. Hilaire.*

Calyx 5-cleft, shortly urceolate at the base. *Petals* 5, 2-parted, inserted on the calyx. *Stamens* 10, perigynous. *Stigmas* 3. *Ovarium*

1-celled, many-seeded. *Ovules* attached to a central axis. *Capsule* with 6 teeth at the end. — *A smooth* herb, *with the habit of* Alsine media. *Dec.*

5. L. *aquatica* St. Hilaire. E. B. 15. 1074.
Stellaria uliginosa *Smith.*
S. aquatica *Pollich.*
S. hypericifolia *Wiggers.*
S. Dilleniana *Leers.*
S. lateriflora *Krock.*
S. fontana *Jacq.*
} according to Smith.
S. Alsine *W.*

In wet places. — Annual. *June.*

16. STELLARIA *Linn.* STITCHWORT.

Calyx 5-parted. *Petals* 5, bifid. *Stamens* 10, or by abortion 3-8. *Stigmas* 3. *Capsule* of one cell, 6 teeth at the apex, and many seeds. *Dec.*

1. S. *nemorum* Linn. E. B. 2. 92.
Lower leaves heart-shaped, stalked; upper ovate, sessile. Panicle repeatedly forked. *Smith.*

In moist woods, and the neighbourhood of shady springs, in the north of England, and lowlands of Scotland. — Perennial. *May, June.*

2. S. *media* Withering. Common Chickweed. E. B. 8. 537.
Leaves ovate. Stems procumbent, with a hairy alternate line on one side. Stamens from 5 to 10. *Smith.*
Alsine media *Linn.*

Common every where in waste and cultivated ground. — Annual. *March—November.*

3. S. *Holostea* Linn. E. B. 8. 511.
Leaves lanceolate, finely serrated. Petals inversely heart-shaped. Calyx without ribs. *Smith.*

In groves, thickets, and dry hedge bottoms, common. — Perennial. *May.*

4. S. *graminea* Linn. E. B. 12. 803.
Leaves linear-lanceolate, entire. Panicle terminal, spreading. Calyx 3-ribbed, nearly as long as the petals. *Smith.*

In heathy pastures, or bushy places, on a gravelly or sandy soil. — Perennial. *May.*

5. S. *glauca* Withering E. B. 12. 825.
Leaves linear-lanceolate, entire, glaucous. Flower-stalks partly scattered, erect. Calyx 3-ribbed, half as long as the petals. *Smith.*
S. palustris *Retz.*
S. media *Sibth.*

In moist meadows, bogs, and the margins of ditches and ponds. — Perennial. *June, July.*

6. S. *scapigera* Willd. E. B. 18. 1269.
Leaves linear-lanceolate, rough-edged. Stem shorter than the flower-stalks. Calyx 3-ribbed, as long as the petals. *Smith.*

By the sides of rivulets on the Scottish mountains. — Perennial. *June.*

7. S. *cerastoides* Linn. E. B. 13. 911.
Leaves elliptic-oblong, bluntish, smooth. Stems with a hairy lateral line, about 2-flowered. Flower-stalks downy all over. Calyx with a single downy rib. *Smith.*
In the Highlands of Scotland. — Perennial. *June.*

Order 15. LINEÆ *Dec.*

Sepals 3-4-5, with an imbricated æstivation, continuous with the peduncle, persistent.

Petals equal in number to the sepals, hypogynous, unguiculate, with a twisted æstivation.

Stamens equal in number to the petals, and alternate with them, united at the base in an hypogynous ring, from which proceed little teeth opposite to the petals, and indicating abortive stamens; *anthers* ovate, innate.

Ovarium with about as many cells as sepals, seldom fewer. *Styles* equal in number to the cells. *Stigmas* capitate.

Capsule generally pointed with the indurated base of the styles, many-celled; each cell partially divided in two by an imperfect spurious dissepiment, and dehiscing with two valves at the apex.

Seeds in each cell single, compressed, inverted; *albumen* usually absent; inner lining of the *testa* tumid. *Embryo* straight, with the radicle pointing towards the hilum; *cotyledons* flat.

Herbaceous plants, or *small shrubs. Leaves* entire, without stipulæ. *Petals* very fugitive.

1. LINUM *Linn.*

Parts of the flower quinary. *Sepals* entire. *Styles* very seldom 3.

* *Leaves alternate.*

1. L. *usitatissimum* Linn. Common Flax. E. B. 19. 1357.
Sepals ovate, acute, with 3 ribs. Petals crenate. Leaves lanceo late, alternate. Stem mostly solitary.
In cultivated fields, frequent. — Annual. *July.*

2. L. *perenne* Linn. E. B. 1. 40.
Sepals obovate, obtuse, obscurely 5-ribbed, naked. Leaves linear-lanceolate. Stems numerous, ascending.
On chalky hills. — Perennial. *June, July.*

3. L. *angustifolium* Huds. E. B. 6. 381.
Sepals elliptical, 3-ribbed, naked, pointed as well as the capsule. Leaves linear-lanceolate, with 3 ribs. Stems numerous.
L. tenuifolium *Withering.*
In sandy or chalky pastures, especially towards the sea. — Perennial. *July.*

* *Leaves opposite.*

4. L. *catharticum* Linn. Mill-mountain. E. B. 6. 382.
   Leaves opposite, obovate-lanceolate. Panicle forked; partly drooping. Petals acute. *Smith.*
   In dry pastures. — Annual. *June—August.*

2. RADIOLA *Gmelin.*

Parts of the flower quaternary. *Sepals* cohering half way, trifid at the points. *Dec.*

1. R. *linoides* Gmel. Dec. E. B. 13. 893.
   R. Millegrana *Smith.*
   Linum Radiola *Linn.*
   In wet sandy ground. — Annual. *July, August.*

Order 16. TILIACEÆ *Juss.*

Sepals 4-5, with a valvular æstivation.
Petals 4-5, entire, with a little pit at their base.
Stamens generally indefinite, hypogynous, distinct; *anthers* 2-celled, dehiscing longitudinally.
Disk formed of glands equal in number to the petals at the foot of which they are placed, adhering to the stalk of the ovarium.
Ovarium single, compressed, of from 4 to 10 carpella; *style* one; *stigmata* as many as the carpella of the ovarium.
Fruit dry, of several cells.
Seeds numerous. *Embryo* erect in the axis of fleshy albumen, with flat foliaceous cotyledons.
Trees or *shrubs*; seldom *herbaceous* plants. *Leaves* simple, stipulate, toothed.

1. TILIA *Linn.* LIME-TREE.

Calyx 5-parted, deciduous. *Petals* 5, with or without a scale on the inside. *Stamens* numerous, with distinct or somewhat polyadelphous filaments. *Ovarium* with 1 style, and 5 2-seeded cells. *Fruit* coriaceous, 1-celled, with 1 or 2 seeds. *Cotyledons* sinuate.—Trees *with a* bark *separating into distinct layers; and light* wood.

1. T. *intermedia* Dec. E. B. 9. 610.
   Leaves twice the length of the footstalks, quite smooth, except a woolly tuft at the origin of each vein beneath. Cymes many-flowered. Fruit coriaceous, downy.
   T. europæa *Linn.*
   In woods and hedges. — Tree. *July.*

2. T. *grandifolia* Ehr.
   Leaves downy, especially beneath; origin of their veins woolly. Branches hairy. Umbels 3-flowered. Fruit woody, downy, turbinate, with 5 prominent angles.

T. platyphylla *Dec.*
T. cordifolia *Besser.*
In woods and hedges. — Tree. *June, July.*

3. T. *rubra* Dec.
Leaves cordate and unequal at the base, with hairs beneath and on the young shoots; the origin of the veins woolly. Fruit globose, smooth. *Dec.*
T. corallina *Smith.*
In woods and hedges. — Tree. *June, July.*

4. T. *parvifolia* Ehr. E. B. 24. 1705.
Leaves smooth above; glaucous beneath, with scattered, as well as axillary, hairy blotches. Umbels compound, many-flowered. Fruit roundish, brittle, nearly smooth.
T. microphylla *Vent.*
In woods. — Tree. *August.*

Order 17. Acerineæ *Juss.*

Calyx divided into 5, or occasionally from 4 to 9 parts.
Petals equal in number to the lobes of the calyx, inserted round an hypogynous disk.
Stamens inserted upon an hypogynous disk, generally 8, not often any other number; always definite.
Ovarium 2-lobed; *style* 1; *stigmas* 2.
Fruit formed of two parts, which are indehiscent and winged; each 1-celled with 1 or 2 seeds.
Seeds erect with a thickened lining to the testa. *Albumen* none. *Embryo* curved, with foliaceous wrinkled cotyledons, and an inferior radicle.
Trees. *Leaves* opposite, simple, without stipulæ. *Flowers* often polygamous, sometimes apetalous, in axillary corymbs or racemes.

ACER *Linn.*

Flowers polygamous. *Calyx* of 5 lobes or parts. *Stamens* seldom 5, generally 7 or 9. Leaves *simple.* Dec.

1. A. *Pseudo-platanus* Linn. Sycamore. E. B. 5. 303.
Leaves 5-lobed, unequally serrated. Clusters pendulous. *Smith.*
In hedges, and about houses, common, but not truly wild. — Tree. *May.*

2. A. *campestre* Linn. Common Maple. E. B. 5. 304.
Leaves 5-lobed, obtuse, somewhat cut. Clusters corymbose, erect. *Smith.*
In hedges and thickets, common; rare in Scotland, and the north of England. *Smith.* — Tree. *May, June.*

Obs. This species requires careful examination. Several curious varieties, some of which have been even considered species, are described by the botanists of Germany, and probably exist in this country.

Order 18. GERANIACEÆ. *Juss.*

Sepals 5, persistent, more or less unequal, with an imbricated æstivation; 1 sometimes saccate or spurred at the base.

Petals 5, seldom 4 in consequence of 1 being abortive, unguiculate, equal or unequal, either hypogynous or perigynous.

Stamens usually monadelphous, hypogynous, or perigynous, twice or thrice as many as the petals; some occasionally abortive.

Ovarium composed of 5 pieces placed round an elevated axis, each 1-celled, 1-seeded; *ovula* pendulous; *styles* 5, cohering round the elongated axis.

Fruit formed of 5 pieces, cohering round a lengthened indurated axis; each piece consisting of 1 cell, containing 1 seed, having a membranous pericarpium, and terminated by an indurated style, which finally curls back from the base upwards, carrying the pericarpium along with it.

Seeds solitary, pendulous, without albumen. *Embryo* curved; *radicle* pointing to the base of the cell; *cotyledons* foliaceous, convolute and plaited.

Herbaceous plants or *shrubs*. *Stems* tumid at the joints. *Leaves* either opposite or alternate; in the latter case opposite the peduncles.

1. GERANIUM *Linn.* CRANE'S-BILL.

Sepals 5, equal. *Petals* 5, equal. *Stamens* 10, fertile, alternately larger. *Nectariferous glands* at the base of the larger stamens. *Indurated styles* glabrous internally, curling back from the axis, from the base to the point. — *Herbaceous plants with palmate lobed* leaves, *and* 1 *or* 2-*flowered* peduncles.

1. G. *phæum* Linn. E. B. 5. 322.
Stalks 2-flowered, panicled, erect. Calyx slightly pointed. Fruit keeled; hairy below; wrinkled at the summit. Stamens hairy. *Smith.*

In mountainous thickets, rare. — Perennial. *May, June.*

2. G. *nodosum* Linn. E. B. 16. 1091.
Stalks 2-flowered. Leaves opposite, 5- or 3-lobed, pointed, serrated. Fruit even, downy all over. *Smith.*

In mountainous thickets, very rare. — Perennial. *May—August.*

3. G. *sylvaticum* Linn. E. B. 2. 121.
Stalks 2-flowered, somewhat corymbose. Leaves about 7-lobed, cut and serrated. Fruit hairy all over. Stamens awl-shaped, fringed. *Smith.*

In woods, thickets, and pastures. — Perennial. *June, July.*

4. G. *pratense* Linn. E. B. 6. 404.
Stalks 2-flowered. Leaves in about 7 deep segments, sharply pinnatifid and serrated. Fruit hairy all over. Stamens smooth much dilated at the base. *Smith.*

In pastures and thickets. — Perennial. *June, July.*

5. G. *Robertianum* Linn. Herb Robert. E. B. 21. 1486.
Stalks 2-flowered. Leaves somewhat pedate, pinnatifid, 5-angled. Calyx with 10 angles. Fruit wrinkled, simply keeled. *Smith.*

In waste ground, on walls, banks, and under hedges, common. — Annual. *May —October.*

6. G. *Raii.*
Stalks 2-flowered, shaggy. Leaves roundish, succulent, lucid, 5-angled, lobed. Calyx with 10 angles, shaggy. Fruit wrinkled. simply keeled.
Geranium lucidum saxatile, foliis Geranii Robertiani *Raii Syn.* 358.

On the sea-coast in the south of England. — Annual. *June, July.*

7. G. *lucidum* Linn. E. B. 2. 75.
Stalks 2-flowered. Leaves 5-lobed, rounded. Calyx pyramidal transversely wrinkled. Fruit wrinkled, triply keeled. *Smith.*

On walls, cottage roofs, and moist rocks. — Annual. *May—August.*

8. G. *molle* Linn. E. B. 11. 778.
Stalks 2-flowered, alternate, opposite to the leaves, which are rounded, many-lobed, notched, and downy. Fruit much wrinkled, smooth. Seeds without dots. *Smith.*

Very common. — Annual. *April—August.*

9. G. *pusillum* Linn. E. B. 6. 385.
Stalks 2-flowered. Leaves kidney-shaped, palmate, cut, downy. Fruit keeled, even, clothed with erect hairs. Seeds without dots. Anthers only 5. *Smith.*
G. parviflorum *Curtis.*
G. malvæfolium *Scopoli.*
β. *humile* Dec.
Leaves more finely cut. Stem dwarf.
G. humile *Cavan.*

In gravelly fields and waste ground, very common. — Annual. *June—September.*

10. G. *pyrenaicum* Linn. E. B. 6. 405.
Stalks 2-flowered. Petals twice the length of the calyx. Leaves kidney-shaped, lobed. Fruit keeled, even, somewhat downy. Seeds without dots. *Smith.*

In meadows and pastures. — Perennial. *June, July.*

11. G. *rotundifolium* Linn. E. B. 3. 157.
Stalks 2-flowered. Petals entire. Leaves kidney-shaped, cut downy. Fruit even, hairy. Seeds reticulated. *Smith.*

In waste places, not very common. — Annual. *June, July.*

12. G. *dissectum* Linn. E. B. 11. 753.
Stalks 2-flowered. Petals cloven. Leaves in 5 deep laciniated segments. Fruit hairy. Seeds reticulated. *Smith.*

In barren gravelly waste ground, hedges, and fallow-fields, frequent. — Annual. *May, June.*

13. G. *columbinum* Linn. E. B. 4. 259.
Stalks 2-flowered, thrice as long as the leaves, which are in 5 very deep, laciniated segments. Fruit quite even and smooth. Seeds reticulated. *Smith.*

In fields, or on dry banks, on a gravelly or limestone soil, sparingly. — Annual. *June, July.*

14. G. *sanguineum* Linn. E. B. 4. 272.
Stems branched, ascending. Stalks single-flowered. Leaves roundish, in 5 or 7 deeply separated, falcate, 3-cleft lobes. Fruit even; bristly at the summit. Seeds minutely wrinkled. *Smith.*

In bushy, stony, rather hilly situations, or uoon limestone rocks. — Perennial. *July—September.*

15. G. *prostratum* Cavanilles.
Stems dwarf, tufted, nearly simple, shaggy. Stalks single-flowered. Leaves roundish, in 3 or 5 2- or 3-lobed ovate divisions. Fruit even, bristly at the summit. Seeds minutely wrinkled.
G. Lancastriense *Withering.*

On the sandy coast of the isle of Walney, in Lancashire. — Perennial. *July—September.*

2. ERODIUM *L'Herit.* Stork's-bill.

Sepals 5, equal, not extended into a nectariferous tube. *Petals* 5, regular, or irregular. *Stamens* 10, monadelphous, of which 5 are sterile. *Glands* at the base of the sterile stamens. *Indurated styles* bearded internally, twisted spirally when ripe. — Herbaceous *plants or* under-shrubs, *with lobed* leaves, *and* peduncles *usually bearing several flowers.*

1. E. *cicutarium Smith.* E. B. 25. 1768.
Stems procumbent, hairy. Stalks many-flowered. Leaves pinnate; leaflets sessile, pinnatifid, cut. Stamens simple. *Smith.*

In waste ground, frequent. — Annual. *June—September.*

2. E. *moschatum* Smith. E. B. 13. 902.
Stems depressed, hairy. Stalks many-flowered. Leaves pinnate; leaflets nearly sessile, elliptical, unequally cut. Perfect stamens toothed at the base. *Smith.*

In mountainous pastures. — Annual. *June, July.*

3. E. *maritimum* Smith. E. B. 9. 646.
Stems depressed, hairy. Stalks barely 3-flowered. Leaves simple, heart-shaped, cut, crenate, rough. *Smith.*

On the sandy or gravelly sea-coasts of Cornwall, Wales, Sussex, &c. — Perennial. *May—September.*

Order 19. Oxalideæ *Dec.*

Sepals 5, sometimes slightly cohering at the base, persistent, equal.

Petals 5, hypogynous, equal, unguiculate, with a spirally twisted æstivation.

Stamens 10, usually more or less monadelphous, those opposite the petals forming an inner series, and longer than the others; *anthers* 2-celled, innate.

Ovarium with 5 angles and 5 cells; *styles* 5, filiform; *stigmata* capitate or somewhat bifid.

Fruit capsular, membranous, with 5 cells, and from 5 to 10 valves.

Seeds few, fixed to the axis, enclosed within a fleshy arillus, which curls back at the maturity of the fruit, and expels the seeds with elasticity. *Albumen* between cartilaginous and fleshy. *Embryo* inverted, the length of the albumen, with a long radicle, and foliaceous cotyledons.

Herbaceous plants, or under-shrubs. *Leaves* alternate, compound, sometimes simple by abortion.

1. OXALIS *Linn.* Wood-sorrel.

Sepals 5, distinct, or united at the base. *Petals* 5. *Stamens* 10; *filaments* slightly monadelphous; the 5 exterior alternately shorter. *Styles* 5. *Stigmata* pencilled, or capitate. *Capsule* 5-cornered, oblong or cylindrical.

1. O. *Acetosella* Linn. E. B. 11. 762.

Stalks radical, single-flowered. Leaves ternate, inversely heart-shaped, hairy. Root of many scaly joints. Stamens all simple. *Smith.*

In groves and shady places, abundantly. — Perennial. *April, May.*

2. O. *corniculata* Linn. E. B. 24. 1726.

Stem branched, procumbent. Flower-stalks in small umbels. Stipules united to the base of the footstalks.

In shady, rather moist, waste ground. — Annual. *May—October.*

Order 20. Balsamineæ *A. Richard.*

Sepals 2, deciduous, with an imbricate æstivation.

Petals 4, hypogynous, cruciate, unequal; the lower elongated at the base into a spur.

Stamens 5, hypogynous; the 3 lower standing opposite the petals, and bearing anthers with two perfect cells; the 2 upper placed in front of the upper petal, bearing anthers with either 2 cells or only 1; *filaments* thickened at the apex; *anthers* partly connate, bursting lengthwise.

Ovarium single; *style* none; *stigmas* 5, either distinct or connate.

Fruit capsular, with 5 elastic valves, and 5 cells formed by membranous projections of the placenta, which occupies the axis of the fruit, and is connected with the apex by 5 slender threads.

Seeds numerous, suspended; *albumen* none; *embryo* straight, with a superior radicle, and plano-convex cotyledons.

Succulent herbaceous plants. Leaves simple, opposite, or alternate, without stipulæ. *Peduncles* axillary.

1. IMPATIENS *Linn.*

Anthers 5, of which 3 are 2-celled, and 2 1-celled. *Stigmas* 5, united. *Capsule* long, taper, the valves rolling back from the base to the apex.

1. I. *Noli-me-tangère* Linn. Touch me not. E. B. 14. 937.
Peduncles with 3 or 4 flowers shorter than the leaves, and spreading beneath them. Flowers pendulous, their spur recurved at the apex. Leaves ovate, coarsely toothed. Joints of the stem tumid. *Dec.*

In watery shady places in the north, but rarely. — Annual. *July, August.*

Order 21. ILLECEBREÆ *R. Brown.*

Sepals 5, seldom 3 or 4, sometimes distinct, sometimes cohering more or less.

Petals minute, inserted upon the calyx between the lobes.

Stamens exactly opposite the sepals, if equal to them in number; sometimes fewer by abortion; *filaments* distinct; *anthers* 2-celled.

Ovarium superior; *styles* 2 or 3, either distinct or partially combined.

Fruit small, dry, 1-celled, either indehiscent, or opening with 3 valves.

Seeds either numerous, upon a free central placenta, or solitary and pendulous from the apex of the cavity of the fruit; *albumen* farinaceous; *embryo* lying on one side of the albumen, curved more or less, with the radicle always pointing to the hilum.

Herbaceous or *half-shrubby* branching plants, with opposite or alternate leaves, and scarious stipulæ. *Flowers* minute, with scarious bracteæ.

1. CORRIGIOLA *Linn.* STRAPWORT.

Sepals 5, slightly cohering at the base. *Petals* 5, equal to the calyx, into which they are inserted alternately with its lobes. *Stamens* 5, opposite the sepals. *Style* short. *Stigmas* 3. *Fruit* 1-seeded, indehiscent, covered by the calyx. *Seed* single, suspended by its cord, which arises from the bottom of the cavity. — *Procumbent* herbs. Leaves *alternate, glaucous.* Flowers *minute, in terminal corymbose racemes.*

1. C. *littoralis* Linn. E. B. 10. 668.
Stems bearing leaves on the part which bears the flowers. *Dec.*

On the southern coast of England. — Annual. *July, August.*

2. HERNIARIA *Linn.* RUPTURE-WORT.

Sepals 5, slightly cohering at the base, somewhat coloured inside. *Petals* 5, filiform, entire, alternate with the sepals, sometimes wanting. *Stamens* 5, or by abortion 2 or 3, opposite the sepals. *Styles* 2, short, distinct, or cohering at the base. *Fruit* 1-seeded, indehiscent, covered by the calyx. — *Little prostrate, densely branching* herbaceous *plants.* Leaves *opposite, minute.* Flowers *in axillary clusters.*

1. H. *glabra* Linn. E. B. 3. 206.
Herbaceous. Leaves and calyx smooth. *Smith.*
In gravelly or sandy ground. — Perennial. *July, August.*

2. H. *hirsuta* Linn. E. B. 20. 1379.
Herbaceous. Leaves and calyx hairy. *Smith.*
In sandy ground, rare. — Perennial. *July, August.*

3. ILLECEBRUM *Linn.*

Sepals but little cohering at the base, hooded, thickened; their back elongated into a horn-like process. *Petals* either wanting, or in the form of 5 subulate scales, alternate with the sepals. *Stamens* from 2 to 5, opposite the sepals, and inserted into their base. *Stigmas* 2, capitate. *Fruit* included in the calyx, with 5 valves, or separable along 5 streaks. *Seed* solitary, inserted into one side of the cavity. *Embryo* nearly straight, on one side of the farinaceous albumen. — *Small* herbs. Leaves *opposite, with scarious stipulæ.* Flowers *axillary,* or *in cymes, with scarious* bracteæ.

1. I. *verticillatum* Linn. Knot-grass. E. B. 13. 895.
Flowers whorled, without bracteæ. Stems procumbent. *Smith.*
In marshy boggy ground, in Cornwall and Devonshire. — Perennial. *July.*

4. POLYCARPON *Linn.*

Sepals slightly cohering at the base, concave, connate, mucronate, with membranous margins. *Petals* 5, emarginate. *Stamens* from 3 to 5. *Styles* 3, very short. *Capsule* of 1 cell, and 3 valves, many-seeded. — Annuals. Leaves *opposite, whorled.* Flowers *in corymbose cymes.* Stipulæ *and* bracteæ *scarious.*

1. P. *tetraphyllum* Linn. All-seed. E. B. 15. 1031.
Triandrous. Petals emarginate. Leaves of the stem whorled in ours, of the branches opposite. *Dec.*
On various parts of the coasts of Devonshire, Dorsetshire, and Portland island. — Annual. *May—August, or later.*

Order 22. TAMARISCINEÆ *Desv.*

Calyx 4- or 5-parted, persistent, with an imbricated æstivation.
Petals inserted into the base of the calyx, withering, with an imbricated æstivation.

Stamens either equal to the petals in number, or twice as many, either distinct or monadelphous.
Ovarium superior; *style* very short; *stigmata* 3.
Capsule 3-valved, 1-celled, many seeded; *placentæ* 3, either at the base of the cavity, or along the middle of the valves.
Seeds erect, or ascending, comose; *albumen* none; *embryo* straight with an inferior radicle.
Shrubs or *herbs* with rod-like branches. *Leaves* alternate, resembling scales, entire. *Flowers* in close spikes or racemes.

1. TAMARIX *Linn.* TAMARISK.

Calyx 4-5 parted. *Petals* 4 or 5. *Stamens* 4 or 5, alternate with the petals, and almost entirely distinct. *Ovarium* tapering much to the point. *Stigmas* 3, long, straggling, glandular and oblique at the apex. *Seeds* inserted into the base of the valves, or almost in the centre of the capsule. *Coma* of the seeds consisting of numerous simple hairs.

1. T. *gallica* Linn. E. B. 19. 1318.
Lateral clusters numerous. Leaves lanceolate, spurred, acute. Branches smooth. *Smith.*
On rocks and cliffs on the south coast of England.—Shrub. *July.*

Order 23. PORTULACEÆ *Juss.*

Sepals 2, seldom 3 or 5, cohering by the base.
Petals generally 5, occasionally 3, 4, or 6, either distinct or cohering in a short tube.
Stamens inserted along with the petals into the base of the calyx, variable in number, all fertile; but opposite to the petals if equal to them in number. *Filaments* distinct; *anthers* versatile, with 2 cells, opening lengthwise.
Ovarium superior, 1-celled; *style* single or none; *stigmata* several, much divided.
Capsule 1-celled, dehiscing either transversely or by 3 valves; occasionally 1-seeded and indehiscent.
Seeds numerous, if the fruit is dehiscent, attached to a central placenta, which is connected with the style by conducting cords; *albumen* farinaceous; *embryo* curved round the circumference of the albumen, with a long radicle.
Succulent *shrubs* or *herbs*. *Leaves* alternate, entire, without stipulæ, sometimes sheathing at the base. *Flowers* axillary or terminal, fugitive.

1. MONTIA *Linn.*

Sepals 2 or 3. *Petals* 5, cohering at the base, 3 rather smaller than the others. *Stamens* inserted into the claws of the petals, generally 3, before the smaller petals, seldom 4 or 5. *Ovarium* sessile. *Style* very short, in 3 parts; with spreading reflexed branches. *Capsule* of 1 cell, with 3 valves and 3 seeds. — *An* annual, *destitute of hairs.* Leaves *opposite.* Flowers *small, axillary.*

1. M. *fontana* Linn. Water Blinks. E. B. 17. 1206.
In watery places, especially on a gravelly soil. — Annual. *April, May.*

Order 24. CRASSULACEÆ *Dec.*

Sepals from 3 to 20, more or less united at the base.

Petals inserted in the bottom of the calyx, either distinct or cohering in a monopetalous corolla.

Stamens inserted with the petals, either equal to them in number, and alternate with them, or twice as many, those opposite the petals being shortest, and arriving at perfection after the others; *filaments* distinct, subulate; *anthers* of 2 cells, bursting lengthwise.

Hypogynous scales several, 1 at the base of each ovarium, sometimes obsolete.

Ovaria of the same number as the petals, opposite to which they are placed around an imaginary axis; 1-celled, tapering into stigmata.

Fruit consisting of several follicles, opening by the suture in their face.

Seeds attached to the margins of the suture, variable in number; *embryo* straight in the axis of the albumen, with the radicle pointing to the hilum.

Succulent *herbs* or *shrubs*. *Flowers* usually in cymes, sessile, arranged unilaterally along the divisions of the cymes.

1. TILLÆA *Linn.*

Sepals 3 or 4, cohering at the base. *Petals* 3 or 4, oblong, taper-pointed. *Stamens* 3 or 4. *Hypogynous scales* obsolete. *Fruit* of 3 or 4 parts, each of which is contracted in the middle and 2-seeded. — *Small* annual *plants, growing in damp exposed places.* Leaves *opposite.* Flowers *minute, axillary.*

1. T. *muscosa* Linn. E. B. 2. 116.
Stems procumbent. Flowers sessile, mostly 3-cleft. *Smith.*
On the most barren sandy heaths. — Annual. *May, June.*

2. UMBILICUS *Dec.* NAVELWORT.

Sepals 5, cohering at the base. *Petals* cohering in a campanulate, 5-cleft corolla. *Stamens* 10, inserted upon the corolla. *Hypogynous scales* 5, obtuse. *Fruit* in 5 parts, which taper to the point, and are tipped each with a subulate style. — Herbaceous *plants.* Leaves *fleshy, alternate, or growing in clusters.* Flowers *whitish or yellow.*

1. U. *pendulinus* Dec. E. B. 5. 325.
Leaves peltate, notched. Flowers clustered, drooping. Bracteas entire. Root tuberous. *Smith.*
Cotyledon umbilicus *Hudson.*
On moist dripping rocks, and old walls, in mountainous countries. — Perennial. *June, July.*

2. U. *erectus* Dec. E. B. 22. 1522.
Leaves deeply toothed; the lowermost slightly peltate. Flowers erect. Root creeping. *Smith.*
Cotyledon lutea *Hudson.*
On moist rocks and old walls, very rare. — Perennial. *July.*

3. SEDUM *Linn.*

Sepals 5, cohering at the base, turgid, and often foliaceous. *Petals* 5, spreading. *Stamens* 10. *Hypogynous* scales entire. *Fruit* in 5 parts. — Herbs *with fleshy* leaves, *many* branches, *and cymose* flowers.

* *Leaves flat.*

1. S. *Telephium* Linn. Orpine. E. B. 19. 1319.
Leaves flattish, serrated. Corymb leafy. Stem erect. *Smith.*
In the borders of fields, hedges, and bushy places, on a gravelly or chalky soil. — Perennial. *August.*

2. S. *Rhodiola* Dec. E. B. 8. 508.
Leaves oblong, serrated at the tip, smooth. Root fleshy. Stem simple. Flowers usually with 4 petals and divisions. *Dec.*
Rhodiola Rosea *Linn.*
Rhodiola odorata *Lam.*
In alpine rocks or cliffs, near the sea. — Perennial. *May, June.*

** *Leaves tumid, or somewhat cylindrical.*

3. S. *dasyphyllum* Linn. E. B. 10. 656.
Leaves ovate, obtuse, fleshy, sessile; the lower ones opposite. Stems flaccid. Panicles glutinous. *Smith.*
On walls and rocks. — Perennial. *June.*

4. S. *anglicum* Hudson. E. B. 3. 171.
Leaves ovate, thick, mostly alternate; spurred at the base. Cyme of two smooth branches. *Smith.*
Sedum annuum *Hudson.*
S. rubens *Lightf.*
On the sandy or rocky sea-coast, as well as on mountains. — Annual. *July.*

5. S. *acre* Linn. E. B. 12. 839.
Leaves alternate, nearly ovate, thick, tumid; spurred at the base. Cyme of three smooth branches, leafy. *Smith.*
On walls, roofs, and dry sandy ground, common. — Perennial. *June.*

6. S. *sexangulare* Linn. E. B. 28. 1946.
Leaves in six or seven rows, nearly cylindrical, obtuse, fleshy,

spreading; spurred at the base. Cyme of three smooth branches, leafy. *Smith.*

On dry sandy ground, and old walls, not common. — Perennial. *July.*

7. S. *villosum* Linn. E. B. 6. 394.

Leaves alternate, linear, flattened, slightly hairy as well as the flower-stalks. Stem erect. *Smith.*

In wet mountainous pastures, and the clefts of moist rocks, in the North. — Perennial. *June, July.*

8. S. *album* Linn. E. B. 22. 1578.

Leaves oblong, cylindrical, obtuse, spreading, smooth. Panicle much branched. *Smith.*

On rocks, walls, and roofs, not common. — Perennial. *July.*

9. S. *reflexum* Linn. E. B. 10. 695.

Leaves awl-shaped, scattered, spurred at the base; the lowermost recurved. Flowers cymose. Segments of the calyx ovate. *Smith.*

On walls and thatched roofs, abundantly. — Perennial. *July.*

10. S. *albescens* Haworth. E. B. 35. 2477.

Leaves glaucous, awl-shaped, scattered; spurred at the base; those of the branches thread-shaped. Flowers cymose. Segments of the calyx lanceolate. *Smith.*

S. glaucum *Smith,* not of others.

On barren sandy ground, or on walls. — Perennial. *July, August.*

11. S. *rupestre* Linn. E. B. 3. 170.

Leaves glaucous, spurred at the base; those of the branches awl-shaped, erect, in 5 close rows. Flowers imperfectly cymose. Segments of the calyx elliptical, obtuse. *Smith.*

On rocks, but rare. — Perennial. *July.*

12. S. *Forsterianum* Smith. E. B. 26. 1802.

Leaves spurred at the base, those of the branches semicylindrical, bluntish, pointed, spreading, in many rows. Flowers cymose. Segments of the calyx elliptical, obtuse. *Smith.*

t the fall of the Rhydoll, near the Devil's bridge, Cardiganshire, and on the rocks of Hisväe, overhanging the little valley of Nant-phrancon. *Smith.* — Perennial. *July.*

4. SEMPERVIVUM *Linn.*

Sepals from 6 to 20, slightly cohering at the base. *Petals* the same number, acuminate. *Stamens* twice as numerous as the petals. *Hypogynous scales* lacerated. *Fruit* of as many parts as there are petals. — *Herbaceous* perennial *plants, or* shrubs; *propagated by* offsets *arising from the axillæ of the leaves.* Leaves *thick, fleshy.* Flowers *in cymes, corymbs, or panicles, white, yellow, or purple.*

1. S. *tectorum* Linn. Common Houseleek. E. B. 19. 1320.

Leaves fringed. Offsets spreading. Edges of the petals hairy, entire. *Smith.*

On walls and cottage roofs, frequent. — Perennial. *July.*

Order 25. SAXIFRAGEÆ *Juss.*

Calyx either superior or inferior, of 4 or 5 sepals, which cohere more or less at their base.

Petals 5, or none, inserted between the lobes of the calyx.

Stamens 5-10, inserted either into the calyx (perigynous), or beneath the ovarium (hypogynous); *anthers* 2-celled, bursting longitudinally.

Disk either hypogynous or perigynous, sometimes nearly obsolete, sometimes annular and notched, rarely consisting of 5 scales.

Ovarium adhering to the calyx or distinct from it, usually consisting of 2 parts, cohering more or less by their face, but distinct at the apex; sometimes 2-celled with a central placenta; sometimes 1-celled with parietal placentæ; rarely 4- or 5-celled. *Styles* none. *Stigmata* sessile on the tips of the lobes of the ovarium.

Fruit generally a membranous 1- or 2-celled capsule with 2 bracteæ; rarely a 4-celled 4-valved capsule; sometimes a 4-celled berry.

Seeds numerous, very minute; usually with long hexagonal reticulations on the sides of a transparent testa. *Embryo* taper, in the axis of fleshy albumen, with the radicle next the hilum.

Herbaceous plants, often growing in patches. *Leaves* simple, either divided or entire, alternate, without stipulæ. *Flower stems* simple, often naked.

ANALYSIS OF THE GENERA.

Petals none
  Fruit a capsule - - - - - - - - 1. CHRYSOSPLENIUM.
  Fruit a berry - - - - - - - - 2. ADOXA.
Petals 5
  Stamens 5 - - - - - - - - - - 3. PARNASSIA.
  Stamens 10
    Calyx spreading or erect
      Petals with a nectariferous furrow at the base - - - - - - - - } 4. HIRCULUS.
      Petals not nectariferous
        Capsule superior - - - - 5. LEIOGYNE.
        Capsule half inferior - - - 6. SAXIFRAGA.
    Calyx reflexed - - - - - - - 7. ROBERTSONIA.

---

1. CHRYSOSPLENIUM *Linn.* GOLDEN SAXIFRAGE.

Calyx 4- or 5-parted, coloured inside. *Petals* none. *Stamens* 8 or 10, short, perigynous. *Disk* annular. *Styles* 2, spreading. *Capsule* inferior, of 1 cell, and 2 valves at the apex. — *Small succulent* herbs, *with reniform notched* leaves, *and green inconspicuous* flowers.

1. Ch. *alternifolium* Linn. E. B. 1. 54.
Leaves alternate.

On Poringland heath, Norfolk; and in many parts of the north of England, and lowlands of Scotland. — Perennial. *May.*

2. Ch. *oppositifolium* Linn. E. B. 7. 490.
Leaves opposite.
In watery shady places. — Perennial. *May.*

2. ADOXA *Linn.*

Sepals 4 or 5, united at the base. *Petals* none. *Stamens* 8 or 10, perigynous. *Disk* obsolete. *Styles* 4 or 5. *Berry* half inferior, of 1 cell, with 4 bordered seeds. — Leaves *compound.* Flowers *terminal, capitate green.*

1. A. *Moschatellina* Linn.
In groves, thickets, and under shady hedges. — Perennial. *April, May.*

3. PARNASSIA *Linn.*

Calyx 5-parted, spreading. *Petals* 5, ribbed, sometimes glandular. *Stamens* 5. *Disk* consisting of 5 fleshy scales, opposite the petals, and often fringed with glands. *Ovarium* superior, 1-celled, with 4 parietal placentæ. *Stigmas* 4, obtuse, opposite the placentæ! *Capsule* superior, or nearly so, of 1 cell and 4 valves. *Seeds* numerous, bordered. — Herbaceous *plants, natives of boggy places.* Roots *fibrous.* Leaves *radical, entire.* Stems *simple, with* 1 *leaf, and one white flower.*

1. P. *palustris* Linn. E. B. 2. 82.
Leaves heart-shaped. Bristles of each scale numerous.
On spongy bogs and commons. — Perennial. *September, October.*

4. HIRCULUS *Haworth.*

Calyx 5-leaved, erect. *Petals* equal, with a 2-valved nectariferous furrow at their base. *Stamens* 10, hypogynous. *Disk* obsolete. *Stigmas* subsessile, capitate. *Capsule* superior, with 2 beaks and 2 cells. — Stems *simple.* Leaves *entire.* Flowers *yellow.*

1. H. *ranunculoides* Haworth. E. B. 15. 1009.
Leaves lanceolate, blunt, naked. Sepals lanceolate, obtuse. Runners none.
Saxifraga Hirculus *Linn.*
n turfy bogs, very rare. — Perennial. *August.*

5. LEIOGYNE *Don.*

Calyx 5-parted, erect. *Petals* equal. *Stamens* 10, perigynous. *Disk* obsolete. *Capsule* superior, with 2 cells. *Seeds* roundish. — Herbaceous *plants, with simple* stems, *producing few branches.* Leaves *either reniform or linear, never with a cartilaginous border.* Flowers *white, rarely yellow.*

§ 1. *Capsule long.*

1. L. *aizoides.* E. B. 1. 39.
Stem decumbent at the base. Leaves alternate, linear, with fringe-like teeth.
Saxifraga aizoides *Linn.*
S. autumnalis *W.*
On Ingleborough hill, Yorkshire, and on most of the Westmoreland and Scottish mountains. — Perennial. *June—September.*

L. *granulata*. E. B. 7. 500.
Leaves kidney-shaped, lobed. Stem panicled, leafy. Root granulated.
Saxifraga granulata *Linn.*
In meadows and pastures. — Perennial. *May.*

3. L. *cernua.* E. B. 10. 664.
Leaves somewhat palmate, stalked. Stem with aggregate, axillary bulbs. Petals obovate.
Saxifraga cernua *Linn.*
About alpine rills, on the loftiest mountains of Scotland. — Perennial. *July.*

4. L. *rivularis.* E. B. 32. 2275.
Leaves palmate, stalked; the uppermost spatulate. Stem with few flowers. Root fibrous.
Saxifraga rivularis *Linn.*
About alpine rivulets, and in wet fissures of rocks, on the mountains of Scotland. — Annual. *June, July.*

§ 2. *Capsule depressed* (Micranthes *Don.*)

5. L. *nivalis.* E. B. 7. 440.
Leaves roundish-obovate, serrated; tapering and entire at the base. Cluster dense, capitate, of few flowers.
Saxifraga nivalis *Linn.*
On the loftiest mountains of Wales and Scotland, in the moist fissures of rocks. — Perennial. *July.*

6. SAXIFRAGA *Linn.*

Calyx 5-lobed, erect. *Petals* equal. *Stamens* 10, perigynous. *Disk* obsolete. *Capsule* half inferior, with 2 cells. — Stems *generally branching and forming tufts, sometimes simple.* Leaves *usually divided more or less.* Flowers *white or purple, seldom yellow.*

1. S. *oppositifolia* Linn. E. B. 1. 9.
Branches single-flowered, clothed with opposite, imbricated, fringed leaves. Petals ovate. *Smith.*
n alpine rocks and precipices. — Perennial. *April.*

2. S. *tridactylites* Linn. E. B. 7. 501.
Leaves wedge-shaped, with 3 or 5 segments; the uppermost undivided. Stem panicled, leafy. Stalks single-flowered, alternate. *Smith.*
On walls, roofs, and dry barren ground, common. — Annual. *April, May.*

3. S. *muscoides* Wulfen.
Leaves linear, obtuse, smooth, triple-ribbed, undivided, or with 2 small lateral lobes. Flowers few, corymbose. Petals nearly linear. Calyx almost naked. *Smith.*
S. cæspitosa *Hudson.*
S. moschata *Withering.*
On rocky mountains, very rare. — Perennial. *May.*

4. S. *pygmæa* Haworth. E. B. 33. 2314.
Leaves linear, abrupt, smooth, undivided. Flowers corymbose. Petals obovate, scarcely longer than the very obtuse glandular calyx. *Smith.*

S. muscoides *Hooker.*
S. moschata *E. Bot.*

In the Highlands of Scotland. — Perennial. *May.*

5. S. *cæspitosa* Linn. E. B. 12. 794.
Radical leaves crowded, 3- or 5-cleft, obtuse, veiny, fringed; lowermost undivided. Flowers from 1 to 5, or more. Fruit hairy. Calyx smoother, obtuse. Petals rounded, triple-ribbed.
S. gronlandica *Linn.*
β. *decipiens* B. 7. 455
S. petræa *Withering.*
S. palmata *Smith.*
S. decipiens *Ehr.*

On the loftiest mountains of Wales and Ireland. — Perennial. *May, June.*

6. S. *hirta* Donn. E. B. 32. 2291.
Radical leaves rather crowded, 5- or 3-cleft, pointed, veiny, fringed. Flowers few, corymbose. Calyx acute. Petals obovate, triple-ribbed.

On alpine rocks in Ireland, Scotland, and Wales. — Perennial. *June, July.*

7. S. *affinis* Don.
Radical leaves 5-cleft; those of the trailing shoots mostly 3-cleft; lobes linear, pointed. Segments of the calyx awl-shaped, channelled, pointed, recurved. Petals oblong, inflexed at the edges. *Smith.*
S. lævis *Donn.*

On the top of Brandon mountain, county of Kerry. — Perennial. *May, June.*

8. S. *platypetala* Smith. E. B. 32. 2276.
Radical leaves 5-cleft; those of the trailing shoots 3-cleft; lobes bristle-pointed. Segments of the calyx ovate, pointed, erect. Petals nearly orbicular, flat, with many lateral veins. *Smith.*

On the mountains of Scotland and Wales. — Perennial. *June.*

9. S. *incurvifolia* Don.
Radical leaves 5-cleft; those of the upright shoots 3-cleft; segments lanceolate, obtuse, incurved. Segments of the calyx ovate, acute. Petals roundish, slightly cloven. *Don.*

On alpine rocks in Ireland. — Perennial.

10. S. *denudata* Don.
Radical leaves 5-cleft; those of the upright shoots 3-cleft; segments linear-awl-shaped, acute, bristle-pointed, smooth. Segments of the calyx lanceolate, minutely pointed. Petals obovate, cloven. *Don.*

On the Grampian hills, in Angus-shire. — Perennial.

11. S. *hypnoides* Linn. E. B. 7. 454.
Radical leaves 3- or 5-cleft; those of the long, procumbent shoots undivided; all bristle-pointed and fringed. Segments of the calyx ovate, pointed. Petals obovate. Stigmas nearly smooth. *Smith.*

β. *condensata.*

Radical leaves 5-parted; those of the short procumbent shoots trifid; all linear, bristle-pointed, and smooth. Segments of the calyx triangular, ovate, acute, not bristle-pointed. Petals oval. *Don.*

S. condensata *Gmelin.*

γ. *elongella.* E. B. 32. 2277

Radical leaves 3- or 5-cleft; those of the upright short shoots undivided or three-cleft; all bristle-pointed, slightly fringed. Primary flower-stalks very long, simple and naked. Calyx pointed. Petals obovate. *Smith.*

S. elongella *Smith.*

On moist rocks. γ. On a rock, by the river near Lintrathen, in Angus-shire. — Perennial. *June.*

12. S. *leptophylla* Persoon.

Radical leaves deeply 5-cleft; those of the very long procumbent shoots deeply 3-cleft or undivided; segments linear-lanceolate, very sharp, widely spreading. Calyx oblong-ovate. Petals spatulate, undivided. *Don.*

On mountains in Wales. — Perennial. *May.*

13. S. *lætevirens* Don.

Leaves deeply 5- or 3-cleft, with linear acute segments. Shoots long and trailing. Calyx lanceolate, pointed. Petals spatulate, slightly cloven. *Don.*

In very elevated situations, upon moist rocks, on the mountains of Angus-shire and Aberdeenshire, and on hills to the north of Loch Lomond. — Perennial. *May.*

14. S. *pedatifida* Ehr. E. B. 32. 2278.

Radical leaves kidney-shaped, divided in a pedate manner into 7 lobes. Panicle cymose, level-topped, many-flowered. Calyx with linear-lanceolate segments, as long as the capsule. *Smith.*

S. quinquefida *Donn.*

In the Highlands of Scotland. — Perennial. *May.*

7. ROBERTSONIA *Haworth.*

(Gymnopera *Don.*)

Calyx 5-leaved, reflexed. *Petals* equal, or nearly so. *Stamens* 10, hypogynous. *Disk* obsolete. *Capsule* superior, with 2 cells. *Seeds* globose. — Stems *branching and forming dense tufts.* Leaves *broad, notched, often cartilaginous at the edge.* Flowers *white* or *pink, rarely pale yellow.*

Obs. I distinguish LIGULARIA of Haworth by the following character: — *Calyx* 5-leaved, spreading. *Petals* very unequal, the 2 lower long and hanging down. *Stamens* 10, perigynous. *Disk* cyathiform, notched. *Capsule* nearly superior.

1. R. *Geum.* E. B. 22. 1561.

Leaves roundish, kidney-shaped, notched, somewhat hairy. Footstalks linear, channelled, much longer than the leaves. Flower-stalk panicled. *Smith.*

Saxifraga Geum *Linn.*

β. *elegans.*

Leaves roundish, cordate, smooth on both sides. *Don.*

S. Geum β. *Don.*

γ. *guttata.*

Thrice the usual size. Leaves smooth on both sides. Panicle more spreading. Petals large, elegantly spotted. *Don.*

S. Geum γ. *Don.*

On the mountains of Ireland. — Perennial. *June.*

2. R. *hirsuta.* E. B. 33. 2322.

Leaves oval, with acute serratures, roundish or cordate at the base, hairy on each side, when full-grown erect. Peduncles very long, taper, villous. Pedicels long, 1-flowered. *Don.*

Saxifraga hirsuta *Linn.*

β. *depilata.*

Leaves roundish, cordate, smooth on each side. *Don.*

On the mountains of Ireland. — Perennial. *June.*

3. R. *umbrosa.* E. B. 10. 663.

Leaves obovate, retuse, quite smooth, with cartilaginous crenatures, when full-grown spreading. Petioles short, dilated. Pedicels few-flowered. *Don.*

Saxifraga umbrosa *Linn.*

β. *punctata.*

Leaves roundish, with 5 sharp serratures, when full-grown erect. Petioles long. *Don.*

S. umbrosa *punctata* Don.

γ. *serratifolia.*

Leaves oblong, ovate, with deep serratures, when full-grown erect. Petioles long. *Don.*

S. umbrosa *serratifolia* Don.

On the mountains of Ireland and Yorkshire. — Perennial. *June.*

4. R. *stellaris.* E. B. 3. 167.

Leaves elliptic, wedge-shaped, coarsely serrated, tapering and entire at the base. Panicle corymbose, of few flowers. *Smith.*

Saxifraga stellaris *Linn.*

Or. in moist places. — Perennial. *June, July.*

Order 26. Salicariæ *Juss.*

Calyx monopetalous; the lobes with a valvate or separate æstivation; their sinuses sometimes lengthened into other lobes.

Petals inserted between the lobes of the calyx, very deciduous.

Stamens inserted into the tube of the calyx below the petals, to which they are sometimes equal in number; sometimes they are twice, or even thrice, and four times as numerous; they are seldom four; *anthers* innate, 2-celled, opening longitudinally.

Ovarium superior, 2- or 4-celled; *style* filiform; *stigma* usually capitate.

Capsule membranous, covered by the calyx, 1-celled, dehiscing either longitudinally or in an irregular manner.

Seeds numerous, small, without *albumen*, adhering to a central placenta; *embryo* straight; *radicle* turned towards the hilum; *cotyledons* flat and leafy.

Herbs, rarely *shrubs*. *Branches* frequently 4-cornered. *Leaves* opposite, seldom alternate, entire, without either stipulæ or glands. *Flowers* axillary, or in spikes or racemes.

1. PEPLIS *Linn.*

Calyx campanulate, with 12 lobes, of which 6 are broader than the rest and erect, the others subulate, spreading. *Petals* 6, minute, fugacious. *Stamens* 6, opposite the broader lobes of the calyx. *Capsule* 2-celled, many-seeded. — Herbaceous, *with opposite or alternate* leaves. Flowers *axillary.*

1. P. *Portula* Linn. E. B. 17. 1211.
Petals wanting, or scarcely visible. Leaves opposite, obovate, stalked. *Smith.*

In watery places, on a gravelly, sandy, or heathy soil, frequent. — Annual. *July, August.*

2. LYTHRUM *Linn.*

Calyx cylindrical, striated, with 8 to 12 teeth, of which from 4 to 6 are broader than the rest and erect, the others smaller and spreading. *Petals* 4 or 6, inserted in the orifice of the calyx, opposite the smaller lobes of the calyx. *Stamens* situated in the middle or at the base of the calyx, twice as numerous as the petals, or occasionally fewer. *Capsule* oblong, 2-celled, many-seeded, included in the calyx. — *Erect* herbaceous *plants.* Leaves *opposite.* Stems *square.* Flowers *purple, axillary.*

1. L. *Salicaria* Linn. E. B. 15. 1061.
Leaves opposite, lanceolate; heart-shaped at the base. Flowers in whorled leafy spikes. Stamens twelve. *Smith.*

In ditches and watery places, especially about the margins of ponds and rivers, abundantly. — Perennial. *July, August.*

2. L. *hyssopifolium* Sibth. E. B. 5. 292.
Leaves alternate. linear-lanceolate. Flowers axillary, solitary. Stamens 6. *Smith.*

In partially dried pits or ditches; or places where water has stagnated during winter. — Annual. *August.*

Order 27. Rhamneæ. *Juss.*

Calyx monophyllous, 4-5 cleft, with a valvate æstivation.

Petals distinct, cucullate, or convolute, inserted into the orifice of the calyx, occasionally wanting.

Stamens opposite the petals.

Disk flesh

Ovarium superior, or half superior, 2, 3, or 4-celled; *ovula* solitary, erect.

Fruit fleshy, indehiscent, or dry, separating in 3 divisions.

Seeds erect; *albumen* fleshy, seldom wanting; *embryo* about as long as the seed, with large flat *cotyledons*, and a short inferior *radicle*.

Trees or *shrubs*, often spiny. *Leaves* simple, alternate, very seldom opposite, with minute stipulæ. *Flowers* axillary or terminal.

1. RHAMNUS *Linn.*

Calyx urceolate, 4-5 cleft. *Petals* 0, or emarginate. *Anthers* ovate, 2-celled. *Disk* thin, overspreading the tube of the calyx. *Ovarium* superior, 3- or 4-celled. *Styles* 3 or 4, distinct or united. *Fruit* fleshy, with 3 or 4, or in consequence of abortion 2, fibrous indehiscent stones. *Brongniart.*

1. R. *catharticus* Linn. Common Buckthorn. E. B. 23. 1629.
Thorns terminal. Flowers four-cleft, diœcious. Leaves ovate, serrated. Stem erect. Berry with four seeds. *Smith.*
In hedges, groves, and thickets. — Shrub. *May.* Fruit in *September.*

2. R. *Frangula* Linn. Berry-bearing Alder. E. B. 4. 250.
Thorns none. Flowers all perfect. Style simple. Leaves entire, smooth. Berry with two seeds. *Smith.*
In woods and thickets. — Shrub. *May.* Fruit in *July.*

Order 28. Ilicineæ *Brongniart.*

Sepals 4 to 6, imbricated in æstivation.

Petals cohering at the base, hypogynous, imbricated in æstivation.

Stamens alternate with the petals, inserted into the corolla; *filaments* erect; *anthers* adnate.

Disk none.

Ovarium fleshy, somewhat truncate, with from 2 to 6 cells; *ovula* solitary, pendulous, from a cup-shaped funiculus; *stigma* subsessile, lobed.

Fruit fleshy, indehiscent, with from 2 to 6 stones.

Seed suspended, nearly sessile; *albumen* large, fleshy; *embryo* small, 2-lobed, lying next the hilum, with minute *cotyledons*, and a superior *radicle*.

Trees or *shrubs*. *Leaves* alternate or opposite, coriaceous. *Flowers* small, axillary, solitary or fascicled.

1. ILEX *Linn.*

Calyx 4- or 5-toothed, persistent. *Petals* 4 or 5, either distinct, or cohering at the base. *Stamens* 4 or 5, alternate with the petals. *Ovarium* 4-celled. *Stigmas* nearly sessile, 4 or 5, either distinct or united in one. *Fruit* fleshy, containing 4 or 5 hard stones, each umbilicate at the apex, and containing 1 seed. *Seed* inverted; *albumen* fleshy; *embryo* in the apex. — *Evergreen* trees *or* shrubs. Leaves *coriaceous.* Flowers *sometimes polygamous.*

1. I. *Aquifolium* Linn. Common Holly. E. B. 7. 496.
Leaves ovate, acute, spinous, and wavy. Flowers axillary, somewhat cymose. *Smith.*

In hedges and bushy places, upon dry hills. — Tree. *May.*

Order 29. CELASTRINEÆ *R.Brown.*

Sepals 4 or 5, imbricated, inserted into the margin of an expanded torus.

Petals inserted by a broad base, under the margin of the disk, with an imbricate æstivation.

Stamens alternate with the petals, inserted into the disk, either at the margin or within it; *anthers* innate.

Disk large, expanded, flat, closely surrounding the ovarium, covering the flat expanded torus.

Ovarium superior, immersed in the disk and adhering to it, with 3 or 4 cells; *cells* 1- or many-seeded; *ovules* ascending from the axis, attached to a short funiculus.

Fruit superior; either a 3- or 4-celled capsule, with 3 or 4 septiferous valves; or a dry drupe with a 1- or 2-celled nut, the cells of which are 1 or many-seeded.

Seeds ascending, seldom inverted by resupination, either provided with an arillus or without one; *albumen* fleshy; *embryo* straight; *cotyledons* flat and thick, with a short inferior radicle.

Shrubs. Leaves simple, alternate or opposite. *Flowers* in axillary cymes.

1. EUONYMUS *Linn.*

Calyx 4-6 lobed, flat, with a peltate disk in the bottom. *Petals* 4-6, spreading, inserted in the disk. *Stamens* 4-6, inserted into glands projecting from the disk, alternate with the petals. *Style* one. *Capsule* 3- or 5-celled, with 3 or 5 angles; dehiscence loculicidal. *Seeds* from 1 to 4, with a fleshy arillus. *Embryo* green, straight, in the axis of a fleshy albumen. — Shrubs *with square branches.* Leaves *generally opposite.* Peduncles *axillary.*

1. E. *europæus* Linn. Common Spindle-tree, or Prickwood.
E. B. 6. 362.
Flowers mostly 4-cleft. Petals acute. Branches smooth and even. *Smith.*

In hedges and thickets. — Shrub, or small tree. *May.*

Order 30. STAPHYLEACEÆ.

(Celastrineæ Sect. 1. *Dec.*)

Sepals 5, connected at the base, coloured, with an imbricated æstivation.
Petals 5, alternate, with an imbricated æstivation.
Stamens 5, alternate with the petals, perigynous.
Disk large, urceolate.
Ovarium 2- or 3-celled, superior; *ovula* erect; *styles* 2 or 3, cohering at the base.
Fruit membranous or fleshy, indehiscent or opening internally, often deformed by the abortion of some of the parts.
Seeds ascending, roundish, with a bony testa; *hilum* large, truncate; *albumen* none; *cotyledons* thick.
Shrubs. Leaves opposite, pinnate, with both common and partial stipulæ. *Flowers* in terminal, stalked racemes.

1. STAPHYLEA *Linn.*

Calyx 5-parted, with an urceolate disk. *Petals* 5. *Ovarium* 2- or 3-lobed. *Styles* 2 or 3, sometimes combined. *Fruit* membranous, or 2 or 3 cells, dehiscing internally. *Seeds* bony, roundish, truncate at the hilum. — Flowers *large, white, in racemose panicles.*

1. S. *pinnata* Linn. Common Bladder-nut. E. B. 22. 1560.
Leaves pinnate. Styles and capsules but 2. *Smith.*
In hedges and thickets, rare. — Shrub. *June.*

Order 31. LEGUMINOSÆ *Juss.*

Calyx 5-parted, toothed, or cleft, inferior, with the odd segment anterior; the segments often unequal, and variously combined.
Petals 5, or by abortion 4, 3, 2, 1, or none, inserted into the base of the calyx, either papilionaceous or regularly spreading; the odd petal posterior.
Stamens definite or indefinite, perigynous, either distinct or monadelphous, or diadelphous; very seldom triadelphous; *anthers* versatile.
Ovarium simple, superior, 1-celled, 1 or many-seeded; *style* simple, proceeding from the upper margin; *stigma* simple.
Fruit either a legume or a drupa.
Seeds attached to the upper suture, solitary or several; occasionally with an arillus; *embryo* destitute of *albumen*, either straight or with the radicle bent upon the cotyledons; *cotyledons* either remaining under ground in germination, or elevated above the ground and becoming green like leaves.

Shrubs, trees, or *herbaceous* plants. *Leaves* compound, with stipulæ at the base of the petiole and of each leaflet. *Petiole* usually tumid at the base. *Flowers* axillary, either solitary, or in racemes or panicles.

ANALYSIS OF THE GENERA.

Pod continuous
- Cotyledons in germination rising above the ground, and becoming green leaves - - - - - - - - § *Loteæ.*
  - Stamens monadelphous
    - Calyx 2-lipped
      - upper lip 3-toothed, lower 2-toothed - - 1. ULEX.
      - upper lip 2-parted, lower 3-toothed - - - 2. GENISTA.
      - upper lip entire, lower 3-toothed, slightly - 3. CYTISUS.
    - Calyx equal
      - 5-toothed, inflated - - - - - - - 4. ANTHYLLIS.
      - 5-cleft - - - - - - - - - 5. ONONIS.
  - Stamens diadelphous
    - Pod 2-celled, or partially so
      - Keel blunt - - - - - - - - - 6. ASTRAGALUS.
      - Keel rostrate - - - - - - - - 7. OXYTROPIS.
    - Pod 1-celled
      - few-seeded, scarcely longer than the calyx
        - Calyx 5-toothed. Flowers racemose - - 8. MELILOTUS.
        - Calyx 5-cleft. Flowers capitate or spiked - 9. TRIFOLIUM.
      - many-seeded, much longer than the calyx
        - straight
          - Keel as long as the wings - - - 10. LOTUS.
          - Keel very short - - - - - 11. TRIGONELLA.
        - falcate or spiral - - - - - 12. MEDICAGO.
- Cotyledons in germination remaining under ground, and not changing to green leaves - - - - - - - § *Vicieæ.*
  - Pod 2- or 4-seeded - - - - - - - - - 13. ERVUM.
  - Pod many-seeded
    - Calyx with 5 foliaceous segments - - - 14. PISUM.
    - Calyx with 5 teeth
      - Style at right angles with the ovarium, villous on both sides - - - - - - - 15. VICIA.
      - Style curved, dilated at the end, villous in front 16. LATHYRUS.
      - Style curved, slender, villous at the end - 17. OROBUS.

Pod, divided internally by transverse articulations - - - § *Hedysareæ.*
- Pod with several joints
  - Joints with parallel margins - - - - - - 18. ORNITHOPUS.
  - Joints with 1 margin plane, and the other concave - 19. HIPPOCREPIS.
- Pod with 1 joint only - - - - - - - - - 20. ONOBRYCHIS.

---

§ *Loteæ* Dec.

Corolla papilionaceous. Stamens either monadelphous or diadelphous. Pod continuous, 1-celled, or occasionally 2-celled, in consequence of the bending inwards of one of the sutures. Cotyledons, in germination, rising above the ground, and acquiring a green colour.

1. ULEX *Linn.*

Calyx with 2 bracteæ, 2-lipped; the upper lip with 3, the lower with 2 teeth. *Stamens* monadelphous. *Pod* oval-oblong, turgid, scarcely longer than the calyx, few-seeded. — *Branching spiny* shrubs. Flowers *solitary, yellow.* Pods *villous.*

1. U. *europæus* Linn. Common Furze, Whin or Gorse. E. B. 11. 742.
Teeth of the calyx obsolete, converging. Bracteas ovate, lax. Branches erect. *Smith*.

On sandy or gravelly heaths and commons abundantly. — Shrub. *May;* and occasionally at all seasons.

2. U. *nanus* Forster. Dwarf Furze. E. B. 11. 743.
Teeth of the calyx lanceolate, spreading. Bracteas minute, close-pressed. Branches reclining. *Smith*.

On dry elevated heaths, less common than the preceding. — Shrub. *August—October*.

2. GENISTA *Linn*.

Calyx 2-lipped; the upper lip 2-parted, the lower 3-toothed. *Vexillum* oblong. *Keel* oblong, straight, not entirely restraining the stamens. *Stamens* monadelphous. *Pod* flat, compressed, or rather turgid, many-seeded, seldom few-seeded, not glandular. — Shrubs *with yellow* flowers. *Dec*.

1. G. *tinctoria* Linn. E. B. 1. 44.
Leaves lanceolate, smooth. Branches round, striated, erect, without thorns. *Smith*.

In pastures, thickets, and the dry borders of fields. — Shrub. *July, August*.

2. G. *pilosa* Linn. E. B. 3. 208.
Leaves obovate-lanceolate, obtuse; hairy beneath. Stem tuberculated, prostrate, without thorns. Calyx and flower-stalks silky. *Smith*.

On dry elevated sandy tracts and rocks. — Shrub. *May*, and again in *September*.

3. G. *anglica*. Linn. Petty Whin. E. B. 2. 132.
Thorns nearly simple. Flowering branches unarmed. Leaves ovate-lanceolate. *Smith*.

On moist boggy heaths, frequent. — Shrub. *May, June*.

3. CYTISUS *Linn*.

Calyx 2-lipped; the upper lip generally entire, the lower slightly 3-toothed. *Vexillum* ovate, large. *Keel* very blunt, enclosing the stamens. *Stamens* monadelphous. *Pod* compressed, flat, many-seeded, without glands. — Shrubs *with yellow* flowers, *and ternate* leaves. *Dec*.

1. C. *scoparius* Link. Common Broom. E. B. 19. 1339.
Leaves ternate, or solitary. Branches angular, without thorns. Legume fringed.
Spartium scoparium *Linn*.

In dry gravelly thickets and fields, abundantly. — Shrub. *May, June*.

4. ANTHYLLIS *Linn*.

Calyx tubular, 5-toothed, persistent, more or less inflated. *Wings, keel*, and *vexillum* nearly equal. *Stamens* monadelphous. *Pod* ovate, 1- or 2-seeded, very seldom many-seeded, always covered by the calyx. — *Herbaceous* plants *or* under hrubs, *of various habits*.

1. A. *vulneraria* Linn. Common Kidney-vetch, or Ladies' finger. E. B. 2. 104.

Herbaceous. Leaves pinnate, unequal. Heads of flowers in pairs. *Smith.*

β. A. *Dillenii* Schultes.

Flowers red or scarlet. Leaves and stems slightly villous or nearly smooth. *Dec.*

A. vulneraria rubriflora *Dec.*

In chalky or limestone countries, where the soil is dry and rather barren. β. In Pembrokeshire. — Perennial. *June—August.*

5. ONONIS *Linn.* Rest-harrow.

Calyx campanulate, 5-cleft, with linear segments. *Vexillum* large, streaked. *Stamens* monadelphous. *Pod* turgid, sessile, few-seeded. — Herbaceous *plants or* under-shrubs. Leaves *ternate, occasionally simple.* Flowers *axillary, yellow or purple.* Pedicels *often bearing a bristle indicating an abortive floral leaf.*

1. O. *procurrens* Wallroth. E. B. 10. 682.

Stem rooting, procumbent, spreading, spinous or unarmed. Flowering branches ascending, downy all over. Leaves simple or ternate, roundish-ovate, glandular on each side, slightly serrated. Lobes of the calyx longer than the pods.

O. arvensis *Linn.*, *Smith*, &c.

O. repens *Linn.*

In fields, common in many places. — Half-shrubby. *June—August.*

2. O. *spinosa* Linn.

Stem nearly erect, spinous, with 1 or 2 separate rows of hairs. Leaves ternate, oblong, wedge-shaped and entire towards the base. Flowers solitary. Lobes of the calyx shorter than the pods.

O. arvensis β. *Smith.*

In fields and neglected pastures. — Half-shrubby. *June—August.*

6. ASTRAGALUS *Linn.*

Calyx 5-toothed. *Corolla* with an obtuse keel. *Stamens* diadelphous. *Pod* 2-celled, or half 2-celled, the lower suture being turned inwards. — Herbaceous *plants or small* shrubs, *with pinnate* leaves. *Dec.*

1. A. *glycyphyllos* Linn. E. B. 3. 203.

Stem prostrate. Legumes obscurely triangular, incurved. Leaves longer than the flower-stalks; leaflets oval. *Smith.*

In woods, thickets, the borders of fields, or at the sides of hills, on a chalky or gravelly soil. — Perennial. *June.*

2. A. *hypoglottis* Linn. E. B. 4. 274.

Stem prostrate. Flowers in round heads. Legumes ovate, deeply channelled along the back, compressed, hairy; hooked at the point. Leaflets blunt. *Smith.*

A. arenarius *Huds.*

A. danicus *Retz.*

A. epiglottis *Dicks.*

On open mountainous heaths, in a chalky or sandy soil; also on the sea-coast. — Perennial. *June, July.*

7. OXYTROPIS *Dec.*

Calyx 5-toothed. *Keel* ending in an exserted point. *Stamens* diadelphous. *Pod* 2-celled, or half 2-celled, the upper suture being turned inwards. — Herbaceous *plants, with pinnated* leaves. Peduncles *axillary or radical.* Flowers *in spikes.* Dec.

1. O. *uralensis* Dec. E. B. 7. 466.
Stem none. Stalk upright, taller than the leaves. Legumes oblong, tumid, pointed, shaggy, erect. Leaflets ovate, acute, all over silky like the calyx. *Smith.*
A. uralensis *Linn.*
Phaca uralensis *Wahl.*
On the Scottish mountains, in a sandy soil. — Perennial. *July.*

2. O. *campestris* Dec. E. B. 36. 2522.
Stem none. Stalk ascending. Legumes ovate, inflated, hairy, erect. Leaflets lanceolate, acute, somewhat hairy. *Smith.*
A. campestris *Linn.*
A. sordidus *Willd.*
Phaca campestris *Wahl.*
Upon a high rock, on one of the mountains at the head of Clova, Angusshire, near the White Water, in great abundance. *Mr. G. Don.* — Perennial. *July.*

8. MELILOTUS *Tournef.* Melilot.

Calyx tubular, 5-toothed. *Keel* simple; wings shorter than the vexillum. *Pod* longer than the calyx, coriaceous, 1 or few-seeded, indehiscent, of various form. — Herbaceous *plants.* Stipules *adnate to the petiole.* Leaves 3-*leaved;* leaflets *often toothed.* Flowers *in loose racemes, either yellow or white.* Dec.

1. M *officinalis* Willd. E. B. 19. 1340.
Clusters unilateral. Legume prominent, acute, transversely wrinkled, hairy, with two seeds. Stem erect. Stipulas awl-shaped. *Smith.*
Trifolium officinale *Smith.*
In thickets, hedges, and the borders of fields; sometimes among corn. — Annual. *June, July.*

9. TRIFOLIUM *Linn.*

Calyx tubular, persistent, 5-cleft, not glandular; with subulate segments. *Keel* shorter than both wings and vexillum. *Stamens* diadelphous. *Pod* small, indehiscent, often ovate with 1 or 2 seeds, shorter than the calyx by which it is covered, seldom oblong, with 3 or 4 seeds, and a little longer than the calyx. — Herbaceous *plants.* Stipules *adhering to the petiole.* Leaves 3- *or 5-leaved.* Flowers *in heads or dense spikes, bracteate, purple, white, or pale yellow.* Petals *in some cohering.* Dec.

* *Seeds several.*

1. T. *repens* Linn. Dutch Clover. E. B. 25. 1769.
Heads globose. Flowers somewhat stalked. Legume within the calyx, 4-seeded. Stems creeping, solid. *Smith.*
In meadows and pastures, very common. — Perennial. *May—September.*

2. T. *suffocatum* Linn. E. B. 15. 1049.
Heads sessile, lateral, roundish. Legume concealed, 2-seeded.

Calyx nearly smooth, with lanceolate, acute, recurved teeth, longer than the corolla. *Smith.*

On the sandy sea-coast. — Annual. *June, July.*

* *Seeds single. Calyx generally hairy.*

3. T. *subterraneum* Linn. E. B. 15. 1048.

Heads hairy, of about 4 flowers. Involucrum central, reflexed, rigid, starry, embracing the fruit. *Smith.*

In dry gravelly pastures, and barren heathy situations. — Annual. *May.*

4. T. *ochroleucum* Linn. E. B. 17. 1224.

Flowers in a solitary, terminal, hairy head. Stem erect, downy. Lower leaflets inversely heart-shaped. Lowest calyx-tooth thrice as long as the rest. *Smith.*

T. squarrosum *Linn.*

In pastures, fields, and thickets, on a dry gravelly, or chalky, soil. — Perennial. *June, July.*

5. T. *pratense* Linn. Common Purple Clover. E. B. 25. 1770.

Spikes dense. Stems ascending. Petals unequal. Calyx hairy; 4 of its teeth equal. Stipulas ovate, bristle-pointed. *Smith.*

In meadows and pastures, especially on limestone or gravelly hills. — Perennial. *May—September.*

6. T. *medium* Linn. E. B. 3. 190.

Spikes lax. Stems zigzag and branching. Petals nearly equal. Stipulas tapering, converging. Two upper calyx-teeth rather the shortest. *Smith.*

T. flexuosum *Jacq.*

T. alpestre *Huds.*

In elevated dry chalky pastures, or in gravelly ones with a clay bottom. — Perennial. *July.*

7. T. *maritimum* Hudson. E. B. 4. 220.

Spikes ovate, somewhat hairy. Stipulas lanceolate, erect. Calyx-teeth after flowering dilated, leafy, and spreading. Leaflets obovate-oblong. *Smith.*

T. stellatum *Hudson.*

In muddy salt-marshes, on the east and south coasts of England, from Norfolk to Somersetshire, in various places. — Annual. *June, July.*

8. T. *stellatum* Linn. E. B. 22. 1545.

Spikes hairy, roundish. Stipulas elliptical. Calyx-teeth longer than the corolla; after flowering dilated, leafy, reticulated and spreading; tube closed. Leaflets inversely heart-shaped. *Smith.*

Between Shoreham harbour, Sussex, and the sea, in great plenty. — Annual. *July, August.*

9. T. *arvense* Linn. E. B. 14. 944.

Spikes cylindrical, very hairy. Stipulas lanceolate, bristle-pointed. Calyx-teeth longer than the corolla, permanently bristle-shaped. Leaflets linear-obovate. *Smith.*

In sandy barren fields, very common. — Annual. *July, August.*

10. T. *scabrum* Linn. E. B. 13. 903.

Heads sessile, axillary, ovate. Calyx-teeth unequal, lanceolate, rigid; finally recurved. Stems procumbent. *Smith.*

In chalky, or dry sandy, fields. — Annual. *May, June.*

11. T. *glomeratum* Linn. E. B. 15. 1063.
Heads sessile, axillary, hemispherical, smooth. Calyx-teeth heart-shaped, reflexed, veiny. Stems prostrate. *Smith.*

In gravelly fields and pastures, chiefly in the east and south of England. — Annual. *June.*

12. T. *striatum* Linn. E. B. 26. 1843.
Heads sessile, axillary and terminal, ovate. Calyx elliptical, furrowed, hairy; with straight bristle-shaped teeth. Stems procumbent. *Smith.*

In dry, barren, sandy fields and pastures. — Annual. *June.*

* *Calyx of the fruit inflated, bladdery.*

13. T. *fragiferum* Linn. E. B. 15. 1050.
Heads roundish. Calyx finally inflated, deflexed, with 2 terminal teeth. Stems creeping. *Smith.*

In moist meadows, pastures, and osier holts, in a black boggy soil. — Perennial. *July, August.*

*** *Standards deflexed, dry and membranous.*

14. T. *procumbens* Linn. E. B. 14. 945.
Heads oval, many-flowered. Standard finally deflexed, furrowed. Stems spreading or procumbent. Common footstalk longest at the base. *Smith.*
T. agrarium *Hudson.*

n dry gravelly fields and pastures. — Annual. *June, July.*

15. T. *minus* Relhan. E. B. 18. 1256.
Heads hemispherical. Flower-stalks straight, rigid. Standard nearly even. Stems prostrate. Common footstalk very short. *Smith.*
T. procumbens *Hudson.*
T. dubium *Sibth.*

In dry gravelly fields and pastures, with the last. — Annual. *June, July.*

16. T. *filiforme* Linn. E. B. 18. 1257.
Clusters lax, of few flowers. Common-stalks capillary, wavy Standard even. Stems prostrate. Leaflets all nearly sessile. *Smith.*

In sandy or gravelly grassy pastures, whether dry or moist. — Annual. *June, July.*

10. LOTUS *Linn.* Bird's-foot-trefoil.

Calyx tubular, 5-cleft; wings about as long as the vexillum; keel beaked. *Pod* cylindrical or compressed, apterous; *style* straight, subulate. — Herbaceous *plants.* Leaves *ternate.* Stipules *leafy.* Peduncles *axillary, from* 1- *to* 6-*flowered, supported by a floral leaf.* Flowers *yellow, rarely white or pink.*

1. L. *corniculatus* Linn. E. B. 30. 2090.
Heads depressed, of few flowers. Stems recumbent, pithy. Legumes spreading, nearly cylindrical. Claw of the standard obovate. Filaments all dilated. *Smith.*

In open grassy pastures, common. — Perennial. *June—September*

2. L. *major* Scopoli. E. B. 30. 2091.
Heads depressed, many-flowered. Stems erect, tubular. Legumes drooping, cylindrical. Claw of the standard linear. Shorter filaments not dilated. *Smith.*

In wet bushy places, osier-holts, and hedges. — Perennial. *July, August.*

3. L. *decumbens* Forster.
Heads of few flowers. Stems recumbent, nearly solid. Legumes somewhat spreading, cylindrical, 2-edged. Calyx hairy; its teeth shorter than the tube. *Smith.*

At Hastings, Sussex, near Bulverhithe; also in meadows near Tonbridge. In fields near Forfar, North Britain. — Perennial. *July.*

4. L. *angustissimus* Linn. E. B. 13. 925.
Flowers solitary, or in pairs. Stems much branched, prostrate, fistular. Legumes 2-edged, very slender, somewhat compressed. Calyx loosely hairy; teeth fringed, twice the length of the tube. *Smith.*
L. diffusus *Smith.*

In meadows towards the sea, on the south and western coasts of Eng.and. — Annual. *May, June.*

11. TRIGONELLA *Linn.*

Calyx campanulate, 5-cleft. *Keel* very small; the wings and vexillum together imitating a tripetalous corolla. *Pod* oblong, compressed, or cylindrical, acuminate, nearly erect, many-seeded. — *Strong-scented* herbaceous *plants.* Leaflets 3, *the terminal one stalked.* Dec.

1. T. *ornithopodioides* Dec. E. B. 15. 1047.
Stems prostrate. Leaflets obcordate, denticulate. Stipules lanceolate, entire, rather membranous, very acute. Flowers from 2 to 4 in number. Legumes rather falcate, compressed, twice as long as the calyx. *Dec.*
Trifolium ornithopodioides *Linn.*

In barren gravelly places. — Annual. *June, July.*

12. MEDICAGO *Linn.*

Calyx somewhat cylindrical, 5-cleft. *Keel* rather distant from the vexillum. *Stamens* diadelphous. *Pod* many-seeded, variable in form, always falcate or spirally twisted. — Herbaceous *plants or* shrubs. Stipulæ *usually cut.* Leaves *stalked, trifoliate; leaflets toothed.* Peduncles *axillary, with* 1, 2, *or many flowers.* Flowers *yellow or purple.*

1. M. *sativa* Linn. Lucerne. E. B. 25. 1749.
Clusters upright. Legumes spiral. Stem erect, smooth. *Smith.*

In hedges, pastures, and the borders of fields, in dry calcareous soils, but scarcely a native. — Perennial. *June, July.*

2. M. *falcata* Linn. E. B. 15. 1016.
Clusters upright. Legumes sickle-shaped. Stem procumbent. *Smith.*

On dry gravelly banks and old walls. — Perennial. *June, July.*

3. M. *lupulina* Linn. Black Nonesuch. E. B. 14. 971.
Spikes ovate, erect. Legumes kidney-shaped, rugged and veiny, single-seeded. Stem procumbent. *Smith.*

In meadows, pastures, and cultivated fields, very common. — Annual. *May—August.*

4. M. *maculata* Sibthorp. E. B. 23. 1616.
Stalks 2- or 3-flowered. Leaflets inversely heart-shaped, spotted. Stipulas dilated, sharply toothed. Legumes spiral, depressed, fringed with long spreading bristles. *Smith.*

M. polymorpha *Linn.*
M. arabica *Withering.*
M. hispida *Gærtn.*

On a gravelly soil in the southern parts of England. — Annual. *May, June.*

5. M. *muricata* Willd.
Stalks barely 3-flowered. Stipulas deeply toothed, hairy as well as the obovate, somewhat rhomboid, leaflets. Legumes even, with short, depressed, radiating teeth, in a single row. *Smith.*

On the sea-coast. — Annual. *June, July.*

6. M. *minima* Willd.
Stalks many-flowered. Stipulas half-ovate, nearly entire. Leaflets obovate, hairy. Legumes orbicular, with a double row of hooked spines..

At Narburgh, Norfolk, and near Newmarket. — Annual. *June, July.*

§ *Vicieæ.* Dec.

Corolla papilionaceous. Stamens always diadelphous. Pod continuous, 1-celled. Cotyledons thick, farinaceous, in germination remaining under ground, and never acquiring a green colour.

13. ERVUM *Linn.* TARE.

Calyx 5-cleft, with linear acute segments as long as the corolla. *Style* smooth. *Pod* oblong, 2 or 4-seeded. *Dec.*

1. E. *tetraspermum* Linn. E. B. 17. 1223.
Flowers mostly in pairs. Legume smooth, with 4 seeds. Leaflets oblong, bluntish. *Smith.*

In corn-fields, hedges, and thickets, particularly such as are rather moist. — Annual. *June, July.*

2. E. *hirsutum* Linn. E. B. 14. 970.
Clusters many-flowered. Legumes hairy, with 2 seeds. Leaflets abrupt. *Smith.*

In corn-fields, and other cultivated ground, as well as in hedges: a very troublesome weed, especially in wet seasons. — Annual. *June—August.*

14. PISUM *Linn.* PEA.

Calyx with foliaceous segments, the 2 upper shortest. *Vexillum* large, reflexed. *Style* compressed, keeled, villous on the upper side. *Pod* oblong, compressed, not winged, many-seeded. *Seeds* roundish, with a roundish hilum. — Annuals. Leaves *abruptly pinnate, of 3 pair, with a tendril in place of a terminal leaflet.* Stipules *large.* Dec.

1. P. *maritimum* Linn. E. B. 15. 1046.
Footstalks flattish on the upper side. Stem angular. Stipulas arrow-shaped. Stalks many-flowered. *Smith.*

On the stony sea-shore, in several parts of the east and south of England. — Perennial. *July.*

15. VICIA *Linn.*

Calyx tubular, 5-cleft or 5-toothed, the 2 upper teeth shorter than the others. *Stamens* diadelphous. *Style* filiform, at nearly right angles with the ovarium, villous on the upper side, and below the apex on the under. *Pod* oblong, 1-celled, many-seeded. *Seeds* with an oval or linear lateral hilum. — *Climbing* herbaceous *plants.* Leaves *abruptly pinnate, with a* tendril *in place of an odd leaflet.* Stipulæ *generally sagittate.* Peduncles *axillary, either long and many-flowered, or short and 1-flowered.* Dec.

* *Stalks elongated, many-flowered.*

1. V. *sylvatica* Linn. E. B. 2. 79.
Stalks many-flowered. Leaflets elliptical. Stipulas crescent-shaped, deeply toothed. *Smith.*

In woods and hedges, chiefly in the more mountainous parts of Britain. — Perennial. *July, August.*

2. V. *Cracca* Linn. E. B. 17. 1168.
Stalks many-flowered. Flowers imbricated. Leaflets lanceolate, downy. Stipulas half-arrow-shaped, mostly entire. *Smith.*

In hedges, thickets, osier-grounds, and bushy low meadows, common. — Perennial. *July, August.*

** *Flowers axillary, nearly sessile.*

3. V. *sativa* Linn. Common Vetch. E. B. 5. 334.
Flowers nearly sessile, mostly in pairs. Leaflets elliptic-oblong; lower ones abrupt. Stipulas with a blackish depression beneath. Seeds orbicular, smooth. *Smith.*
β. V. *angustifolia* Willd.
Leaflets lanceolate and more acute. Flowers paler.

In corn-fields, and other cultivated ground. β. Among grass or bushes, on more barren or sandy ground. — Annual. *May, June.*

4. V. *angustifolia* Sibthorp.
Flowers solitary, nearly sessile. Leaflets linear; lower ones inversely heart-shaped. Stipulas with a pale depression beneath. Seeds orbicular, smooth. *Smith.*
V. lathyroides *Hudson.*

In grassy pastures, on a chalky or gravelly soil. — Annual? *June.*

5. V. *lathyroides* Linn. E. B. 1. 30.
Flowers solitary, nearly sessile. Leaflets elliptic-oblong; lower ones inversely heart-shaped. Tendrils simple, shorter than the leaflets. Seeds cubic, warty. *Smith.*
Ervum soloniense *L.*

In fallow fields on a gravelly soil, in chalky pastures, or on dry banks. — Annual. *April, May.*

6. V. *lutea* Linn. E. B. 7. 481.
Flowers solitary, very nearly sessile. Standard smooth. Legumes reflexed, hairy. Stems diffuse. Stipulas ovate, pointed, coloured. *Smith.*

On stony ground, chiefly near the sea. — Perennial. *August.*

7. V. *hybrida* Linn. E. B. 7. 482.
Flowers solitary, almost sessile. Standard hairy. Legumes reflexed, hairy. Stems ascending. Leaflets abrupt. Stipulas ovate, unstained. *Smith.*

On Glastonbury Tor-hill. — Perennial. *June, July.*

8. V. *lævigata* Smith. E. B. 7. 483.
Flowers solitary, nearly sessile. Legumes reflexed, smooth. Stems ascending. Stipulas cloven, unstained. Leaflets bluntish, very smooth. *Smith.*
V. hybrida *Hudson.*

At Weymouth, Dorsetshire. — Perennial. *July, August.*

9. V. *sepium* Linn. E. B. 22. 1515.
Flowers about 4 together, in short axillary clusters. Legumes upright, smooth. Leaflets ovate, obtuse; the upper ones gradually smaller. *Smith.*

In thickets and under hedges, common. — Perennial. *May, June.*

10. V. *bithynica* Linn. E. B. 26. 1842.
Flowers stalked, mostly solitary. Legumes upright, rough. Leaflets 2 pair, lanceolate. Stipulas with lateral teeth. *Smith.*

In bushy places, on a gravelly soil, often near the sea. — Perennial. *July, August.*

16. LATHYRUS *Linn.*

Calyx campanulate, 5-cleft, the 2 upper lobes the shortest. *Stamens* diadelphous. *Style* flat, dilated at the end, villous or pubescent in front. *Pod* oblong, many-seeded, 2-valved, 1-celled. *Seeds* round, or angular. — *Climbing* herbaceous *plants.* Stipules *half sagittate.* Leaves *abruptly pinnate, of from* 1 *to* 3 *pairs, with a tendril in place of the terminal leaflet.* Peduncles *axillary.* Dec.

* Perennials. Peduncles many-flowered.

† *Leaves conjugate.*

1. L. *sylvestris* Linn. E. B. 12. 805.
Quite smooth. Stems prostrate, winged. Leaflets linear-lanceolate, acuminate, coriaceous. Stipules half-sagittate, shorter than the leaf. Peduncles 3 to 6-flowered, the length of the leaf. Pods reticulated longitudinally. Seeds roundish, scabrous.

In groves, thickets, and moist hedges. — Perennial. *July, August.*

2. L. *latifolius* Linn. E. B. 16. 1108.
Quite smooth. Stems prostrate, winged. Leaflets elliptical, somewhat glaucous, obtuse and mucronate, with 3 or 5 prominent veins. Stipules broad, half-sagittate. Peduncles many-flowered, longer than the leaves. Pods reticulated longitudinally.

In woods; rare. — Perennial. *July, August.*

3. L. *pratensis* Linn. E. B. 10. 670.
Nearly smooth. Stems rather erect, 4-cornered. Leaflets oblong or linear-lanceolate, with 3 prominent veins. Stipules sagittate, ovate, shorter than the leaflets. Peduncles many-flowered, twice as long as the leaves. Calyx ribbed, with nearly equal teeth, the length of the tube. Pods obliquely reticulated. Seeds round, polished.

In meadows, pastures, and thickets, common. — Perennial. *July, August.*

†† *Leaves of several pair.*

4. L. *palustris* Linn. E. B. 3. 169.
Quite smooth. Stems winged, rather erect. Leaves of 3 pairs; petiole subulate; leaflets oblong, mucronulate. Stipules half-sagittate, acute, small. Peduncles with 3 to 5 flowers, scarcely longer than the leaves. Segments of the calyx unequal, somewhat linear, the length of the tube.

In meadows and thickets. — Perennial. *July, August.*

** Annuals. Peduncles from 1 to 3-flowered.

† *Leafless.*

5. L. *Aphaca* Linn. E. B. 17. 1167.
Stem erect. Petioles cylindrical, filiform, usually leafless, occasionally with 2 leaflets. Stipules very large, sagittate. Peduncles 1-flowered, with a minute bractea at the apex. Segments of the calyx twice as long as the tube. Pods few-seeded, broad. Seeds compressed.

In the borders of sandy or gravelly fields, rare. — Annual. *June—August.*

6. L. *Nissolia* Linn. E. B. 2. 112.
Stem erect. Petioles dilated, with from 3 to 5 prominent veins. Stipules minute, subulate. Flowers solitary, on long stalks. Peduncles without a bractea at the apex. Pods narrow, with projecting veins, reflexed.

In bushy places, and the grassy borders of fields. — Annual. *May.*

†† *Leaves conjugate.*

7. L. *hirsutus* Linn. E. B. 18. 1255.
Stems diffuse, winged. Leaflets linear, oblong. Stipules linear, half-sagittate, about as long as the petiole. Peduncles with from 1 to 3 flowers, scarcely longer than the leaves. Segments of the calyx ovate, the length of the tube. Pods oblong, hairy. Seeds round, scabrous.

In fields, rare. — Annual. *July.*

17. OROBUS *Linn.*

Calyx campanulate, 5-cleft, the 2 upper lobes the shortest. *Stamens* diadelphous. *Style* slender, linear, villous at the end. *Pod* cylindrical, oblong, 1-celled, 2-valved, many-seeded. *Seeds* with a linear hilum. — *Erect* herbaceous *plants.* Stipules *half-sagittate.* Leaves *abruptly pinnate, with a short simple seta in place of the terminal leaflet.* Racemes *axillary, stalked.* Dec.

1. O. *sylvaticus* Linn. E. B. 8. 518.
Stem branching, decumbent, hairy. Leaves hairy, of many pairs: leaflets ovate-lanceolate, acuminate. Stipules half-sagittate. Peduncles many-flowered, scarcely so long as the leaves. Teeth of the calyx unequal, short. Pods ovate, stalked.

In mountainous woods and thickets. — Perennial. *May, June.*

2. O. *tuberosus* Linn. E. B. 17. 1153.
Quite smooth. Roots tuberous. Stems prostrate. Leaves of 2 or 4 pairs: leaflets elliptical, mucronate, dotted, with nearly parallel veins. Stipules half-sagittate. Peduncles bearing few flowers, and scarcely longer than the leaves. Teeth of the calyx unequal, ovate, obtuse, shorter than the tube. Pods compressed, with obsoletely-reticulated veins. Seeds spherical.

β. O. *tenuifolius* Roth.
Leaflets linear-lanceolate.

In mountainous pastures, thickets, and woods. — Perennial. *May, June.*

3. O. *niger* Linn.
Leaves pinnate, of from 4 to 6 pair of elliptic-lanceolate leaflets. Stipulas linear-awl-shaped, simple, entire. Stem branched, angular, erect. *Smith.*

In the den of Airly, 12 miles west of Forfar, sparingly. — Perennial. *June, July.*

§ *Hedysareæ* Dec.

Corolla papilionaceous. Stamens usually monadelphous; sometimes diadelphous, 1 and 9, or 5 and 5. Legume divided transversely into 1-seeded joints. Cotyledons thin, in germination rising above the ground, and acquiring a green colour.

18. ORNITHOPUS *Linn.*

Calyx with bracteæ, tubular, nearly equally 5-toothed. *Keel* very small and compressed. *Stamens* diadelphous. *Pod* compressed, consisting of numerous 1-seeded, indehiscent joints, truncate equally on each side, with parallel margins. — *Hairy* annuals. Leaves *pinnate.* Stipules *small, adhering to the petiole.* Peduncles *axillary, few-flowered.* Flowers *small, white or rose-colour. A leafy pinnated* bractea *under each head.* Dec.

1. O. *perpusillus* Linn. Bird's-foot. E. B. 6. 369.
Leaves pinnate. Flowers capitate, accompanied by a leaf. Legumes incurved, beaded. *Smith.*

In sandy or gravelly pastures. — Annual. *May*

19. HIPPOCREPIS *Linn.*

Calyx 5-cleft, with equal, acute lobes. *Keel* 2-lobed. *Stamens* diadelphous. *Style* filiform, acute. *Pod* of several joints, each containing 1 seed, and curved like a horse-shoe; whence the upper edge of the pod appears as if cut into several rounded recesses. *Seeds* cylindrical or compressed, oblong, attached to the middle part of each curvature. — Herbaceous *plants or* shrubs. Leaves *pinnated.* Flowers *yellow, either axillary, solitary, and sessile, or in umbels at the end of an axillary peduncle.* Dec.

1. H. *comosa* Linn. E. B. 1. 31.
Legumes umbellate, rough; their joints neither dilated nor bordered. *Smith.*

On dry chalky banks; sometimes on limestone. — Perennial. *May—August.*

20. ONOBRYCHIS *Tournef.*

Calyx 5-cleft, with subulate equal divisions. *Keel* truncate obliquely; *wings* short. *Stamens* diadelphous. *Pod* sessile, of 1 joint, compressed, indehiscent, coriaceous, prickly, crested or winged; the upper side thick and straight; the lower convex and thinner. — Herbaceous *plants*. Leaves *pinnated*. Peduncles *long, bearing spikes of flowers at the end.* Flowers *red or white.*

1. O. *sativa* Lam. Saintfoin. E. B. 2. 96.
Leaves pinnate, nearly smooth. Legumes single-seeded, toothed at the margin and ribs. Wings of the corolla not longer than the calyx. Stem elongated. *Smith.*
Hedysarum Onobrychis *Linn.*

On dry chalky hills and open downs, in various parts of England. — Perennial. *June, July.*

Order 32. ROSACEÆ *Juss.*

Calyx 4- or 5-lobed, with a disk either lining the tube or surrounding the orifice; the fifth lobe uppermost.

Petals 5, perigynous, equal.

Stamens either definite or indefinite, inserted on the calyx, just within the petals, in æstivation curved inwards; *anthers* innate, 2-celled, bursting longitudinally.

Ovaries superior, either solitary or several, 1-celled; *ovula* 2 or more, suspended, very rarely erect; *styles* lateral; *stigmata* usually simple, and emarginate on one side.

Fruit either 1-seeded nuts or small drupes, or follicles containing several seeds.

Seeds suspended, rarely ascending. *Embryo* straight, with a taper short radicle pointing to the hilum, and flat cotyledons. *Albumen* in small quantity, fleshy, usually almost obliterated when the seeds are ripe.

Herbaceous plants or *trees*. *Leaves* simple or compound, alternate, with 2 stipulæ at their base.

ANALYSIS OF THE GENERA.

Petals 4 or 5
  Fruit consisting of 1 or more follicles surrounded by a dry calyx - - - - - - - - - - - - 1. SPIRÆA.
  Fruit a naked drupe
    Vernation convolute. Drupe covered with bloom - 2. PRUNUS.
    Vernation conduplicate. Drupe glabrous - - - 3. CERASUS.

Fruit consisting of numerous small nuts or drupes invested with a dry calyx
- Fruit succulent, receptacle dry - - - - - - 4. RUBUS.
- Fruit dry, receptacle succulent - - - - - - 5. FRAGARIA.
- Fruit and receptacle both dry
  - Calyx with external bracteæ
    - Styles withering
      - Stamens indefinite - - - - - - 6. POTENTILLA.
      - Stamens definite - - - - - - 7. SIBBALDIA.
    - Styles persistent, indurated - - - - - 8. GEUM.
  - Calyx without external bracteæ
    - 8 or 9 parted, naked - - - - - - 9. DRYAS.
    - 5-cleft, lappaceous - - - - - - - 10. AGRIMONIA.

Fruit consisting of numerous nuts enclosed in a fleshy calyx, contracted at the orifice of its tube - - } 11. ROSA.

Petals none.
- Stamens 1 to 4
  - Calyx 8-parted - - - - - - - - - - - 12. ALCHEMILLA.
  - Calyx 4-parted - - - - - - - - - - - 13. SANGUISORBA.
- Stamens 20 to 30 - - - - - - - - - - - - 14. POTERIUM.

---

§ 1. *Spiræaceæ* Dec.

Follicles several, invested by the calyx. *Seeds* from 1 to 6, suspended from the inner edges of the follicle. — Shrubs, *or* herbaceous *plants.*

1. SPIRÆA *Linn.*

Calyx 5-cleft, persistent. *Stamens* from 10 to 50, inserted along with the petals upon a disk adhering to the calyx. *Follicles* 1 or several, distinct, or occasionally cohering by the base. *Seeds* from 2 to 6.

1. S. *salicifolia* Linn. E. B. 21. 1468.
Leaves elliptic-lanceolate, unequally serrated, smooth. Clusters terminal, compound. *Smith.*

In swampy, shady, mountainous situations. — Shrub. *July.*

2. S. *Filipendula* Linn. Common Dropwort. E. B. 4. 284.
Leaves interruptedly pinnate; leaflets uniform, serrated, smooth. Stem herbaceous. Flowers cymose, with many styles. *Smith.*

In open elevated pastures, on a chalky or gravelly soil. — Perennial. *July.*

3. S. *Ulmaria* Linn. Meadow-sweet. Queen of the Meadows. E. B. 14. 960.
Leaves interruptedly pinnate; downy beneath; the terminal leaflet largest and lobed. Stem herbaceous. Flowers cymose, with many styles. *Smith.*

In moist meadows, and about the banks of rivers and ditches. — Perennial. *June, July.*

§ 2. *Drupaceæ* Dec. (*Amygdaleæ* Juss. & Dec.)

Fruit a solitary drupe, containing 1 or 2 seeds, hanging from the top of their cell. *Calyx* deciduous. — Trees *or* shrubs, *with simple stalked* leaves, *glandular* petioles, *and distinct* stipules. *All the parts abound in prussic acid.*

2. PRUNUS *Tournefort.*

Vernation convolute. *Drupe* covered with bloom, with a smooth stem deeply furrowed at its inner edge.

1. P. *domestica* Linn. Wild Plum. E. B. 25. 1783.
Flower-stalks solitary or in pairs. Leaves lanceolate-ovate. Branches without thorns.

In woods and hedges, probably escaped from gardens. — Tree. *May.*

2. P. *spinosa* Linn. Sloe, or Blackthorn. E. B. 12. 842.
Flower-stalks solitary. Leaves lanceolate, smooth. Branches thorny at the end.

P. *insititia* Linn. Bullace. E. B. 12. 841.
Flower-stalks in pairs. Leaves lanceolate-ovate; downy beneath. Branches thorny at the end.

In hedges and groves. — Tree. *April.*

3. CERASUS *Tournefort.*

Vernation conduplicate. *Drupe* not covered with bloom, with a smooth stone not furrowed at its inner edge.

1. C. *avium* Mœnch. Wild Cherry. E. B. 10. 706.
Flowers in nearly sessile umbels. Leaves ovate-lanceolate, or obovate with a point, with 2 glands at the base.
Prunus avium *Linn.*

In woods and hedges. — Tree. *May.*

2. C. *Padus* Dec. Bird Cherry. E. B. 20. 1383.
Racemes long, pendulous. Leaves ovate-lanceolate, acuminate, thin; smooth beneath, with spreading serratures. Fruit round, bitter.
Prunus Padus *Linn.*

In woods and hedges. — Tree. *May.*

§ 3. *Fragariaceæ* Richard. (*Dryadeæ* Vent.)

Fruit consisting either of small dry nuts or succulent drupeolæ, inserted upon a common receptacle, and invested with a dry permanent calyx. *Calyx* either 4- or 5-cleft, sometimes bearing bracteolæ on its tube, equal in number to the segments, and alternate with them. *Petals* 5. *Seed* solitary, erect, or inverted. — *Mostly* herbaceous *plants, very seldom* shrubs; leaves *usually compound;* stipulæ *adhering to the petiole.*

* 4. RUBUS *Linn.* Bramble.

Calyx somewhat campanulate, 5-lobed, without external bracteolæ. *Petals* 5. *Stamens* indefinite. *Fruit* consisting of numerous suc-

* In the species of this very difficult genus, I follow the truly excellent monograph of the German Rubi by Drs. Weihe and Nees von Esenbeck. Those species only of that work which are here enumerated have hitherto come before me as wild British plants; but it is probable that they are all to be dis-

culent drupeolæ, placed upon an elevated dry receptacle. *Seed* inverted. — Shrubs *or* herbaceous *plants.* Stems *usually long and procumbent, sterile the first year, bearing flowers and fruit the second, and then perishing.* Leaves *either simple, ternate, 5-nate, pedate, or pinnate, always more or less divided at the margin.*

Note. In describing the stem, the barren root shoot is always referred to; the leaves described are those of the root shoot, and in speaking of the form of the leaflets, that of the middle leaflet of the stem-leaves is to be understood. It is important to bear this in mind.

Division 1. *Leaves digitate; stem destitute of hairs, not covered with bloom; the prickles alike in form.*

A. *Corylifolii.* Leaves quinate, pubescent beneath. Panicle lax, not much longer than the leaves.

1. R. *plicatus* Weihe & Nees.
Stem nearly erect, angular, prickly, smooth. Leaves quinate, cordate-ovate, cuspidate, plaited; smoothish above; pubescent beneath. Panicle nearly simple. Calyx naked at the base, reflexed. *W. & N.*

In hedges and woods. — Shrub. *July, August.*

2. R. *fastigiatus* Weihe & Nees.
Stem arched, angular, sparingly prickly, smooth. Leaves quinate, ovate, cordate, with a long point; smooth and flat above; pubescent beneath. Flowering branches horizontal, simply panicled, narrow. Calyxes unarmed at the base, reflexed from the fruit. *W. & N.*

In thickets and hedges. — Shrub. *July, August.*

3. R. *suberectus* Anderson. E. B. 36. 2572.
Stems short, somewhat erect, taper, covered with a few scattered slender unequal prickles. Leaves ternate or quinate; leaflets overlapping each other, shining, plaited, cordate, acuminate, simply serrate, smooth, green on each side. Panicles simple, horizontal, very smooth, with few prickles.
R. nessensis *Hall.*

In thickets and woods in the north. — Shrub. *July, August.*

Obs. Certainly not R. fastigiatus of Weihe and Nees, which has rounder and more prickly shoots; a much more erect and coarse habit, with an obvious tendency to divide the central leaflet into 3, so as to become septenate.

4. R. *affinis* Weihe & Nees.
Stem arched, angular, prickly, smooth. Leaves quinate, cordate-ovate, cuspidate, flat at the base, somewhat wavy towards the

---

covered by diligent research. Sir James Smith's English Flora is my authority for Rubus nitidus being British; for the introduction of the remainder I am alone responsible.

points, between pubescent and downy beneath. Panicle compound, with cymose branches. Calyxes naked, reflexed. *W. & N.* R. collinus *Dec.*

In hedges and woods. — Shrub. *July, August.*

5. R. *nitidus* Weihe & Nees.

Stem nearly erect, angular, prickly, smooth. Leaves quinate, ovate, acute, flat, smooth and shining above, pubescent beneath. Panicle prickly, somewhat compound. Calyxes reflexed, prickly at base. *W. & N.*

In thickets and woods, not rare. *Smith.* — Shrub. *July, August.*

Obs. — W. & N. refer R. suberectus of E. B. to this; but its branches are angular, not taper.

B. *Fastigiati.* Leaves quinate, downy and grey beneath. Panicle dense, long.

6. R. *cordifolius* Weihe & Nees.

Stem arched, angular, furrowed, prickly, smooth. Leaves quinate, roundish, cordate, cuspidate, covered with hoary shining down on the under side. Panicle decompound, straggling at the base. Calyxes spreading, prickly at the base. *W. & N.*

In woods and hedges, not uncommon. — Shrub. *July, August.*

7. R. *rhamnifolius* Weihe & Nees.

Stem arched, angular, furrowed, prickly, smooth. Leaves quinate, roundish, cuspidate, hoary beneath. Panicle compound and decompound, close, with divaricate branches. Calyxes spreading, prickly at the base. *W. & N.*

In woods and hedges, not uncommon. — Shrub. *July, August.*

Obs. E. B. 715., is cited by W. & N. to this with an ? Sir James Smith continues the reference to the true fruticosus.

8. R. *fruticosus* Linn.

Stem arched, angular, furrowed, aculeate, smooth. Leaflets quinate, ovate-oblong, acute, white with down beneath. Panicle decompound, narrow, straight. Calyxes reflexed, almost unarmed. *W. & N.*

In hedges and woods. — Shrub. *July, August.*

9. R. *abruptus.*

Stem arched, strong, angular, very glaucous, with very broad, equal, hooked prickles. Leaflets 3- or 5-nate, small, distant, undulate, shining, obovate, truncate, with an inflexed cuspidate point, simply serrated, very hoary beneath. Panicles long, downy, armed with strong hooked prickles; the branches spreading and corymbose.

In Scotland. — Shrub. *July, August.*

Division 2. *Leaves digitate. Stem hairy, with neither bloom nor glands. The prickles alike in form.*

A. *Corylifolii.* Inflorescence lax, short, either umbelled or panicled. Leaves usually green and pubescent beneath.

10. R. *saxatilis* Linn. E. B. 32. 2233.
Stem nearly herbaceous, veiny, prostrate, unarmed, hairy. Leaves ternate, obovate, coarsely cut and serrate, pubescent. Flowers in contracted umbels. Petals oblong. *W. & N.*

11. R. *macrophyllus* Weihe & Nees.
Stem arched, angular, prickly, with few hairs. Leaves somewhat quinate, roundish, cordate, taper-pointed, with nearly simple serratures, not very hairy above, but velvety beneath. Panicle compound; branches erect, few-flowered.

Hedges in North Wales. — Shrub. *July, August.*

12. R. *vulgaris* Weihe & Nees.
Stem arched, angular, prickly, somewhat hairy. Leaves quinate, elliptical or roundish, obtusely wedge-shaped or subcordate at the base, pubescent beneath. Panicle compound, spreading, few-flowered. *W. & N.*

In hedges and thickets, common. — Shrub. *July, August.*

B. *Fastigiati.* Inflorescence long, in panicled racemes. Leaves quinate, beneath pubescent or hoary.

13. R. *leucostachys* Smith.
Stem arched, angular, hairy, as well as the scattered straight prickles. Leaves ternate or quinate, cordate, with a very long point, coarsely and unequally serrated, smooth above, green and velvety beneath. Panicle villous, contracted, very prickly, with corymbose branches.

In hedges and thickets. — Shrub. *July, August.*

14. R. *diversifolius.*
Stem arched, angular, hairy, as well as the scattered straight prickles. Leaflets ternate, orbicular, cordate, acute, rugose, equally serrated, velvety and rather hoary beneath. Panicle villous, contracted, with corymbose branches, and roundish, cordate, entire, or 3-lobed leaves at the base.

In hedges and thickets. — Shrub. *July, August.*

Obs. This is next R. macroacanthus of *W. & N.* I suspect it to be the R. cordifolius of the English Flora.

15. R. *discolor* Weihe & Nees.
Stem arched, angular, aculeate, silky, brownish-purple. Flowering branch hairy. Leaves quinate, roundish, acuminate, finely and doubly serrate, naked above, hoary beneath, with reticulated veins. Raceme dense, decompound. *W. & N.*

Hedges and woods. — Shrub. *July, August.*

Division 3. *Leaves digitate. Stem covered with unequal prickles, setæ, and glands, almost always also hairy, either naked, or covered with bloom. The flowering branches always glandular.*

* Calyx of the fruit reflexed.

A. *Corylifolii.* Inflorescence loose, short, corymbose, or panicled. Leaves always green beneath, more or less pubescent.

16. R. *fusco-ater* Weihe & Nees.
Stem procumbent, angular, hairy, with many glands and setæ; prickles unequal, recurved, numerous. Leaves quinate, cordate, ovate, acuminate, densely pubescent beneath. Panicle downy, with many setæ and glands; the lower branches racemose. Bracteæ lanceolate, bifid or trifid. Prickles of the peduncles long and setaceous, with numerous setæ. *W. & N.*

In thickets and hedges. — Shrub. *July, August.*

17. R. *pallidus* Weihe & Nees.
Stem procumbent, somewhat angular, hairy, with scattered, equal, recurved prickles and setæ. Leaves quinate, cordate-ovate, acuminate, pale green on the veins beneath, which are pilose. Panicle downy, with scattered setæ; branches in corymbose racemes. Bracteæ linear. Prickles of the peduncles short and close set, without setæ. *W. & N.*

In hedges. — Shrub. *July, August.*

18. R. *Köhleri* Weihe & Nees.
Stem procumbent, nearly taper, with scattered hairs and glands; prickles numerous, unequal, straight. Leaves quinate, obovate, cordate, acuminate, downy beneath, with coarse unequal serratures. Panicle somewhat downy at the end, with scattered glands; its branches corymbose. Bracteæ foliaceous. Peduncles with very long dense prickles, and few setæ. *W. & N.*

Hedges near Bury St. Edmunds. *Mr. John Denson.* — Shrub. *July, August.*

B. *Fastigiati.* Inflorescence more dense and panicled, and more naked at the end. Leaves in some species hoary beneath.

19. R. *rudis* Weihe & Nees.
Stem somewhat arched, angular, slightly furrowed, without hairs, but with abundance of glands and setæ; prickles equal, recurved, scattered. Leaves quinate, ovate or obovate-acuminate, coarsely and unequally serrated, between hoary and pubescent beneath. Panicle open, somewhat downy, setose, with corymbose racemose branches. Bracteæ lanceolate, trifid. Prickles and setæ of the peduncles not numerous. *W. & N.*

In hedges. — Shrub. *July, August.*

20. R. *echinatus.*
Stem arched, angular, densely covered with glands and setæ, mixed with short, numerous, nearly straight prickles. Leaflets 5, roundish, cordate, coarsely and unequally serrated, taper-pointed, green, and velvety beneath. Panicle spreading, prickly, and glandular, leafy at the base; the branches corymbose. Bracteæ 3-toothed and entire. Prickles of the peduncle scattered, with very few setæ.

In hedges and woods. — Shrub. *July, August.*

** Calyx of the fruit erect.

21. R. *hirtus* Waldst. & Kitaibel.
Stem procumbent, taper, densely hairy, covered with purple glands and setæ; prickles long, numerous, straight, spreading. Leaves

ternate or quinate-pedate, cordate, ovate, covered beneath with shining hairs. Stipules leafy. Panicle corymbose, with red setæ. Bracteæ long, trifid. *W. & N.*

In hedges and woods.—Shrub. *July, August.*

22. R. *dumetorum* Weihe & Nees.
Stem procumbent, somewhat angular, rather hairy, with a slight coat of bloom; setæ and glands very few; prickles unequal, recurved. Leaves generally quinate; the lateral leaflets sessile. Flowers in corymbs. Fruit black. *W. & N.*
R. nemorosus *Willd.*

In hedges and woods.—Shrub. *July, August.*

Obs. This is said by *W. & N.* to be R. corylifolius *Smith.* E. B. 12. 827.; but I believe that figure represents R. vulgaris, which is the plant usually called R. corylifolius by British botanists.

23. R. *cæsius* Linn. Dewberry. E. B. 12. 826.
Stem procumbent, taper, rather hairy, and covered with bloom; prickles variable, unequal. Leaves ternate or pinnate-quinate; the lateral leaflets sessile, often lobed on the outer edge. Flowers in corymbose panicles. Fruit cæsious. *W. & N.*

In hedges and woods.—Shrub. *June, July, August.*

Division 4. *Leaves pinnate.*

24. R. *idæus* Linn. Raspberry. E. B. 34. 2442.
Stem round, erect, smooth, with downy branches; their prickles straight and slender. Leaves pinnate, of 5 or 3 ovate, rather angular leaflets, very downy beneath. Clusters prickly, somewhat compound. Flowers pendulous. *Smith.*

In mountainous woods and thickets.—Shrub. *May, June.*

Division 5. *Stem herbaceous. Flowers usually solitary.*

25. R. *arcticus* Linn. E. B. 23. 1585.
Leaflets 3, bluntly serrated. Stem without prickles, bearing 1 or 2 solitary flowers. Petals roundish. *Smith.*

On stony mountainous moors, rare.—Perennial. *May, June.*

26. R. *Chamæmorus* Linn. Cloudberry. E. B. 10. 716.
Leaves simple, plaited, lobed. Stem without prickles, simple, single-flowered. Segments of the calyx ovate. *Smith.*

In turfy alpine bogs.—Perennial. *June.*

5. FRAGARIA *Linn.*

Calyx concave, 5-cleft, with 5 external bracteolæ. *Petals* 5. *Stamens* indefinite. *Fruit* consisting of numerous small nuts, placed upon a succulent receptacle. *Seed* inverted.—Herbaceous *plants, propagating themselves by runners.* Leaves *ternate* or *simple.*

1. F. *vesca* Linn. Wood Strawberry. E. B. 22. 1524.
Calyx of the fruit reflexed. Hairs of the footstalks widely spreading; those of the partial flower-stalks close-pressed, silky. *Smith.*

In groves and thickets, common.—Perennial. *May, June.*

2. F. *moschata* Duchesne. Hautboy Strawberry. E. B. 31. 2197.
Leaflets oblong, plaited, coarsely toothed, hairy, as well as the flower-stalks. Calyx smaller than the corolla. Flowers polygamous. Fruit perfumed.

F. elatior *Smith*, not of *Ehrhart*, which is what the gardeners call the Green Pine Strawberry.

In groves in the south of England. — Perennial. *June—September*. A doubtful native.

3. F. *calycina* Loiseleur.
Leaflets sessile, hairy, roundish, wedge-shaped, coarsely toothed. Peduncles longer than the scapes. Calyx as large as the corolla.

In Northumberland. — Perennial. *June—September*.

6. POTENTILLA *Linn.* Cinquefoil.

Calyx concave, 4 or 5-cleft, with 4 or 5 external bracteolæ. *Petals* 5. *Stamens* indefinite. *Fruit* consisting of numerous small nuts, placed upon a dry elevated receptacle. *Seed* inverted. — Herbaceous *plants*, or shrubs. Leaves *compound*. Stipules *adhering to the petiole*. Flowers *white, yellow, or purple*.

§ Petals 5, obtuse.

* *Leaves pinnate.*

1. P. *fruticosa* Linn. E. B. 2. 88.
Leaves pinnate, entire, hairy. Stem shrubby. *Smith.*

In mountainous thickets. — Shrub. *June.*

2. P. *anserina* Linn. E. B. 12. 861.
Leaves interruptedly pinnate, serrated, silky. Stem creeping. Stalks axillary, solitary, single-flowered. *Smith.*

In osier holts and spongy meadows. — Perennial. *June, July.*

3. P. *rupestris* Linn. E. B. 29. 2058.
Leaves lyrate-pinnate; leaflets 7, 5, or 3, ovate, serrated, hairy. Stem erect. *Smith.*

On shady alpine rocks in Wales. — Perennial. *June.*

** *Leaves digitate.*

4. P. *argentea* Linn. E. B. 2. 89.
Leaflets 5, wedge-shaped, jagged; downy beneath. Stem ascending. *Smith.*

In pastures on a gravelly soil. — Perennial. *June, July.*

5. P. *alpestris* Haller. E. B. 8. 561.
Radical leaves of 5 wedge-shaped, somewhat hairy leaflets; deeply cut in their upper half. Upper stipulas ovate. Petals heart-shaped. Stems ascending. *Smith.*

P. salisburgensis *Hænke.*
P. aurea *Smith.*
P. verna β. *Wahlenb.*
P. verna γ. *Nestler.*

On mountains in Scotland, the north of England, and Wales. — Perennial. *July.*

6. P. *verna* Linn. E. B. 1. 37.

Radical leaves of 5 or 7, obovate-wedge-shaped, partly serrated, furrowed leaflets; bristly at the margins and ribs beneath. Upper stipulas dilated. Calyx-segments partly 3-ribbed. Stems procumbent. *Smith.*

In elevated pastures, and barren hilly ground. — Perennial. *April, May.*

7. P. *opaca* Linn. E. B. 35. 2449.

Radical leaves of 7, hairy, linear-wedge-shaped leaflets, deeply serrated throughout; stem-leaves ternate, mostly opposite. Stems recumbent. *Smith.*

On the mountains of Scotland. — Perennial. *June.*

8. P. *alba* Linn. E. B. 20. 1384.

Radical leaves of 5, elliptic-oblong leaflets; silky beneath; tipped with converging serratures. Stems thread-shaped, procumbent. Receptacle very hairy. *Smith.*

In mountainous woods. — Perennial. *June, July.*

9. P. *reptans* Linn. E. B. 12. 862.

Leaflets 5, obovate, serrated. Stem creeping. Stalks axillary, single-flowered. *Smith.*

In meadows, pastures, and by way sides, common. — Perennial. *June—August.*

*** *Leaves ternate.*

10. P. *tridentata* Solander. E. B. 34. 2389.

Leaflets 3, wedge-shaped; smooth above; hairy beneath; with about 3 terminal teeth. Stem panicled, erect. Fruit even. *Smith.*

On a mountain called Werron, and some other hills in Angus-shire, to the westward. — Perennial. *May, June.*

11. P. *Fragaria* Dec. E. B. 25. 1785.

Leaflets 3, roundish-obovate, serrated, hairy. Stems prostrate. Fruit corrugated, hairy at the scar. *Smith.*

P. Fragariastrum *Ehrhart.*

P. Fragarioides *Villars.*

Fragaria sterilis *Linn.*

In dry gravelly pastures, common. — Perennial. *March, April.*

§ 2. Petals 5, acuminate.

12. P. *Comarum* Scopoli. E. B. 3. 172.

Leaves with close lobes. Stipules broad, coriaceous. Petals less than the calyx. Fruit ovate, compressed, smooth. *Dec.*

Comarum palustre *Smith.*

Potentilla palustris *Lehmann.*

Potentilla rubra *Haller.*

In bogs and ditches. — Perennial. *June, July*

§ 3. Petals 4.

13. P. *Tormentilla* Sibth. E. B. 12. 863.

Stem ascending, branched. Leaves almost sessile. Stipulas cut. *Smith.*

Tormentilla erecta *Linn.*

T. officinalis *Smith.*
P. tetrapetala *Haller*

In barren pastures, heaths, and bushy places. — Perennial. *June, July.*

14. P. *reptans* Dec. E. B. 12. 864.
Stem prostrate, scarcely branched. Leaves stalked Stipulas undivided. *Smith.*
Tormentilla reptans *Linn.*
Potentilla procumbens *Sibth.*
P. nemoralis *Nestler.*

About hedges and the borders of fields. — Perennial. *June, July.*

7. SIBBALDIA *Linn.*

Calyx concave, 5-cleft, with 5 external bracteolæ. *Petals* 5. *Stamens* 5. *Fruit* consisting of 5 small nuts, placed upon a dry receptacle. *Seed* inverted. — Herbaceous *plants with compound* leaves *and yellow* flowers.

1. S. *procumbens* Linn. E. B. 13. 897.
Leaflets wedge-shaped, with 3 terminal teeth. *Smith.*

On the summits of the Highland mountains of Scotland. — Perennial. *July.*

8. GEUM *Linn.*

Calyx concave, 5-cleft, with 5 external bracteolæ. *Petals* 5. *Stamens* indefinite. *Fruit* consisting of numerous small nuts, tipped with the indurated persistent naked styles, and placed upon a dry receptacle. *Seed* ascending. — Herbaceous *plants with compound* leaves. Flowers *white or yellow.*

1. G. *urbanum* Linn. Common Avens. Herb Bennet.
E. B. 20. 1400.
Leaves ternate; radical ones somewhat lyrate. Stipulas rounded, cut. Flowers nearly upright. Styles naked. *Smith.*

In woods, shady dry hedges, and thickets. — Perennial. *May—August.*

2. G. *rivale* Linn. E. B. 2. 106.
Radical leaves interruptedly pinnate, somewhat lyrate. Stipulas ovate, acute, cut. Flowers drooping. Styles hairy above the curvature. *Smith.*

intermedium Seringe.
Leaves hairy. Lobes of the upper leaves narrower. Peduncles more slender. *Ser.*
G. intermedium *Ehrh.*

γ. *luxurians* Trattennich.
Flowers semi-double. Sepals distinct, transformed into leaves. *Ser.*
G. hybridum *Wulfen.*

In moist meadows and woods. — Perennial. *June, July.*

9. DRYAS *Linn.*

Calyx 8- or 9-parted, without external bracteolæ. *Petals* 8 or 9. *Stamens* indefinite. *Fruit* consisting of numerous small nuts, tipped with the persistent feathery styles, and placed on a dry receptacle.

Seed ascending. — Herbaceous *plants, with simple* leaves, *hoary beneath, and white* flowers.

1. D. *octopetala* Linn. E. B. 7. 451.
Petals 8. Leaves oblong, notched, downy beneath.
On stony alpine heaths. — Perennial. *July, August.*

10. AGRIMONIA *Linn.*

Calyx turbinate, 5-cleft, without external bracteolæ, covered with rigid hooked bristles. *Petals* 5. *Stamens* 15. *Fruit* consisting of 2 membranous small nuts inclosed in the indurated tube of the calyx. *Seed* suspended. — Herbaceous *plants, with compound* leaves. Flowers *in spikes, yellow.* Bracteæ *trifid.*

1. A. *Eupatoria* Linn. Agrimony. E. B. 19. 1335.
Stem-leaves pinnate; leaflets elliptic-oblong; terminal one-stalked. Calyx encompassed with bristles. Spikes elongated. *Smith.*
In bushy places, by road sides, and about the borders of fields. — Perennial. *June, July.*

§ 4. *Roseæ* Dec.

Nuts numerous, hairy, terminated by the persistent style, and enclosed within the fleshy tube of the calyx, which is contracted at the orifice, where it is surrounded by a fleshy disk. *Seed* suspended. *Sepals* 5. *Petals* 5. *Stamens* indefinite. — Shrubs *with prickly or naked* stems. Leaves *pinnate.* Flowers *red, white, or yellow, usually fragrant.*

11. ROSA *Linn.* Rose.

The characters the same as those of the section, no other genus being included in it.

i. Cinnamomeæ. *Branches setigerous, sometimes unarmed. Flowers with bracteæ. Leaflets lanceolate or oblong, without glands. Disk thin.*

1. R. *Dicksoniana* Lindley.
Branches flexuose, armed with a few slender scattered prickles. Leaflets folded together, unequal, with coarse double serratures. Stipules, petioles, and sepals glandular; the latter equal. Fruit naked.
In Ireland. — Shrub. *May, June.*

Obs. R. cinnamomea must be excluded from the British Flora, there being no good evidence of its having been found wild in this country.

ii. Pimpinellifoliæ. *Branches setigerous, their prickles being all of nearly the same figure; sometimes unarmed. Flowers without bracteæ. Leaflets usually ovate. Sepals converging, not separating from the fruit. Disk very thin.*

2. R. *rubella* Smith. E. B. 36. 2521.
Branches covered with equal-sized prickles and setæ. Fruit long, pendulous.
On the sea-coast in the north. — Shrub. *July.*

3. R. *spinosissima* Linn. E. B. 3. 187.
Prickles very unequal. Leaflets flat, smooth, simply serrated. Fruit globose, erect.

β. *pilosa* Lindley.
Very dwarf. Leaves acute, hairy on the under surface.

On sandy heaths, and hedges near the sea, and upon mountains. β. in Ireland. — Shrub. *May, June.*

4. R. *hibernica* Smith. E. B. 31. 2196.
Prickles unequal; the setæ very few. Leaflets ovate, acute, simply serrated, hairy beneath. Sepals compound, reflexed. Flowers with bracteæ.

Ireland, near Belfast. — Shrub. *May, June.*

5. R. *involuta* Smith. E. B. 29. 2068.
Prickles very unequal and dense. Leaflets doubly serrated, downy. Petals rolled up. Sepals simple. Fruit aculeate.
R. nivalis *Donn.*

In the western part of Scotland. — Shrub. *June.*

6. R. *Sabini* Woods.
Setæ scarce. Prickles unequal, scattered. Leaflets doubly serrate, tomentose. Sepals compound.
R. involuta *Winch.*
R. gracilis *Woods.*

β. *Doniana* Lindley.
Setæ almost wholly wanting. Prickles nearly straight.
R. Doniana *Woods.*

In woods and thickets in several parts of England, chiefly in the north. — Shrub. *June.*

iii. Villosæ. *Prickles nearly straight, and equal. Setæ none. Leaflets with diverging serratures, and turpentine glands. Sepals remaining upon the fruit. Disk thick, closing up the orifice of the tube.*

7. R. *tomentosa* Smith. E. B. 14. 990.
Root-shoots arched. Sepals compound, diverging. Leaflets oblong, downy on both sides. Fruit hispid, or naked.

β. *fœtida.*
Leaflets nearly smooth on the upper side.
R. fœtida *Batard.*
R. scabriuscula *E. Bot.*
R. subglobosa *Smith.*

In hedges and thickets. — Shrub. *June, July.*

8. R. *mollis* Smith. E. B. 35. 2459.
Root-shoots erect, coloured. Sepals nearly simple, converging. Leaflets ovate, downy on both sides. Fruit hispid or naked.
R. villosa *Smith*, not of *Linnæus.*
R. pulchella *Woods.*
R. heterophylla *Woods.*

β. *resinosa.*
More dwarf. Leaflets narrower, with very numerous glands. Flowers deep red.

R. tomentosa γ. *Lindl.*

In hedges and thickets, especially in the north. — Shrub. *June, July.*

9. R. *sylvestris*.

Shoots erect, coloured, flexuose. Prickles hooked. Leaflets oblong, acute, hoary on each side. Sepals diverging, deciduous before the fruit is ripe. Fruit elliptical, setose.

R. tomentosa sylvestris *Woods.*

In Oxfordshire, in hedges. — Shrub. *June, July.*

v. Rubiginosæ. *Prickles very unequal, sometimes tipped with glands, very rarely absent. Leaflets ovate or oblong, usually fragrant and glandular, with diverging serratures. Sepals persistent. Disk thick, closing up the orifice of the tube.*

10. R. *rubiginosa* Linn. E. B. 14. 991.

Prickles much hooked. Leaflets rugose, not lucid, roundish-ovate, with fragrant brown glands at their margin, and on the underside. Calyxes and peduncles hispid.

R. eglanteria *Hudson.*

β. *micrantha* Lindley. E. B. 35. 2490.

Prickles more equal, and less numerous. Sepals deciduous before the fruit is quite ripe. Fruit oblong or obovate.

R. micrantha *Smith.*

γ. *umbellata* Lindley.

Branches of the inflorescence extremely prickly. Fruit taper.

R. umbellata *Leers.*

δ. *inodora* Lindley. E. B. 36. 2579.

Prickles very much hooked, and nearly equal. Leaflets less glandular. Sepals deciduous before the fruit is quite ripe, and very compound.

R. inodora *Agardh.*

R. Borreri *Woods.*

R. dumetorum *E. Bot.*

Subvar. parvifolia. *Leaflets very small and acute.* Fl. Lond. t. 117.

In hedges and thickets. — Shrub. *June, July.*

11. R. *sepium* Thuill.

Prickles slender. Branches flexuose. Leaflets shining, acute at each end. Flowers nearly solitary. Fruit polished. Segments of the sepals very narrow.

R. rubiginosa η. *Lindley.*

Near Bridford, in Warwickshire (Rev. Mr. Bree). — Shrub. *June, July.*

v. Caninæ. *Prickles equal, hooked. Leaflets ovate, without glands, with converging serratures. Sepals deciduous before the fruit is ripe. Disk thick, closing up the orifice of the tube.*

12. R. *canina* Linn. E. B. 14. 992.

Leaflets ovate, acute, smooth on both sides. Prickles falcate.

R. surculosa *Woods.*

β. sarmentacea.
Leaflets doubly se ate.
R. glaucophylla *Winch.*
R. sarmentacea *Woods.*
Hedges and thickets. — Shrub. *June, July.*

13. R. *collina* Jacquin.
Leaflets ovate, simply serrated, smooth above, or very slightly hairy; more or less downy beneath. Prickles falcate.
R. Forsteri *Smith.*
Hedges and thickets. — Shrub. *June, July.*

14. R. *dumetorum* Thuill.
Leaflets ovate, simply serrated, hairy on both sides, and not shining on the upper. Prickles falcate.
R. bractescens *Woods.*
In hedges and thickets chiefly in the north. — Shrub. *June, July.*

15. R. *cæsia* Smith. E. B. 33. 2367.
Leaflets ovate, hoary on both sides. Prickles very numerous, strongly uncinate. Fruit elliptical, smooth.
In the highlands. — Shrub. *July.*

vi. Systylæ. *Styles cohering in the form of a column, protruded beyond the orifice of the tube of the calyx.*

16. R. *systyla* Batard. E. B. 27. 1895.
Root-shoots nearly erect, arched. Prickles strong, hooked.

a. ovata Lindley.
Leaflets ovate. Fruit oblong.
R. collina *Smith.*
R. stylosa *Desvaux.*

β. lanceolata Lindley.
Leaflets ovate-lanceolate. Fruit spherical.
In hedges and thickets. β in the south of Ireland. — Shrub. *June, July.*

Obs. I omit the R. systyla Monsoniæ of my monograph of Roses, because it and some other similar things are perhaps accidental productions owing their origin to garden plants, rather than really native species.

17. R. *arvensis* Hudson. E. B. 3. 188.
Root-shoots long, trailing. Prickles unequal, falcate. Leaves deciduous, glaucous beneath.
R. repens *Ehrh.*
In hedges in the south. — Shrub. *July.*

§ 5. *Sanguisorbeæ* Juss.

Nuts 1 or 2, inclosed within the dry tube of the calyx, which is contracted at the orifice. *Calyx* 3- or 5-cleft, the divisions with a valvular æstivation. *Petals* usually wanting, sometimes 4, cohering at the base into a monopetalous corolla. *Seeds* suspended, very rarely erect. — Herbs *or* shrubs. Leaves *often compound.* Flowers *minute.*

12. ALCHEMILLA *Linn.* LADIES MANTLE.

Calyx 4-toothed, with 4 external bracteolæ. *Petals* 0. *Stamens* 1 to 4. *Nuts* 1 or 2. *Stigmas* capitate. *Seed* suspended. — Herbaceous *plants*. Leaves *palmate, lobed, or cut*. Leaves *corymbose, herbaceous*.

1. A. *vulgaris* Linn. E. B. 9. 597.
Leaves lobed, plaited. *Smith*.
β. A. minor *Hudson*.
In dry, rather mountainous, pastures. — Perennial. *June—August.*

2. A. *alpina* Linn. E. B. 4. 244.
Leaves fingered, serrated, silky at the back. *Smith*.
On alpine rocks. — Perennial. *July.*

3. A. *arvensis* Smith. Parsley Piert. E. B. 15. 1011.
Leaves flat, 3-lobed, cut. *Smith*.
A. aphanes *Leers*.
Aphanes arvensis *Linn*.
In sandy or gravelly fields. — Annual. *May—October.*

13. SANGUISORBA *Linn.*

Flowers hermaphrodite. *Calyx* 4-cleft, with 2 external seales at the base. *Petals* 0. *Stamens* 4. *Nuts* 2. *Stigma* pencil-formed. *Seed* suspended. — Herbaceous *plants*. Leaves *unequally pinnate*. Flowers *in dense spikes*.

1. S. *officinalis* Linn. Great Burnet. E. B. 19. 1312.
Spikes ovate.
In meadows and pastures, on a calcareous soil, that are rather moist; chiefly in the north of England; more sparingly in the lowlands of Scotland. — Perennial. *June, July.*

2. S. *media* Linn.
Spikes cylindrical.
In pastures, in the west of Scotland. — Perennial. *July.*

14. POTERIUM *Linn.*

Flowers monœcious or polygamous. *Calyx* 4-toothed, with 3 scales on the outside at the base. *Petals* 0. *Stamens* 20 to 30. *Nuts* 2. *Stigma* pencil-shaped. *Seed* suspended. — Herbaceous *plants*. Leaves *unequally pinnate*. Flowers *in dense spikes*.

1. P. *Sanguisorba* Linn. E. B. 12. 860
Thorns none. Stem somewhat angular. *Smith*.
On chalky hills, or about limestone rocks, abundantly. — Perennial. *July.*

Order 33. POMACEÆ *Juss.*

Calyx inferior, 5-toothed; the odd segment posterior.
Petals 5, unguiculate, inserted in the throat of the calyx; the odd one anterior.

Stamens indefinite, inserted in a ring in the throat of the calyx.
Disk thin, clothing the sides of the limb of the calyx.
Ovarium from 1 to 5-celled, seldom spuriously 10-celled; *ovules* usually 2, collateral, ascending, very rarely solitary; *styles* from 1 to 5; *stigmata* simple.
Fruit a pome, 1- to 5-celled, seldom spuriously 10-celled; the endocarpium either cartilaginous, spongy, or bony.
Seeds ascending, solitary. *Albumen* none; *embryo* erect, with flat *cotyledons*, and a short conical *radicle*.
Trees or *shrubs*. *Leaves* alternate, stipulate, simple, or compound. *Flowers* in terminal cymes, white or pink.

1. MESPILUS *Linn.*

Segments of the *calyx* foliaceous. *Petals* roundish. *Disk* large, secreting much honey. *Styles* 2 to 5, smooth. *Fruit* turbinate, with the upper ends of the carpella exposed. *Endocarpium* bony. — Trees *with serrated undivided* leaves. Flowers *solitary*.

1. M. *germanica* Linn. Common Medlar. E. B. 22. 1523.
Leaves lanceolate, a little downy. Flowers solitary, nearly sessile, terminal. Styles 5. *Smith.*

In hedges, rare. — Tree. *May.*

2. CRATÆGUS *Linn.*

Segments of the *calyx* acute. *Petals* roundish. *Styles* 2 to 5. *Fruit* oval or round, concealing the upper ends of the carpella. *Endocarpium* bony. — Trees *with lobed* leaves. Flowers *corymbose*.

1. C. *oxyacantha* Linn. Hawthorn, or Whitethorn. E. B. 35. 2504.
Leaves obovate, wedge-shaped, either entire, trifid, or cut, quite smooth and rather lucid. Flowers in corymbs, with from 1 to 3 styles. Calyx destitute of glands.
Mespilus oxyacantha *Smith.*

β. *eriocarpa.*
Leaves oblong, 3- or 5-cleft, slightly serrated. Tube of the calyx densely hoary.

In woods and hedges. — Small tree. *May, June.*

3. COTONEASTER *Lindley.*

Flowers polygamous. *Calyx* turbinate, with 5 short teeth. *Petals* 5 small, erect. *Stamens* erect, the length of the teeth of the calyx. *Fruit* turbinate, with its nuts adhering to the side of the calyx, but not cohering in the centre. — Bushes, *with entire* leaves, *and corymbose* flowers.

1 C. *vulgaris* Lindley.
Leaves ovate, rounded at the base. Calyxes and peduncles smooth. *Dec.*
Mespilus Cotoneaster *Linn.*

On the cliffs of the Great Ormshead, in Carnarvonshire. — Shrub. *July.*

4. PYRUS *Linn.*

Calyx 5-toothed. *Petals* roundish, spreading. *Styles* 2, 3, or 5. *Fruit* fleshy, with 5 distinct cells. *Endocarpium* cartilaginous. *Seeds* 2 in each cell. *Testa* cartilaginous. — Trees, *with serrated, undivided, or pinnated* leaves, and *cymose* flowers. Bracteæ *deciduous.*

* *Leaves simple.*

. P. *communis* Linn. Wild Pear-tree. E. B. 25. 1784.
Leaves simple, ovate, serrated. Flower-stalks corymbose. Fruit turbinate.

In woods and hedges. — Tree. *April, May.*

2. P. *Malus* Linn. Crab-tree. E. B. 3. 179.
Leaves simple serrated, rugose. Flowers in a simple sessile umbel. Fruit round.

In woods, hedges, and parks. — Tree. *May.*

3. P. *torminalis* Smith. E. B. 5. 298.
Leaves simple, somewhat heart-shaped, serrated, 7-lobed; the lower lobes spreading. Flower-stalks corymbose, branched. *Smith.*

In woods and hedges, chiefly in the midland and southern counties. — Tree. *April, May.*

4. P. *Aria* Smith. Beam Tree.
Leaves ovate, doubly serrated, hoary beneath. Flowers in dense flat corymbs. Fruit pisiform.
Cratægus Aria *Linn.*

In mountainous woods. — Tree. *May.*

5. P. *intermedia* Ehrhart. E. B. 26. 1858.
Leaves ovate, doubly-serrated, lobed, hoary beneath. Flowers in dense flat corymbs. Fruit pisiform.
Sorbus hybrida *Hudson.*
Cratægus scandica *Wahlenberg.*
Pyrus aria *β. Smith.*
Sorbus scandica *Fries.*

On the walls of Castle Dinas y Brān, in Denbighshire. *Sm.* — Tree. *May.*

6. P. *pinnatifida* Ehr. E. B. 33. 2331.
Leaves deeply pinnatifid, or half pinnate; downy beneath. Flowers corymbose. Styles about 3. *Smith.*
Sorbus hybrida *Linn.*

On mountains in the western isles of Scotland. — Tree. *May.*

** *Leaves pinnate.*

7. P. *domestica* Smith. True Service-tree. E. B. 5. 350.
Leaves pinnate; leaflets uniform, downy beneath, serrated towards the point. Flowers panicled. Fruit obovate. *Smith.*
Sorbus domestica *Linn.*

In mountainous woods, very rare. — Tree. *May.*

8. P. *aucuparia* Gærtner. Quicken-tree, or Mountain Ash. Roan-tree. E. B. 5. 337.
Leaves pinnate; leaflets uniform, serrated, smooth. Flowers corymbose. Styles about 3. Fruit globular. *Smith.*
Sorbus aucuparia *Linn.*

In mountainous woods, and hedges. — Tree. *May.*

Order 34. Grossulaceæ.

(Grossularieæ *Dec.*)

Calyx inferior, 4- or 5-parted, regular, coloured.
Petals 5, inserted in the throat of the calyx.
Stamens 5, inserted alternately with the petals.
Ovarium 1-celled, with 2 opposite parietal placentæ; *ovules* numerous; *style* 2-3-4-cleft.
Berry crowned with the remains of the flower, 1-celled; the cell filled with pulp.
Seeds numerous, suspended among the pulp by long filiform funiculi; *testa* externally gelatinous, adhering firmly to the *albumen,* which is horny; *embryo* minute, excentrical, with the *radicula* next the hilum.
Shrubs, either unarmed or spiny. *Leaves* alternate, lobed, with a plaited vernation. *Flowers* in axillary racemes, with bracteæ at their base.

1. RIBES *Linn.*

The character the same as that of the order, there being no other genus.

* *Without prickles.* Currants.

1. R. *rubrum* Linn. Common Currant. E. B. 18. 1289.
No prickles. Clusters smooth, pendulous. Flowers but slightly concave. Petals inversely heart-shaped. *Smith.*

In mountainous woods, especially about the banks of rivers, in the north of England, and in Scotland. — Shrub. *May.*

2. R. *petræum* Wulfen. E. B. 10. 705.
No prickles. Clusters somewhat hairy; in flower upright; in fruit pendulous. Flowers slightly concave. Petals bluntish. Bracteas shorter than the flower-stalks. Stem erect. *Smith.*

In the mountainous woods of Durham and Scotland. — Shrub. *May, June.*

3. R. *spicatum* Robson. E. B. 18. 1290.
No prickles. Spikes upright. Flowers nearly sessile. Petals oblong. Bracteas shorter than the flowers. *Smith.*

In woods in the north of England. — Shrub. *May.*

4. R. *alpinum* Linn. E. B. 10. 704.
No prickles. Clusters upright, both in flower and fruit. Bracteas longer than the flowers. Leaves polished at the back. Stem erect. Berries smooth. *Smith.*
In woods in the north. — Shrub. *May.*

5. R. *nigrum* Linn. E. B. 18. 1291.
No prickles. Clusters hairy, pendulous, with a separate flower-stalk at the base of each. Flowers oblong. *Smith.*
In sandy swamps and thickets, about the banks of rivers. — Shrub. *May.*

** *Branches prickly.* Gooseberries.

6. R. *Grossularia* Linn. Common Gooseberry. E. B. 18. 1292.
Prickles 1, 2, or 3 under each bud. Branches otherwise smooth, spreading. Stalks single-flowered. Bracteas close together. Segments of the calyx reflexed, shorter than the tube. *Smith.*
R. Uva crispa *Linn.*

β. *reclinatum.* Lancashire Gooseberry.
Branches arched, recurved.
R. reclinatum *Linn.*
In hedges, thickets, waste ground. — Shrub. *April.*

Order 35. ONAGRARIÆ *Juss.*

Calyx superior, tubular, with the limb either 4- or 5-lobed; the lobes cohering in various degrees, with a valvate æstivation.
Petals generally equal in number to the lobes of the calyx, into the throat of which they are inserted, regular, with a twisted æstivation.
Stamens definite, inserted into the calyx; *filaments* distinct; pollen triangular, usually cohering by filaments.
Ovarium of several cells, generally crowned by a disk; *style* filiform; *stigma* either capitate or 4-lobed.
Fruit baccate or capsular, many-seeded, with from 2 to 4-cells.
Seeds numerous, without *albumen*; *embryo* straight; *radicle* long and taper; *cotyledons* very short.
Herbaceous plants or *shrubs.* *Leaves* alternate or opposite, simple, entire, or toothed. *Flowers* red, purple, white, blue, or yellow, axillary, or in terminal spikes.

1. EPILOBIUM *Linn.*

Calyx tubular, with a 4-parted limb, which falls off after flowering. *Petals* 4. *Stamens* 8. *Capsule* linear, bluntly 4-cornered, with 4 cells, 4 valves, and many seeds. *Seeds* pappose. — Herbaceous *plants.* Leaves *opposite or alternate.* Flowers *axillary and solitary, or terminal in spikes, purple or rose colour.*

* *Flowers irregular.*

1. E. *angustifolium* Linn. Persian, or French Willow. E. B. 28. 1947.
Leaves scattered, linear-lanceolate, veiny, smooth. Petals unequal. Stamens declining. *Smith.*

In meadows and moist shady places, chiefly in the north of England. — Perennial. *July, August.*

** *Flowers regular. Stigma deeply 4-cleft.*

2. E. *hirsutum* Linn. Codlings and Cream. E. B. 12. 838.
Leaves half clasping the stem, ovate, lanceolate, hairy. Stem copiously branched. Root creeping. *Smith.*
E. ramosum *Hudson.*

In watery places, ditches, and margins of rivers, common. — Perennial. *July.*

3. E. *parviflorum* Schreb. E. B. 12. 795.
Leaves sessile, lanceolate, downy. Stem nearly simple, woolly. Root fibrous. *Smith.*
E. pubescens *Willd.*
E. hirsutum *Hudson.*
E. villosum *Curtis.*

Frequent in watery places and about the banks of rivers. — Perennial. *July.*

4. E. *montanum* Linn. E. B. 17. 1177.
Leaves stalked, ovate, toothed. Stem round. Stigma in 4 deep segments. *Smith.*

In dry, shady, hilly, or stony places. — Perennial. *July.*

*** *Fl. regular. Stigma undivided.*

5. E. *roseum* Schreber. E. B. 10. 693.
Leaves stalked, ovate, toothed. Stem erect, with 4 obsolete angles. Stigma undivided. *Smith.*

In waste boggy ground, or watery places. — Perennial. *July.*

6. E. *tetragonum* Linn. E. B. 28. 1948.
Leaves lanceolate, sessile, minutely toothed. Stem erect, unequally quadrangular. Stigma undivided. *Smith.*

In ditches, and watery marshy places. — Perennial. *July.*

7. E. *palustre Linn.* E. B. 5. 346.
Leaves sessile, linear-lanceolate, slightly toothed. Stem round. Stigma undivided. *Smith.*

In boggy turfy ground. — Perennial. *July.*

8 E. *alsinifolium* Villars. E. B. 28. 2000.
Leaves stalked, ovate, acute, toothed. Stigma undivided. Root creeping. Stem decumbent, obtusely quadrangular. *Smith.*

On the margins of mountain rivulets. — Perennial. *July.*

9. E. *alpinum* Linn. E. B. 28. 2001.
Leaves slightly stalked, elliptic-lanceolate, obtuse, mostly entire. Stem decumbent 2- or 3-flowered.

By the sides of alpine rivulets. — Perennial. *June, July.*

2. ŒNOTHERA *Linn.*

Calyx tubular, deciduous, with a reflexed 4-parted limb; the segments of which cohere irregularly. *Petals* 4. *Stamens* 8. *Pollen* cohering by threads. *Stigma* 4-lobed. *Capsule* linear or winged, with 4 cells, 4 valves, and many seeds. *Seeds* naked. — Herbaceous *plants.* Leaves *alternate, toothed,* or *pinnatifid.* Flowers *sessile, axillary, solitary,* or *in terminal spikes,* blue, red, yellow, or white.

1. Œ. *biennis* Linn. Evening-primrose. E. B. 22. 1534.
Leaves ovate-lanceolate, flat. Stem rough, somewhat hairy. Stamens equal. Petals undivided. *Smith.*

On sandy banks, on the west coast of England. — Biennial. *July—September.*

3. ISNARDIA *Linn.*

Calyx with a 4-parted limb. *Petals* 4, or none. *Stamens* 4, alternate with the petals. *Style* deciduous. *Stigma* capitate. *Capsule* 4-cornered, 4-celled, with 4 valves, and a loculicidal dehiscence. *Seeds* many. — Aquatic *or* marsh *plants.* Leaves *entire, alternate, or opposite.* Flowers *axillary, sessile.* Dec.

1. I. *palustris* Linn.
Stem procumbent, rooting, smooth. Leaves opposite, tapering into the petiole, ovate, acute. Flowers apetalous.

Found in a pool at Buxted, Sussex, in 1827, by Mr. W. Borrer. — Annual. *July.*

Order 36. CIRCÆACEÆ.

Calyx superior, deciduous, tubular, with a 2-parted limb.
Petals 2, alternate with the lobes of the calyx.
Stamens 2, alternate with the petals, inserted into the calyx.
Disk large, cup-shaped, filling up the whole of the tube of the calyx, and projecting beyond it.
Ovarium 2-celled, with an erect ovulum in each cell; *style* simple, arising out of the disk; *stigma* emarginate.
Fruit 2-celled, 2-valved, 2-seeded.
Seeds solitary, erect; *albumen* none; *embryo* erect; *radicle* short, inferior.
Herbaceous plants. *Leaves* opposite, toothed, stalked. *Flowers* in terminal and lateral racemes, covered with uncinate hairs.

Obs. This order differs from Onagrariæ in its large fleshy disk which fills up the tube of the calyx, in its solitary erect ovula, and in the binary division of the flower: it is connected with that order through Lopezia, with which it cannot however be absolutely associated, and bears about the same relation to Onagrariæ as is borne by Halorageæ.

1. CIRCÆA *Linn.* ENCHANTER'S NIGHTSHADE.

There being no other genus, the character is the same as that of the order

1. C. *lutetiana* Linn. E. B. 15. 1056.
Stem erect. Leaves ovate, slightly toothed, opaque and downy. *Smith.*
In moist shady places. — Perennial. *June, July.*

2. C. *alpina* Linn. E. B. 15. 1057.
Stem ascending. Leaves heart-shaped, serrated, shining. Calyx membranous. *Smith.*
In moist, shady, stony places. — Perennial. *July, August.*

Order 37. HALORAGEÆ *R. Brown.*

Calyx superior, with a minute limb.
Petals minute, inserted into the summit of the calyx, or wanting.
Stamens inserted in the same place, equal in number to the petals, or occasionally fewer.
Ovarium adhering inseparably to the calyx, with 1 or more cells. *Style* none; *stigmata* equal in number to the cells, papulose, or pencil-formed; *ovula* pendulous.
Fruit dry, indehiscent, membranous, or bony, with 1 or more cells.
Seeds solitary, pendulous; *albumen* fleshy; *embryo* straight, in the axis; *radicle* superior, long and taper; *cotyledons* minute.
Herbaceous plants or *under-shrubs*, often growing in wet places. *Leaves* either alternate, opposite, or whorled. *Flowers* axillary, sessile, occasionally monœcious or diœcious.

1. MYRIOPHYLLUM *Linn.* WATER-MILFOIL.

Flower generally bisexual. — Male. *Calyx* 4-parted. *Petals* 4, fugitive. *Stamens* 4, 6, or 8. — Female. *Calyx* with a 4-lobed limb. *Petals* 0. *Fruit* separable into 4 hard nuts. — *Aquatic* floating *plants, rising to flower.* Leaves *finely cut.* Flowers *minute, in whorls.*

1. M. *spicatum* Linn. E. B. 2. 83.
Flowers in whorled, interrupted, leafless spikes. *Smith.*
In ditches and pools frequent. — Perennial. *July, August.*

2. M. *verticillatum* Linn. E. B. 4. 218
Flowers all axillary. *Smith.*
In ponds and ditches. — Perennial. *July.*

2. HIPPURIS *Linn.*

Calyx with an entire limb. *Petals* none. *Stamens* 1. *Style* filiform, lying in a channel of the anther. *Ovarium* 1-celled, with a single pendulous ovulum. *Fruit* nucamentaceous, 1-celled, 1-seeded. — *Aquatic or mud* herbaceous *plants.* Leaves *linear, whorled.* Flowers *sessile, axillary, minute, sometimes polygamous.*

1. H. *vulgaris* Linn. Common Mare's-tail. E. B. 11. 763.
Leaves linear, many in each whorl. *Smith.*
In ditches. — Perennial. *May, June.*

Order 38. UMBELLIFERÆ *Juss.*

Calyx superior, either entire, or 5-toothed.

Petals 5, inserted on the outside of a fleshy disk; usually inflexed at the point; æstivation generally valvate, very rarely imbricate.

Stamens 5, alternate with the petals, incurved in æstivation.

Ovarium inferior, 2-celled, with solitary pendulous ovula; crowned by a double fleshy disk; *styles* 2, distinct; *stigmata* simple.

Fruit consisting of 2 carpella, separable from a common axis, to which they adhere by their face (*the commissure*); each carpellum traversed by elevated *ridges*, of which 5 are primary, and 4, alternating with them, secondary; the ridges are separated by *channels*, below which are often placed, in the substance of the testa, certain linear receptacles of coloured oily matter, called *vittæ*.

Seed pendulous, usually adhering inseparably to the pericarpium, rarely loose; *embryo* minute, at the base of abundant horny *albumen*; *radicle* pointing to the hilum.

Herbaceous plants, with fistular furrowed stems. *Leaves* usually compound, sometimes simple, sheathing at the base. *Flowers* in umbels, white, pink, yellow, or blue, generally surrounded by an involucrum.

ANALYSIS OF THE TRIBES AND GENERA.

TRIBES.

Albumen involute, or deeply channelled next the axis	
The primary lateral ridges placed on the plane of the commissure - - - - - - -	ii. CAUCALINEÆ.
The primary lateral ridges forming a margin	
Fruit turgid - - - - - - - -	x. SMYRNIEÆ.
Fruit long - - - - - - - - -	ix. SCANDICINEÆ.
Albumen solid	
Fruit with both primary and secondary ridges	
The lateral primary ridges on the plane of the commissure - - - - - - -	i. DAUCINEÆ.
The lateral primary ridges placed within an accessory border - - - - - -	iii. CORIANDREÆ.
Fruit with primary juga only, or with scarcely any, compressed at the back	
with two wings on each side - - -	vi. ANGELICEÆ.
with one wing on each side	
Border knotty, or plaited and thickened	iv. TORDYLINEÆ.
Border smooth, flattened - - -	v. SELINEÆ.
taper	
Umbels compressed - - - - -	vii. SESELINEÆ.
Umbels simple - - - - - -	xi. SANICULEÆ.
compressed at the sides	
Umbels compressed - - - - -	viii. AMMINEÆ.
Umbels simple - - - - - -	xii. HYDROCOTYLINEÆ.

GENERA.

- i. DAUCINEÆ - - - - 1. DAUCUS.
- ii. CAUCALINEÆ
  - Secondary ridges prickly - - - 2. CAUCALIS.
  - Secondary ridges obscure - - - 3. TORILIS.
- iii. CORIANDREÆ - - - - 4. CORIANDRUM.
- iv. TORDYLINEÆ
  - Vittæ triple - - - - 5. CONDYLOCARPUS.
  - Vittæ single - - - - 6. TORDYLIUM.
- v. SELINEÆ
  - Vittæ single
    - filiform - - - - 7. PASTINACA.
    - clavate - - - - 8. HERACLEUM.
  - Vittæ 1 to 3 - - - - 9. PEUCEDANUM.
- vi. ANGELICEÆ
  - Seed covered all over with vittæ - - - 10. ARCHANGELICA.
  - Vittæ single in each channel - - - 11. ANGELICA.
- vii. SESELINEÆ
  - Seed loose in the pericarpium when dry - - - 12. CRITHMUM.
  - Seed adhering closely to the pericarpium
    - Vittæ numerous
      - Petals unguiculate - - - 13. LIGUSTICUM.
      - Petals sessile
        - acute at each end - - - 14. MEUM.
        - obovate - - - 15. SILAUS.
    - Vittæ single
      - Carpella adhering to an axis
        - Petals roundish, entire - - - 16. FŒNICULUM
        - Petals obovate, emarginate
          - Fruit roundish, ovate - - - 17. ÆTHUSA.
          - Fruit oval, crowned by the reflexed styles - - - 18. SESELI.
      - Carpella with no real axis - - - 19. ŒNANTHE.
- viii. AMMINEÆ
  - Leaves simple - - - - 20. BUPLEURUM.
  - Leaves compound
    - Vittæ numerous
      - Fruit crowned with a conical disk and straight styles - - - 21. CONOPODIUM.
      - Fruit crowned by a flattened disk and reflexed styles
        - ovate - - - - 22. PIMPINELLA.
        - nearly double - - - 23. SIUM.
    - Vittæ single
      - Fruit apparently single
        - Petals ovate, entire, acute, or bluntish 24. HELOSCIADIUM.
        - Petals obovate, emarginate - - 25. CARUM.
        - Petals roundish, deeply emarginate - 26. SISON.
      - Fruit apparently double
        - Petals roundish, entire
          - All the vittæ single - - - 27. PETROSELINUM.
          - Outer vittæ triple - - - 28. APIUM.
        - Petals obovate, emarginate - - 29. CICUTA.
    - Vittæ wanting
      - Flowers hermaphrodite - - - 30. ÆGOPODIUM.
      - Flowers polygamous - - - 31. TRINIA.
- ix. SCANDICINEÆ
  - Fruit beaked
    - Carpella without ridges - - - 32. ANTHRISCUS.
    - Carpella with 5 ridges - - - 33. SCANDIX.
  - Fruit not beaked
    - Pericarpium solid - - - 34. CHÆROPHYLLUM.
    - Pericarpium hollow - - - 35. MYRRHIS.

x. Smyrnieæ
 Vittæ single
  Flowers hermaphrodite - - - - - - 36. Physospermum.
  Flowers monœcious - - - - - - 37. Echinophora.
 Vittæ numerous - - - - - - - - 38. Smyrnium.
 Vittæ none - - - - - - - - - - 39. Conium.

xi. Saniculeæ
 Fruit covered with scales - - - - - - 40. Eryngium.
 Fruit covered with prickles - - - - - 41. Sanicula.

xii. Hydrocotylineæ - - - - - - - - - 42. Hydrocotyle.

Tribe i. *Daucineæ* Koch.

Fruit compressed at the back, or somewhat taper. Carpella with the 5 primary ridges filiform and bristly, the lateral being placed on the plane of the commissure, and with 4 secondary ridges which are more prominent and prickly; the prickles sometimes running together into a sort of wing. Seed flat or half taper flattish in front.

1. DAUCUS *Linn.*

Calyx 5-toothed. *Petals* obovate, emarginate, inflexed; the outer often radiant and deeply bifid. *Fruit* compressed from the back. *Carpella* with the 5 primary ridges filiform and bristly, of which the 3 intermediate ones occupy the back, and the 2 lateral ones the plane of the commissure; the 4 secondary ridges equal, more prominent, with the prickles in a single row. *Channels* under the secondary ridges with single vittæ. *Seed* flat in front. — Involucrum *universal and partial, many-leaved.* Flowers *white or pink.*

1. D. *Carota* Linn. Wild Carrot. E. B. 17. 1174.
Bristles of the fruit slender. Leaflets pinnatifid, with linear-lanceolate acute segments. Umbels with a solitary, coloured, abortive flower; when in fruit concave.

In pastures, and the borders of fields, common. — Biennial. *June, July.*

2. D. *maritimus* Withering. E. B. 36. 2560.
Bristles of the fruit flattened. Leaflets dilated, fleshy, pinnatifid, with rounded segments. Umbels convex when in fruit.

On the sea-coast of the south of England. — Biennial. *July, August.*

Tribe ii. *Caucalineæ* Koch.

Fruit contracted at the side or nearly taper. Carpella with the 5 primary ridges filiform, bristly or prickly, the lateral being placed on the plane of the commissure, and with 4 secondary ridges, which are more prominent and prickly, or obliterated by the multitude of their prickles, which fill up the channels. Seed involute or inflexed at the margin.

2. CAUCALIS *Linn.*

Calyx 5-toothed. *Petals* obovate, emarginate, inflexed; the outer radiant and deeply bifid. *Fruit* laterally compressed. *Carpella* with the 5 primary ridges filiform, bristly, or with little prickles; of these the 3 middle are dorsal, the 2 lateral on the plane of the commissure;

the 4 secondary ridges more prominent, armed with prickles, in one or two rows. *Channels* under the secondary ridges with single vittæ. *Seed* involute, or inflexed at the edge. — Involucrum *both universal and partial, of many leaflets.* Flowers *white or pink.*

1. C. *daucoides* Linn. E. B. 3. 197.
Umbels 3-cleft, without general involucra; partial involucra 3-leaved, and ripening about 3 fruits. Leaves repeatedly subdivided.
Caucalis leptophylla *Huds.*
In corn fields, on a chalky soil. — Annual. *June*

2. C. *latifolia* Linn. E. B. 3. 198.
Umbels 3-cleft, with membranous involucra; partial ones ripening about 5 fruits. Leaves pinnate, serrated.
Tordylium latifolium *Linn*
In fields on a chalky soil. — Annual. *July.*

3. TORILIS *Adanson.*

Calyx 5-toothed. *Petals* obovate, emarginate, inflexed; the outer larger than the others and bifid. *Fruit* contracted at the side. *Carpella* with the 5 primary ridges bristly, of which the 3 middle are dorsal, and the 2 lateral ones in the plane of the commissure; the secondary ridges obliterated by the multitude of prickles which cover the channels. *Channels* under the prickles with single vittæ. *Seed* inflexed at the edge. — Involucrum *variable; the partial many-leaved.* Flowers *white or pink.*

1. T. *Anthriscus* Gærtn. E. B. 14. 987.
Umbels of many close rays. General involucrum many-leaved. Leaflets pinnatifid. Branches nearly upright.
Caucalis Anthriscus *Hudson.*
In hedges, and the borders of fields. — Annual. *July.*

2. T. *infesta* Spreng. E. B. 19. 1314.
Umbels of many close rays. General involucrum scarcely any. Leaflets pinnatifid. Branches spreading.
Caucalis infesta *Curtis.*
C. arvensis *Hudson.*
C. helvetica *Jacq.*
Scandix infesta *Linn.*
In fields and by waysides. — Annual. *July.*

3. T. *nodosa* Gærtn. E. B. 3. 199.
Umbels lateral, simple, nearly sessile. Stem prostrate. Fruit partly granulated. *Smith.*
Caucalis nodosa *Hudson.*
On banks, and about the borders of fields. — Annual. *May, June.*

Tribe iii. *Coriandreæ* Koch.

Fruit globose, or double, being formed of two globose carpella. Carpella with 5 primary ridges which are depressed and wavy, or in the form only of obscure furrows, the lateral being placed within an accessory margin; and with 4 secondary ridges, which are more prominent, but not winged. Seed rolled inwards from the base to the apex, or curved and hollowed out in front.

4. CORIANDRUM *Linn.*

Calyx 5-toothed. *Petals* obovate, emarginate, inflexed; the outer radiant and bifid. *Fruit* globose. *Carpella* with the 5 primary ridges depressed, and wavy; the lateral ones placed next an accessory margin; the 4 secondary more prominent and carinate. *Channels* without vittæ; the commissure with 2 vittæ. *Seed* hollowed in front, covered by a loose membrane. — Universal involucrum *wanting*; partial *halved*. Flowers *white*.

1. C. *sativum* Linn. Common Coriander. E. B. 1. 67.
The 2 hemispherical carpella making 1 uniform globe. Leaflets of the lower leaves wedge-shaped.
In fields and on dunghills; not really wild. — Annual. *June.*

Tribe iv. *Tordylineæ* Koch.

Fruit pressed flat at the back, surrounded by a thick dilated knotty or plaited border. Carpella with 5 primary, very fine or almost obsolete ridges, of which the lateral are contiguous to the dilated border, or make part with it; and without secondary ridges. Seed flattened.

5. CONDYLOCARPUS *Hoffmann.*

Calyx 5-toothed. *Petals* obovate, emarginate, inflexed; the outer radiant and bifid. *Fruit* flattened at the back, surrounded by an accessory thick knotted margin. *Carpella* with very fine ridges; the 3 dorsal equidistant, the 2 lateral contiguous to the thickened margin. *Channels* with 3 vittæ, each of which is separated by a furrow. *Seed* flat. — Involucrum *both universal and partial, of many leaves.* Flowers *white.*

C. *officinalis* Koch. E. B. 34. 2440.
Partial involucrum about as long as the flowers. Leaflets ovate, cut, crenate. Radiant petals in pairs, with very unequal lobes.
Tordylium officinale *Linn.*
Said to be British; but a very doubtful native. — Annual. *June, July.*

6. TORDYLIUM *Linn.*

Calyx 5-toothed. *Petals* obovate, emarginate, inflexed; the exterior radiant and bifid. *Fruit* compressed from the back, surrounded by an accessory thickened warted margin. *Carpella* with extremely obscure ridges; the 3 dorsal equidistant, the 2 lateral contiguous to the thickened margin, or even covered over by it. *Channels* with single filiform vittæ. *Seed* flat. — Involucrum *both universal and partial, of many leaves.* Flowers *white.*

1. T. *maximum* Linn. E. B. 17. 1173.
Leaflets lanceolate, deeply serrated and notched. Flowers somewhat radiant; the outermost petal of 2 equal lobes. Stem rough, with close deflexed bristles. *Smith.*
On banks and waste ground, but rare. — Annual. *June, July.*

Tribe v. *Selineæ* Koch.

Fruit pressed flat at the back, with a smooth, dilated, winged, flat or convex border. Carpella with 5 primary, filiform, occasionally very fine ridges, of which the lateral are contiguous to the border or make part of it; secondary ridges wanting. Seed flattened or rather convex at the back.

7. PASTINACA *Linn.*

Calyx almost obsolete. *Petals* roundish, entire, involute, with a broad, inflexed, blunt middle segment. *Fruit* flattened at the back, surrounded by a dilated flat margin. *Carpella* with very fine ridges; the 3 dorsal equidistant, the 2 lateral contiguous to the dilated margin. *Channels* with single vittæ. *Seed* flattened. — Involucrum *neither universal nor partial, or with very few leaves.* Flowers *yellow.*

1. P. *sativa* Linn. Common Parsnep. E. B. 8. *556.*
Leaves simply pinnate; downy beneath. *Smith.*
About the borders of fields in a chalky soil. — Biennial. *July.*

8. HERACLEUM *Linn.*

Calyx 5-toothed. *Petals* obovate, emarginate, inflexed; the outer often radiant and bifid. *Fruit* flattened at the back, surrounded by a flat dilated margin. *Carpella* with very minute ridges; the 3 dorsal equidistant, the 2 lateral contiguous to the dilated margin. *Channels* with single clavate vittæ. *Seed* flattened. — Universal involucrum *deciduous;* partial *many-leaved.* Flowers *large, white.*

1. H. *Sphondylium* Linn. Common Cow-Parsnep. Hogweed. E. B. 14. 939.
Leaves pinnate; leaflets pinnatifid, cut and serrated. *Smith.*
β. angustifolium.
H. angustifolium *Smith.*
In hedges and meadows, very common. — Biennial. *July.*

9. PEUCEDANUM *Linn.*

Calyx slightly 5-toothed, sometimes obsolete. *Petals* obovate, emarginate, or nearly entire, contracted into an inflexed segment. *Fruit* flattened at the back, surrounded by a flat dilated margin. *Carpella* with equidistant ridges; the 3 middle filiform, the 2 lateral more obsolete, contiguous to the dilated margin, or confounded with it. *Seed* flat in front. *Channels* with from 1 to 3 vittæ. — Universal involucrum *various;* partial *many-leaved.* Flowers *small, yellow or white.*

1. P. *Ostruthium* Koch. Great Masterwort. E. B. 20. 1380.
Leaves twice ternate, undivided or 3-lobed, rough-edged. Flower-stalks alternate. *Smith.*
Imperatoria Ostruthium *Linn.*
In rather moist meadows in Scotland. — Perennial. *June.*

2. P. *palustre* Mœnch. E. B. 4. 229.
Milky. Root generally single. Leaves triply pinnate; leaflets

pinnatifid, with elliptic-lanceolate segments. Rays of the umbel rough. Ridges of the fruit broad and obtuse. *Smith.*
Thysselinum palustre *Tourn.*
Th. Plinii *Spreng.*
Selinum palustre *Linn.*

In marshes and meadows. — Perennial. *July.*

3. P. *officinale* Linn. Sea Sulphur-wort. E. B. 25. 1767.
Leaves five times deeply 3-cleft; leaflets linear, undivided, flat. Bracteas linear, almost capillary. *Smith.*

In salt marshes; very rare. — Perennial. *July—September*

Tribe vi. *Angeliceæ* Koch.

Fruit compressed at the back, with a winged dilated margin, which, on account of the commissure being almost central, is double, so that the fruit has two wings on each side. Carpella with 5 primary ridges, of which the 3 dorsal are filiform or winged, and the lateral always winged and much wider than the dorsal ridges; secondary ridges wanting. Seed plano-convex.

10. ARCHANGELICA *Hoffmann.*

Calyx 5-toothed. *Petals* elliptical, entire, acuminate, with the point curved inwards. *Fruit* somewhat compressed from the back, with 2 wings on each side. *Carpella* with thick carinate ridges; the 3 dorsal elevated, the 2 lateral dilated into a wing twice as broad as the rest. *Seed* a loose kernel covered all over with numerous vittæ. — Universal involucrum *scarcely any*; partial *halved, many-leaved.* Flowers *white.*

1. A. *officinalis* Hoffm. Garden Angelica. E. B. 36. 2561.
Angelica archangelica *Linn.*

In watery places, rare, apparently a naturalized plant. — Biennial. *June—September.*

11. ANGELICA *Linn.*

Calyx an obsolete margin. *Petals* lanceolate, entire, acuminate, either incurved or not. *Fruit* compressed at the back, with 2 wings on each side. *Carpella* with 3 dorsal filiform raised ridges; the 2 lateral dilated into a wing twice as broad as the rest. *Channels* with single vittæ. *Seed* rather taper. — Universal involucrum *few-leaved or none*; partial *many-leaved.* Flowers *white.*

1. A. *sylvestris* Linn. E. B. 16. 1128.
Leaflets equal, ovate, serrated. *Smith.*

In watery places in thickets. — Perennial. *July.*

Tribe vii. *Seselineæ* Koch.

Fruit taper, or but little compressed either way. Carpella with 5 primary filiform or winged ridges, of which the lateral form a border; all equal, or the lateral a little the widest; secondary ridges wanting. Seed plano-convex. Commissure occupying the whole face of the seed.

12. CRITHMUM *Linn.*

Calyx an obsolete margin. *Petals* roundish, entire, involute, with an obovate end. *Fruit* taper. *Carpella* with 5, raised, sharp, somewhat winged ridges, of which the lateral are rather larger than the rest, and form a margin. *Seed* half-taper, forming a loose kernel, covered with numerous vittæ. — Universal and partial involucra *many-leaved.* Pericarpium *spongy, cellular.* Flowers *greenish-white.*

1. C. *maritimum* Linn. Samphire. E. B. 12. 819.
Leaflets lanceolate, fleshy. Bracteas ovate. *Smith.*
On rocky sea shores and cliffs. — Perennial. *August.*

13. LIGUSTICUM *Linn.*

Calyx 5-toothed or obsolete. *Petals* obovate, acutely emarginate, inflexed, with a very short unguis. *Fruit* taper, or a little compressed at the sides. *Carpella* with 5 sharp, winged, equal ridges; the lateral of which form a margin. *Channels* with many vittæ. *Seed* nearly half-taper. — Universal involucrum *various;* partial *many-leaved.* Flowers *white.*

1. L. *scoticum* Linn. E. B. 17. 1207.
Leaves twice ternate.
On the sea coasts of Scotland, and the north of England. — Perennial. *July.*

14. MEUM *Tournefort.*

Calyx an obsolete margin. *Petals* entire, elliptical, acute at each end. *Fruit* nearly taper. *Carpella* with 5, projecting, acutely carinate, equal ridges, of which the lateral form the margin. *Channels* with many vittæ. *Seed* nearly half-taper. — Universal involucrum *few-leaved or none;* partial *many-leaved.*

1. M. *athamanticum* Jacq. Spignel, Meu, or Bald-money. E. B. 32. 2249.
Leaflets all in numerous, deep, bristle-like segments. Involucra both general and partial.
Athamanta Meum *Linn.*
Æthusa Meum *Linn.*
Ligusticum Meum *Crantz.*
In mountainous pastures in the north. — Perennial. *May, June.*

15. SILAUS *Besser.*

Calyx an obsolete margin. *Petals* obovate-oblong, contracted into an inflexed lobe; entire or somewhat emarginate, either sessile, or with an appendage at the base. *Fruit* nearly taper. *Carpella* with 5 sharp, winged, equal ridges, of which the lateral form a margin. *Channels* with many vittæ. *Seed* nearly taper. — Universal involucrum *few-leaved or none;* partial *many-leaved.*

1. S. *pratensis* Besser. E. B. 30. 2142.
Leaflets deeply pinnatifid; their segments opposite, decurrent.
Peucedanum Silaus *Linn.*
Cnidium Silaus *Spreng.*
In rather moist meadows and pastures. — Perennial. *August, September.*

16. FÆNICULUM *Hoffmann.*

Calyx obsolete. *Petals* roundish, entire, with a nearly square, retuse, involute segment. *Fruit* nearly taper. *Carpella* with 5, prominent, obtusely-keeled ridges, of which the lateral form a margin, and are rather broader than the others. *Channels* with 1 vitta. *Seed* nearly half-taper. — Involucra *none.* Flowers *yellow.*

1. F. *vulgare* Hoffmann. E. B. 17. 1208.
Anethum Fæniculum *Linn.*
Meum Fæniculum *Spreng.*
On chalky cliffs and old banks. — Biennial. *July, August.*

17. ÆTHUSA *Linn.*

Calyx obsolete. *Petals* obovate, emarginate, inflexed. *Fruit* roundish-ovate. *Carpella* with 5 elevated, thick, acutely-keeled ridges, of which the lateral form a margin, and are rather wider than the others, surrounded by a somewhat winged keel. *Channels* with 1 vitta. *Seed* half globose.—Universal involucrum *wanting;* partial *3-leaved, pendulous.* Flowers *white.*

1. Æ. *Cynapium* Linn. Fool's-parsley. E. B. 17. 1192.
Leaves uniform; leaflets wedge-shaped, decurrent, with lanceolate segments. *Smith.*
In gardens and cultivated fields, a common weed. — Annual. *July, August.*

18. SESELI *Linn.*

Calyx 5-toothed. *Petals* obovate, contracted into an inflexed segment, emarginate or almost entire. *Fruit* oval or oblong, nearly taper, crowned by the reflexed styles. *Carpella* with 5 prominent, filiform, or thick ridges; the lateral of which form a margin, and are usually rather broader than the others. *Channels* with 1 vitta. *Seed* almost half round. — Universal involucrum *various;* partial *many-leaved.* Flowers *white.*

S. *Libanotis* Koch. E. B. 2. 138.
Leaves doubly pinnate, cut. Umbels hemispherical. *Smith.*
Athamanta Libanotis *Linn.*
In elevated chalky pastures. — Perennial. *August.*

19. ŒNANTHE *Linn.*

Calyx 5-toothed. *Petals* obovate, emarginate, inflexed. *Fruit* nearly taper, crowned by the erect styles. *Fruit* with 5, rather convex, obtuse ridges, of which the lateral form a margin, and are rather broader than the others. *Channels* with single vittæ. *Seed* taper, convex. *Axis* wanting. — Universal involucrum *wanting;* partial *many-leaved.* Flowers *white.*

1. Œ. *fistulosa* Linn. Water-dropwort. E. B. 6. 363.
Root sending forth runners. Stem-leaves pinnate, cylindrical, tubular. Universal involucrum mostly wanting.
In ditches, and other watery places. — Perennial. *July, August.*

2. Œ. *pimpinelloides* Linn. E. B. 5. 347.
Leaflets of the radical leaves wedge-shaped, cloven; of the

rest entire, flat, linear, elongated. Universal involucra many leaved.

In salt marshes. — Perennial. *July.*

3. Œ. *peucedanifolia* Pollich. E. B. 5. 348.
Leaflets all linear. Universal involucrum none. Knobs of the root sessile, elliptical.

In fresh-water ditches and bogs. — Perennial. *June.*

4. Œ. *crocata* Linn. E. B. 33. 2313.
Leaflets all wedge-shaped, many-cleft, nearly uniform. Fruit linear-oblong, with slender ridges.

In rivers and ditches. — Biennial. *June, July.*

5. Œ. *Phellandrium* Spreng. E. B. 10. 684.
Leaflets all uniform, with narrow, wedge-shaped, cut, divaricated segments. Fruit ovate, with 5 broad ridges, and narrow channels.
Œ. aquatica *Lam.*
Phellandrium aquaticum *Linn.*

In watery places. — Perennial. *July.*

Tribe viii. *Ammineæ* Koch.

Fruit evidently compressed at the sides, and generally double. Carpella with 5 primary, filiform, sometimes winged ridges, of which the lateral forming the edge are of the same size as the others. Seed taper or prominent on one side. *Koch.*

20. BUPLEURUM *Linn.*

Calyx an obsolete margin. *Petals* roundish, entire, closely involute, with a broad retuse segment. *Fruit* compressed at the side, crowned by the depressed disk. *Carpella* with 5 equal winged ridges, either sharp and filiform, or very slight and obsolete; the lateral ones forming a margin. *Channels* with or without vittæ. *Seed* taper, convex, flattish in front. — Involucra *various.* Flowers *greenish-yellow.* Leaves *entire.*

1. B. *rotundifolium* Linn. Common Hare's-ear. Thorow-wax. E. B. 2. 99.
Universal involucrum wanting. Leaves perfoliate.

In corn-fields. — Annual. *July.*

2. B. *Odontites* Linn. E. B. 35. 2468.
Leaves of the partial involucrum 5, ovate, acute, 3-ribbed; of the general 3 or 4. Branches widely spreading. Umbels all stalked. Leaves lanceolate.
Odontites lutea *Sprengel.*

On rocks in Devonshire. — Annual. *July.*

3. B. *tenuissimum* Linn. E. B. 7. 478.
Umbels simple, alternate, of about 3 flowers, with 5 awl-shaped bracteas. *Smith.*
Odontites tenuissima *Spreng.*

In muddy salt-marshes. — Annual. *August, September.*

21. CONOPODIUM *Koch.*

Calyx an obsolete margin. *Petals* obovate, emarginate, or nearly so, inflexed. *Fruit* contracted at the side, linear-oblong, crowned by a conical unedged disk and straight styles. *Carpella* with 5, equal, filiform, obtuse ridges, and many vittæ. *Seed* taper, convex, flat in front.—Universal involucrum *none*, partial *with few leaves*. Flowers *white*.

1. C. *flexuosum*. Common Earth-nut. Kipper- or Pig-nut. E. B. 14. 988.
General bracteas scarcely 3. Stem tapering and zigzag at the base. Fruit somewhat beaked. Styles nearly upright. *Smith.*
Bunium flexuosum *Withering.*
In grassy pastures, common.—Perennial. *May, June.*

22. PIMPINELLA *Linn.*

Calyx an obsolete margin. *Petals* obovate, emarginate, inflexed. *Fruit* contracted at the side, ovate, crowned by a convex disk and reflexed styles. *Carpella* with 5, equal, filiform ridges, of which the lateral form a margin. *Channels* with many vittæ. *Seed* convex, flattish in front. — Involucrum *none*. Flowers *white*.

1. P. *saxifraga* Linn. Common Burnet-saxifrage. E. B. 6. 407.
Leaves pinnate; leaflets of the radical ones roundish; of the uppermost in various linear segments. *Smith.*
P. dissecta *Retz.*
In dry gravelly or chalky pastures.—Perennial. *July, August.*

2. P. *magna* Linn. E. B. 6. 408.
Leaves pinnate; leaflets all ovate, serrated, somewhat cut; the terminal one 3-lobed. *Smith.*
P. major *Hudson.*
In shady places and under hedges.—Perennial. *July, August.*

23. SIUM *Linn.* Water-parsnep.

Calyx 5-toothed or obsolete. *Petals* obovate, emarginate, with an inflexed segment. *Fruit* compressed at the sides, or contracted and nearly double, crowned by the disk and reflexed styles. *Carpella* with 5, equal, filiform, bluntish ridges and numerous vittæ. *Seed* nearly taper. — Universal involucrum *various*; partial *many-leaved*.

1. S. *latifolium* Linn. E. B. 3. 204.
Leaves pinnate; leaflets oblong-lanceolate, equally serrated. *Smith.*
In rivers, ditches, and fens.—Perennial. *July, August.*

2. S. *angustifolium* Linn. E. B. 2. 139.
Leaves pinnate; leaflets unequally lobed and serrated. Umbels stalked, opposite to the leaves. Stem erect. *Smith.*
S. erectum *Hudson.*
In ditches and rivulets.—Perennial. *July, August.*

24. HELOSCIADIUM *Koch.*

Calyx a 5-toothed margin, occasionally obsolete. *Petals* ovate, entire, acute, or bluntish, with a point; the point either straight or inflexed.

Fruit compressed at the side, ovate, or oblong. *Carpella* with 5, filiform, prominent, equal ridges, of which the lateral form a margin. *Channels* with single vittæ. *Seed* more or less convex, flattish in front. — Involucra *various*. Flowers *white*.

1. H. *nodiflorum* Koch. E. B. 9. 689.
Leaves pinnate; leaflets ovate, equally serrated. Umbels nearly sessile, opposite the leaves. Stem procumbent. *Smith*.
Sium nodiflorum *Linn*.
In ditches, common. — Perennial. *July, August*.

2. H. *repens* Koch. E. B. 20. 1431.
Leaves pinnate; leaflets roundish, deeply toothed. Umbels stalked, opposite the leaves. Stem creeping. *Smith*.
Sium repens *Linn*.
In wet places. — Perennial. *August*.

3. H. *inundatum* Koch. E. B. 3. 227.
Leaves pinnate, cut; the lowermost in many compound capillary segments. Umbels 5-flowered, in pairs. *Smith*.
Sium inundatum *Wiggers*.
Sison inundatum *Linn*.
Meum inundatum *Spreng*.
Hydrocotyle inundata *Smith*.
In ditches and wet places. — Perennial. *May*.

25. CARUM *Linn*.

Calyx an obsolete margin. *Petals* obovate, emarginate, regular, inflexed. *Fruit* compressed at the sides, oblong. *Carpella* with 5, filiform, equal ridges, of which the lateral form a margin; commissure flat. *Channels* with single vittæ. *Seed* taper, convex, flattish in front. — Involucra *various*. Flowers *white*.

1. C. *Carui* Linn. Common Caraway. E. B. 21. 1503.
Stem branched. Partial involucrum none.
In meadows and pastures; a naturalized plant. — Biennial. *June*.

2. C. *verticillatum* Koch. E. B. 6. 395.
Leaflets in numerous, linear, capillary, almost whorled segments. Involucra many-leaved.
Sison verticillatum *Linn*.
Sium verticillatum *Lam*.
In salt marshes, rare. — Perennial. *July, August*.

26. SISON *Linn*.

Calyx an obsolete margin. *Petals* roundish, curved, deeply emarginate, inflexed. *Fruit* compressed at the sides, ovate. *Carpella* with 5, equal, filiform ridges, of which the lateral form a margin. *Channels* with single, short, clavate vittæ. *Seed* very convex, flattish in front. — Involucra *few-leaved*. Flowers *white*.

1. S. *Amomum* Linn. Honewort. E. B. 14. 954.
Leaves pinnate; the upper ones ternate. Umbels erect, of about 4 general rays. Disk globose.
In marly or chalky, rather moist, ground, under hedges. — Annual or Biennial. *August*.

27. PETROSELINUM *Hoffm.*

Calyx an obsolete margin. *Petals* roundish, incurved, entire, scarcely emarginate, contracted into an inflexed lobe. *Fruit* ovate, contracted at the side, nearly double. *Carpella* with 5, equal, filiform ridges, of which the lateral form a margin. *Channels* with single vittæ. *Seed* gibbous, convex, flattish in front. — Universal involucrum *few-leaved*; partial *many-leaved*. Flowers *white*.

1. P. *segetum* Koch. E. B. 4. 228.
Leaves pinnate; leaflets roundish-ovate, numerous. Umbels drooping, irregularly aggregate. Disks subulate, depressed.
Sison segetum *Linn.*
In moist fields. — Annual or Biennial. *August.*

28. APIUM *Linn.*

Calyx an obsolete margin. *Petals* roundish, entire, with an involute point. *Fruit* roundish, contracted at the sides, double. *Carpella* with 5, filiform, equal ridges, of which the lateral form a margin. *Channels* with single vittæ, except the outermost, which have sometimes 2 or 3. *Seed* very convex, flattish in front. — Involucra *none*. Flowers *white*.

1. A. *graveolens* Linn. Celery. E. B. 17. 1210.
Leaflets of the stem-leaves wedge-shaped. Stem furrowed.
In ditches and marshy ground. — Biennial. *August, September.*

29. CICUTA *Linn.*

Calyx a 5-toothed, somewhat leafy margin. *Petals* obovate, emarginate, inflexed. *Fruit* roundish, contracted at the sides, double. *Carpella* with 5, flattish, equal ridges, of which the lateral form a margin. *Channels* with single vittæ, which in the ripe fruit are more elevated than the ridges. *Seed* taper.—Universal involucrum *few-leaved or wanting*; partial *many-leaved*. Flowers *white*.

1. C. *virosa* Linn. Water Hemlock. E. B. 7. 479.
Leaves twice ternate; leaflets linear-lanceolate, decurrent. *Smith.*
In ditches, and about the margins of rivers. — Perennial. *August.*

30. ÆGOPODIUM *Linn.*

Calyx an obsolete margin. *Petals* obovate, emarginate, inflexed. *Fruit* compressed at the side, oblong. *Carpella* with 5, filiform ridges, of which the lateral form a margin. *Channels* without vittæ. *Seed* taper, convex, flattish in front. — Involucra *none*. Flowers *white*.

1. Æ. *Podagraria* Linn. Common Gout-weed. Herb Gerarde. E. B. 14. 940.
Sison Podagraria *Spreng.*
In shady waste places, and under hedges. — Perennial. *May, June.*

31. TRINIA *Hoffm.*

Calyx an obsolete margin. *Petals* of the male plant lanceolate, contracted into a lanceolate segment, somewhat emarginate; of the female or hermapnrodite plant ovate, with a short inflexed point. *Fruit* compressed at the side, ovate. *Carpella* with 5, prominent, equal, filiform ridges, of which the lateral form a margin. *Channels* without vittæ, or with scarcely any appearance of them, but a distinct channel under each ridge. *Seed* convex, flattish in front. — Involucra *various*. Flowers *white*.

1. T. *glaberrima* Hoffm. E. B. 17. 1209.
Root fusiform. Stems weak, nearly smooth. Leaflets multifid, with linear segments. Umbels numerous, simple, and compound. Involucrum none, or 1-leaved. *Dec.*
Pimpinella dioica *Linn.*
P. pumila *Jacq.*
On limestone rocks, in warm situations. — Perennial. *May, June.*

Tribe ix. *Scandicineæ* Koch.

Fruit evidently compressed at the side, linear, and usually beaked. Carpella with 5, primary, filiform, sometimes winged ridges, of which the lateral, forming a margin, are the same size as the others; secondary ridges wanting; or sometimes the ridges are altogether obliterated except in the beak. Seed taper, with a deep cleft in front, or with an involute margin.

32. ANTHRISCUS *Sprengel.*

Calyx an obsolete margin. *Petals* obovate, truncate, or emarginate, inflexed, often very short. *Fruit* contracted at the side, beaked. *Carpella* almost taper, without ridges, the beak only having 5. *Seed* taper, deeply furrowed in front. — Universal involucrum *none;* partial *many-leaved.* Flowers *white.*

1. A. *vulgaris* Spreng. E. B. 12. 818.
Fruit ovate, twice the length of its beak. Leaves triply pinnate, pinnatifid. *Smith.*
Scandix Anthriscus *Linn.*
Caucalis scandicina *Wiggers.*
In banks and waste ground, chiefly near large towns. — Annual. *May.*

2. A. *Cerefolium* Hoffm. Chervil. E. B. 18. 1268.
Umbels sessile, lateral. Leaflets of the involucra lanceolate. Leaves doubly pinnate, cut.
Scandix Cerefolium *Linn.*
Chærophyllum sativum *Spreng.*
In waste ground. — Annual. *June.*

3. A. *sylvestris* Hoffm. E. B. 11. 752.
Umbels terminal, stalked. Leaflets of the involucra ovate, membranous. Leaves triply pinnate; leaflets ovate, pinnatifid, rough-edged.
Chærophyllum sylvestre *Linn.*
In hedges, very common. — Perennial. *April, May.*

33. SCANDIX *Linn.*

Calyx an obsolete margin. *Petals* obovate, truncate, inflexed. *Fruit* compressed at the side, with a very long beak. *Carpella* with 5, obtuse, equal ridges, of which the lateral form a margin. *Channels* without vittæ, or with scarcely any. *Seed* taper, with a deep furrow in front. — Universal involucrum *none, or few-leaved; partial 5- or 7-leaved.* Flowers *white.*

1. S. *Pecten-Veneris* Linn. Shepherd's-needle. Venus's Comb. E. B. 20. 1397.
   Fruit nearly smooth, with a bristly-edged beak. Umbels simple; solitary or in pairs. Leaflets of the involucra jagged. Petals inflexed at the point.
   In waste fields, common. — Annual. *June—September.*

34. CHÆROPHYLLUM *Linn.*

Calyx an obsolete margin. *Petals* obovate, emarginate, inflexed. *Fruit* compressed or contracted at the sides. *Carpella* with 5, obtuse, equal ridges, of which the lateral form a margin; the commissure with a deep furrow. *Channels* with a single vitta. *Seed* taper, its transverse section lunate. — Universal involucrum *wanting, or few-leaved; partial of several leaflets.* Flowers *white.*

1. C. *aromaticum* Linn.
   Fruit smooth. Styles long. Leaflets ovate, acute, serrated, undivided.
   Myrrhis aromatica *Spreng.*
   Near Guthrie, by the road leading from Forfar to Arbroath. — Perennial. *June.*

2. C. *aureum* Linn. E. B. 30. 2103.
   Fruit nearly smooth, coloured. Stem slightly swelling, angular, hairy. Leaflets pointed, sharply pinnatifid, or cut. Universal involucrum few-leaved; partial deflexed.
   Myrrhis aurea *Spreng.*
   Borders of fields in Scotland. — Perennial. *June.*

3. C. *temulum* Linn. E. B. 22. 1521
   Fruit nearly smooth. Stem rough, swollen under each joint.
   Myrrhis temula *Spreng.*
   In bushy places. — Biennial. *June, July.*

35. MYRRHIS *Scopoli.*

Calyx an obsolete margin. *Petals* obovate, emarginate, inflexed. *Fruit* compressed at the side. *Pericarpium* hollow, externally with 5 elevated, sharp, keeled ridges, internally adhering closely to the seed. *Seed* involute. *Vittæ* wanting. — Universal involucrum *none; partial many-leaved.* Flowers *white.*

1. M. *odorata* Scopoli. Sweet Cicely. E. B. 10. 697.
   Scandix odorata *Linn.*
   Chærophyllum odoratum *Hooker.*
   In mountainous pastures. — Perennial. *May.*

Tribe x. *Smyrnieæ* Koch.

Fruit compressed or contracted at the sides, turgid. Carpella with 5 primary ridges, of which the lateral either occupy the margin, or are placed within the margin; secondary ridges wanting; sometimes the ridges are almost obliterated when the fruit is polished. Seed involute, or with a deep incision in front.

36. PHYSOSPERMUM *Cusson.*

Calyx a 5-toothed margin. *Petals* obovate, somewhat emarginate, inflexed. *Fruit* contracted at the side, double. *Carpella* roundish, uniform, with 5 fine equal ridges, of which the lateral are placed within the margin. *Channels* with single vittæ. *Seed* involute, lunate. — Universal *and* partial involucra *many-leaved.* Flowers *white.*

1. P. *commutatum* Sprengel. E. B. 10. 683.
Radical leaves twice or thrice pinnate, rough-edged, cut; stem-leaves ternate, lanceolate, entire. Ridges of the fruit bluntish.
Ligusticum cornubiense *Linn.*
In fields about Bodmin, in Cornwall. — Perennial. *July.*
Obs. Good *ripe* fruit of this plant requires to be examined.

37. ECHINOPHORA *Linn.*

Calyx a 5-toothed margin. *Petals* obovate, emarginate, inflexed, or the external larger than the rest and bifid. *Flowers* of the ray male with long stalks, of the centre a solitary female. *Fruit* ovate, nearly taper, included in a hollow receptacle, with a short projecting beak. *Carpella* with 5, equal, depressed, wavy, streaked ridges. *Channels* with single vittæ, which are covered by an arachnoid membrane. — Involucra *both of many leaves.*

1. E. *spinosa* Linn. Prickly Samphire. E. B. 34. 2413.
Leaves with spinous, awl-shaped, entire segments. *Smith.*
On the sea-coast — a doubtful native. — Perennial. *July.*

38. SMYRNIUM *Linn.*

Calyx an obsolete margin. *Petals* lanceolate, or elliptical, entire, with a long inflexed point. *Fruit* contracted at the sides, double. *Carpella* roundish, reniform, with the 3 dorsal ridges prominent and sharp; the 2 lateral forming an obscure margin. *Channels* with many vittæ. *Seed* involute. — Involucra *various.*

1. S. *Olusatrum* Linn. Common Alexanders. E. B. 4. 230.
Stem-leaves ternate, stalked, serrated. *Smith.*
In waste ground; often on rocks and cliffs near the sea. — Biennial. *May.*

39. CONIUM *Linn.*

Calyx an obsolete margin. *Petals* obovate, emarginate, inflexed. *Fruit* compressed at the side, ovate. *Carpella* with 5, prominent, wavy, crenated, equal ridges, of which the lateral form a margin. *Channels* with many streaks, but no vittæ. *Seed* with a deep narrow incision in the face. — Universal involucrum *few-leaved;* partial *3-leaved, halved.*

1. C. *maculatum* Linn. Common Hemlock. E. B. 17. 1191.
Stem polished and spotted, much branched. *Smith.*
In hedges and waste ground, frequent. — Biennial. *June, July.*

Tribe xi. *Saniculeæ* Koch.

Fruit nearly taper. Carpella with 5, primary, equal ridges, and no secondary ones; or without ridges, in place of which the fruit is covered with scales or prickles. Seed half-taper, flat in front. Umbels fascicled or capitate. Petals erect, broken inwards in the middle.

40. ERYNGIUM *Linn.*

Calyx a 5-toothed leafy margin. *Petals* erect, converging, oblong-obovate, emarginate, with an abruptly incurved segment the length of the petal. *Fruit* nearly taper, obovate. *Carpella* covered with scales, with neither ridges nor vittæ. *Seed* about half-taper. — Umbels *simple.* Involucrum *many-leaved.* Flowers *usually blue.* Leaves *simple.*

1 E. *maritimum* Linn. Sea Holly. E. B. 10. 718.
Radical leaves roundish, plaited, spinous. Heads stalked. Bracteæ 3-cleft.

On the sandy sea-shore. — Perennial. *July, August.*

2. E. *campestre* Linn. E. B. 1. 57.
Leaves clasping the stem; radical ones twice or thrice pinnatifid. Bracteæ undivided.

In waste ground, especially near the sea, but rare. — Perennial. *July, August.*

41. SANICULA *Linn.*

Calyx a 5-toothed leafy margin. *Petals* erect, converging, obovate, emarginate, with an abruptly inflexed segment, the length of the petal. *Fruit* taper, nearly round. *Carpella* densely covered with hooked prickles, no ridges, but many vittæ. *Seed* half round. — Universal *and* partial involucra *of several leaves.*

1. S. *europæa* Linn. Wood Sanicle. E. B. 2. 98.
Radical leaves simple, deeply lobed. Flowers all nearly sessile. *Smith.*

In woods and groves abundantly. — Perennial. *May.*

Tribe xii. *Hydrocotylineæ* Koch.

Fruit very much flattened laterally. Carpella with 5 primary ridges, of which the lateral either occupy the margin, or the place of the commissure; secondary ridges wanting. Seed flattish in front. Umbels simple or imperfect, the flowers being capitate or whorled. Petals spreading, entire, acute, with a straight or inflexed end.

42. HYDROCOTYLE *Linn.*

Calyx an obsolete margin. *Petals* ovate, entire, acute, with a straight point. *Fruit* compressed at the side, so as to form 2 little shields. *Carpella* with 5 filiform ridges, those of the keel and sides nearly obsolete, the intermediate arched, without vittæ. *Seed* carinate, compressed. — *Creeping* herbs, *with simple* leaves, *and green obscure* flowers.

1. H. *vulgaris* Linn. Common White-rot. Marsh Pennywort. E. B. 11. 751.
Leaves orbicular, peltate, smooth; cloven at the base. Umbels somewhat aggregate. Flowers nearly sessile. *Smith.*
On boggy commons, and the margins of little clear rivulets, very frequent. — Perennial. *May, June.*

Order 39. Stellatæ.

Calyx superior, 4- 5- or 6-lobed.
Corolla monopetalous, rotate or tubular, regular, inserted into the calyx; the number of its divisions equal to those of the calyx.
Stamens equal in number to the lobes of the corolla, and alternate with them.
Ovarium simple, 2-celled; *ovules* solitary, erect; *style* simple; *stigmata* 2.
Fruit a dry indehiscent pericarpium, with 2 cells, and 2 seeds.
Seeds erect, solitary; *embryo* straight in the axis of horny albumen; *radicle* inferior; *cotyledons* leafy.
Herbaceous plants; with whorled *leaves,* destitute of *stipulæ;* square *stems; roots* staining red; *flowers* minute.

1. GALIUM *Linn.*

Corolla rotate, or campanulate, 5-cleft. *Fruit* dry, not crowned by the calyx.

* *Fruit smooth.*

1. G. *cruciatum.* Linn. Mug-weed. E. B. 2. 143.
Leaves ovate, hoary, 4 in a whorl. Stem hairy, simple above. Flower-stalks axillary, corymbose, with 2 leaves. *Smith.*
In thickets and hedges, common. — Perennial. *May.*

2. G. *palustre* Linn. E. B. 26. 1857.
Leaves obovate, obtuse; the upper ones 4 in a whorl, unequal in size. Stem weak; branched in the upper part. *Smith.*
In moist meadows, and ditches. — Perennial. *July.*

3. G. *Witheringii* Smith. E. B. 31. 2206.
Leaves about 5 in a whorl, widely spreading, lanceolate, fringed with bristles. Stem upright, slightly branched, rough with reversed hooks. *Smith.*
G. montanum *Withering.*
In moist, heathy spots. — Perennial. *July.*

4. G. *saxatile* Linn. E. B. 12. 815.
Leaves 6 in a whorl, obovate, obtuse, with a small point. Stem much branched, prostrate, smooth. Fruit granulated. *Smith.*
G. montanum *Hudson.*

G. procumbens *Withering.*
G. hercynicum *Weig.*
On heaths and hilly ground. — Perennial. *June—August.*

5. G. *uliginosum* Linn. E. B. 28. 1972.
Leaves 6 in a whorl, obovate-lanceolate, rigid, bristle-pointed; their edges rough, like the stem, with recurved prickles. Fruit smooth, smaller than the corolla. *Smith.*
In watery places. — Perennial. *August.*

6. G. *erectum* Hudson. E. B. 29. 2067.
Leaves about 8 in a whorl, lanceolate, bristle-pointed, with marginal prickles all pointing forward. Stem weak, slightly hairy under each joint. Fruit smooth and even. Corolla taper-pointed. *Smith.*
In hedges and pastures — Perennial. *June, July.*

7. G. *cinereum* Allioni.
Leaves 6 or 8 in a whorl, linear, bristle-pointed, with marginal prickles all pointing forward. Stem weak, much-branched, smooth. Fruit smooth. Corolla taper-pointed. *Smith.*
G. diffusum *Hooker.*
In the lowlands of Scotland. — Perennial. *August.*

8. G. *aristatum* Linn.
Leaves 6 in a whorl, stalked, lanceolate, flat, reticulated with veins, bristle-pointed, with minute marginal prickles pointing forward. Stem much-branched, spreading, smooth. Fruit smooth, reniform; its lobes distinct. Corolla taper-pointed.
In Angusshire. — Perennial. *July, August.*

9. G. *verrucosum* Smith. E. B. 31. 2173.
Leaves 6 in a whorl, lanceolate, with marginal prickles all pointing forward. Stalks axillary, 3-flowered. Fruit warty, drooping. *Smith.*
G. tricorne *G. Don.*
Valantia aparine *Linn.*
In corn fields, rare. — Annual. *June—August.*

10. G. *tricorne* Withering. E. B. 23. 1641.
Leaves about 8 in a whorl, lanceolate, with reflexed marginal prickles, like those on the stem. Stalks axillary, 3-flowered. Fruit sharply granulated, drooping. *Smith.*
G. spurium *Hudson.*
In dry chalky fields. — Annual. *July.*

11. G. *spurium* Linn. E. B. 26. 1871.
Leaves about 8 in a whorl, lanceolate, with reflexed marginal prickles, like those on the stem. Stalks axillary, many-flowered, cymose. Fruit smooth, erect. *Smith.*
About Forfar, sparingly. — Annual. *June, July.*

12. G. *pusillum* Linn. E. B. 2. 74.
Leaves 8 in a whorl, linear-lanceolate, hair-pointed, entire, somewhat hairy. Panicles terminal, forked. Fruit very smooth. *Smith.*

G. scabrum *Jacq.*
G. obliquum *Villars.*
On limestone hills. — Perennial. *July, August.*

13. G. *verum* Linn. E. B. 10. 660.
Leaves 8 in a whorl, linear, channelled, entire, rough. Flowers in dense panicles. Fruit smooth. *Smith.*
In hilly, bushy places. — Perennial. *July, August.*

14. G. *Mollugo* Linn. Hedge Bed-straw. Whip-tongue. E. B. 24. 1673.
Leaves 8 in a whorl, elliptical, bluntish, bristle-pointed, rough-edged. Flowers in loose spreading panicles. Corolla thick-tipped. Fruit smooth, globular. *Smith.*
In hedges and thickets. — Perennial. *July, August.*

15. G. *anglicum* Hudson. E. B. 6. 384.
Leaves about 6 in a whorl, lanceolate, pointed, fringed with prickles. Stems straggling, rough. Flower-stalks cloven. Fruit granulated, without hairs. *Smith.*
On walls, and dry sandy ground. — Annual. *June, July.*

** *Fruit bristly.*

16. G. *boreale* Linn. E. B. 2. 105
Leaves 4 in a whorl, ovate-lanceolate, 3-ribbed, smooth, with rough edges. Stem erect. Fruit rough with hooked bristles. *Smith.*
In rocky shady places. — Perennial. *July.*

17. G. *Aparine* Linn. Goose-grass, or Cleavers. E. B. 12. 816.
Leaves 8 in a whorl, lanceolate, keeled, rough, fringed with reflexed prickles. Stem weak. Fruit bristly. *Smith.*
In hedges, every where. — Annual. *May—August.*

2. ASPERULA *Linn.*

Corolla funnel-shaped, with 3 or 4 segments. *Fruit* dry, not crowned by the calyx.

1. A. *odorata* Linn. Sweet Woodruff. E. B. 11. 755.
Leaves 8 in a whorl, lanceolate. Panicles stalked, of few flowers. *Smith.*
In dry mountainous woods. — Perennial. *May.*

2. A. *cynanchica* Linn. Squinancy-wort. E. B. 1. 33.
Leaves linear, 4 in a whorl; the upper ones very unequal. Flowers all 4-cleft. Fruit smooth. *Smith.*
On dry chalky sunny banks. — Perennial. *June, July.*

3. SHERARDIA. *Linn.*

Corolla funnel-shaped, 4-cleft. *Fruit* dry, crowned with the persistent teeth of the calyx.

1. S. *arvensis* Linn. E. B. 13. 891.
All the leaves whorled. Flowers terminal. *Smith.*
In fields, on a light soil. — Annual. *June—August.*

4. PUBIA *Linn.*

Corolla campanulate, spreading, 4- or 5-lobed. *Stamens* 4 or 5. *Fruit* succulent, double, smooth.

1. R. *peregrina* Linn. Wild Madder. E. B. 12. 851.
Leaves 4, or more, in a whorl, elliptical, shining and smooth on the upper side. Flowers 5-cleft. *Smith.*
In thickets, and on stony or sandy ground, in the west of Britain. — Perennial. *June—August.*

Order 40. CAPRIFOLIACEÆ *Juss.*

Calyx superior, usually with 2 or more bracteæ at its base; entire or lobed.

Corolla superior, monopetalous or polypetalous, rotate or tubular, regular or irregular.

Stamens equal in number to the lobes of the corolla, and alternate with them.

Ovarium with from 1 to 5 cells, one of which is often monospermous, the others polyspermous; in the former the ovulum is pendulous; *style* 1; *stigmas* 1 or 3.

Fruit indehiscent; 1 or more celled, either dry, fleshy, or succulent, crowned by the persistent lobes of the calyx.

Seeds either solitary and pendulous, or numerous and attached to the axis; *testa* often bony; *embryo* straight at the top of the fleshy albumen; *radicle* superior.

Shrubs or *herbaceous* plants; with opposite or alternate *leaves*, destitute of *stipulæ*. *Flowers* usually corymbose and often sweet-scented.

§ *Lonicereæ.*

Corolla monopetalous.

1. CAPRIFOLIUM *Tourn.*

Calyx 5-toothed, persistent. *Corolla* tubular, 2-lipped, usually saccate at the base. *Stamens* 5. *Ovarium* 3-celled, with the cells equally many-seeded. *Berry* 1-celled, 1-seeded. — *Twining* shrubs, *with simple* leaves, *and capitate fragrant* flowers.

1. C. *perfoliatum.* E. B. 12. 799.
Flowers ringent, whorled, terminal. Leaves deciduous; the upper most confluent and perfoliate. *Smith.*
Lonicera Caprifolium *Linn.*
In woods and thickets. — Shrub. *May, June.*

2. C. *Periclymenum.* Common Honeysuckle, or Woodbine. E. B. 12. 800.
Heads of flowers ovate, imbricated, terminal. Leaves all separate, deciduous. Flowers ringent. *Smith.*
Lonicera Periclymenum *Linn.*
In hedges, groves, and thickets. — Shrub. *June, July, to October.*

2. LONICERA *Linn.*

Calyx 5-toothed, deciduous. *Corolla* funnel-shaped, saccate at the base, with an erect 2-lipped limb. *Stamens* 5. *Ovarium* 3-celled, with the cells equally many-seeded. *Berry* 2-celled, 2-seeded. — *Upright deciduous* shrubs, *with simple* leaves, *and twin inodorous* flowers.

1. L. *Xylosteum* Linn. Fly Honeysuckle. E. B. 13. 916.
Stalks 2-flowered. Berries distinct. Leaves entire, downy. *Smith.*
In thickets and rocky places. — Shrub. *July.*

3. LINNÆA *Gronovius.*

Calyx 5-cleft, with 4 connate bracteæ at the base. *Corolla* campanulate, 5-lobed. *Stamens* 4, of which 2 are shorter than the others. *Ovarium* 3-celled; 2 of the cells many-seeded, 1 few-seeded. *Fruit* dry, 3-celled, with 2-seeded cells. — *A* creeping *plant, with cernuous twin pink* flowers.

1. L. *borealis* Gronovius. E. B. 7. 433.
In dry stony shady fir woods in the north. — Perennial. *May, June.*

4. VIBURNUM *Linn.*

Calyx 5-cleft. *Corolla* campanulate, 5-lobed. *Stamens* 5. *Fruit* succulent, 3-seeded. — *Upright deciduous* shrubs, *with cymose* flowers *and simple* leaves.

1. V. *Lantana* Linn. Way-faring Tree. E. B. 5. 331.
Leaves heart-shaped, serrated, veiny; downy beneath. *Smith.*
In woods and hedges. — Shrub. *May.*

2. V. *Opulus* Linn. Common Guelder-rose. E. B. 5. 332.
Leaves lobed. Foot-stalks beset with glands. *Smith.*
In watery hedges and thickets. — Shrub. *June.*

5. SAMBUCUS *Linn.*

Calyx 5-cleft. *Corolla* rotate, 5-lobed. *Stamens* 5. *Berry* 3-seeded. — *Upright deciduous* shrubs, *with pinnated* leaves, *and cymose* flowers.

1. S. *Ebulus* Linn. Dwarf Elder. Danewort. E. B. 7. 475.
Cymes with 3 main branches. Stipulas leafy. Stem herbaceous. *Smith.*
In waste ground, and about hedges. — Perennial. *July.*

2. S. *nigra* Linn. Common Elder. E. B. 7. 476.
Cymes with 5 main branches. Stipulas obsolete. Leaflets ovate. Stem arboreous. *Smith.*
In hedges, coppices, and woods. — A small tree. *June.*

§ *Hederaceæ* Ach. Rich.

Corolla polypetalous.

6. CORNUS *Linn.*

Calyx 4-toothed, deciduous. *Petals* 4. *Stamens* 4. *Drupe* with a

2-celled nut. — *Erect deciduous* shrubs *or* herbaceous *plants, with simple* leaves, *and cymose or umbellate* flowers.

§. *Flowers naked.*

1. C. *sanguinea* Linn. Dog-wood. E. B. 4. 249.
Branches straight. Leaves green on both sides. Cymes naked, flat. *Smith.*
In hedges and thickets. — Shrub. *June.*

§§. *Flowers in an involucrum.* Thæmatia.

2. C. *suecica* Linn. E. B. 5. 310.
Herbaceous. Umbel between 2 branches, stalked, with an involucrum. Ribs of the leaves but slightly combined. *Smith.*
In moist alpine pastures. — Perennial. *June, July.*

7. HEDERA *Linn.*

Calyx 5-toothed. *Petals* 5. *Stamens* 5. *Anthers* forked at the base. *Berry* succulent, 5-celled, 5-seeded. — *Climbing evergreen* shrubs, *with umbellate* flowers *and simple* leaves.

1. H. *Helix* Linn. Common Ivy. E. B. 18. 1267.
Leaves some ovate, some lobed. *Smith.*
In woods, and on old buildings. — Shrub. *October.*

Order 41. Loranthеæ *Rich. & Juss.*

Calyx superior, with 2 bracteæ at the base.
Corolla with 4 or 8 petals, more or less united at the base.
Stamens equal in number to the petals, and opposite to them.
Ovarium 1-celled; *ovulum* pendulous; *style* 1 or none; *stigma* simple.
Fruit succulent, 1-celled.
Seed solitary, pendulous; *testa* membranous; *embryo* cylindrical, longer than the fleshy *albumen*; *radicle* naked, clavate, superior.
Parasitical herbaceous plants. *Leaves* opposite, veinless, fleshy, without stipulæ. *Flowers* often monœcious, axillary or terminal, solitary, corymbose, or spiked.

1. VISCUM. *Linn.* Misseltoe.

Diœcious. *Calyx* an entire margin. *Corolla* deeply 4-cleft, fleshy. Male. *Anthers* sessile, in the middle of the petals. Female. *Style* very small. *Stigma* capitate. *Berry* 1-seeded, crowned with the calyx.

1. V. *album* Linn. E. B. 21. 1470.
Leaves obovate-lanceolate, obtuse. Stem forked, with sessile intermediate heads, of about 5 flowers. *Smith.*
Parasitical on trees. — Shrub. *May.*

Order 42. VACCINIEÆ *Dec.*

Calyx superior, entire, or with from 4 to 6 lobes.
Corolla monopetalous, lobed as often as the calyx.
Stamens distinct, double the number of the lobes of the corolla, inserted into an epigynous disk; *anthers* with 2 horns and 2 cells.
Ovarium inferior, 4- or 5-celled, many-seeded; *style* simple; *stigma* simple.
Berry crowned by the persistent limb of the calyx, succulent, 4- or 5-celled, many-seeded.
Seeds minute; *embryo* straight, in the axis of a fleshy albumen; *cotyledons* very short; *radicle* long, inferior.
Shrubs with alternate coriaceous leaves.

1. VACCINIUM *Linn.*

Calyx entire or toothed. *Corolla* 4-cleft, with erect segments. *Stamens* 8. *Berry* crowned with the persistent calyx, many-seeded. — *Evergreen or deciduous* shrubs. Leaves *simple.* Flowers *axillary or racemose.*

* *Leaves deciduous.*

1. V. *Myrtillus* Linn. Bilberry. E. B. 7. 456.
Stalks solitary, single-flowered. Leaves ovate, serrated, membranous, smooth, deciduous. Stem acutely angular. Calyx wavy, nearly entire. *Smith.*
On stony heaths, and in woods where the soil is turfy. — Shrub. *May.*

2. V. *uliginosum* Linn. Great Bilberry. E. B. 9. 581.
Stalks somewhat aggregate, single-flowered. Leaves obovate, entire, smooth, deciduous. Branches round. *Smith.*
On boggy mountainous heaths. — Shrub. *May.*

** *Leaves evergreen.*

3. V. *Vitis Idæa.* Linn. Cow-berry. E. B. 9. 598.
Clusters terminal, drooping, with ovate concave bracteas, longer than the flower-stalks. Leaves obovate, revolute, minutely toothed; dotted beneath. Corolla bell-shaped. *Smith.*
On dry, stony, turfy heaths in Scotland, Wales, and the north of England. — Shrub. *June.*

2. OXYCOCCUS. *Rich.*

Calyx toothed. *Corolla* 4-cleft, with reflexed segments. *Berry* crowned with the persistent calyx, many-seeded. — *Evergreen trailing* shrubs.

1. O. *palustris Rich.* Cranberry. E. B. 5. 319.
Leaves ovate, entire, smooth, revolute, acute. Flowers terminal.
Vaccinium oxycoccus *Linn.*
In turfy bogs, among running waters. — Shrub. *June.*

Order 43. Campanulaceæ *Juss.*

Calyx superior, 5-lobed.

Corolla monopetalous, inserted into the top of the calyx, 5-lobed, withering on the fruit; regular or irregular.

Stamens 5, inserted into the calyx, alternately with the lobes of the corolla. *Anthers* distinct. *Pollen* spherical.

Ovarium inferior, with 2 or more polyspermous cells. *Style* simple; *stigma* with from 2 to 5 lobes.

Fruit dry, crowned by the withered calyx and corolla, with from 3 to 5 cells, dehiscing by lateral irregular apertures.

Seeds numerous, attached to a placenta in the axis; *embryo* straight in fleshy albumen; *radicle* inferior.

Herbaceous plants or *under* shrubs. *Leaves* alternate. *Flowers* single, or in heads; usually purple.

1. PHYTEUMA *Linn.*

Calyx 5-cleft. *Corolla* rotate, with a very short tube, and 5 long linear segments. *Stamens* 5. *Stigma* 3-parted. *Capsule* 3-celled, opening by lateral perforations. — Flowers *in spikes or heads.*

1. Ph. *orbiculare* Linn. E. B. 2. 142.
Flowers in a roundish head. Leaves crenate; radical ones heart-shaped, or elliptic-lanceolate. *Smith.*
In pastures, and by road sides, on a chalky soil. — Perennial. *August.*

2. P. *spicatum* Linn.
Radical leaves blunt, cordate-ovate, doubly toothed, with a winged foot-stalk; cauline, linear-lanceolate, toothed, sessile. Bracteæ few, linear-lanceolate, acuminate, 4 times as short as the long spike. *Dec.*
In hedges, very rare. — Perennial. *August.*

2. PRISMATOCARPUS *L'Heritier.*

Corolla rotate, with a flat limb. *Capsule* prismatical, 2- or 3-celled, dehiscing towards the top. *Dec.*

1. P. *hybridus* L'Heritier. E. B. 6. 375.
Stem upright, a little branched at the base. Leaves oblong, somewhat crenated. Flowers solitary. Segments of the calyx shorter than the corolla. *Dec.*
Campanula hybrida *Linn.*
In corn-fields. — Annual. *August.*

3. CAMPANULA *Linn.*

Calyx 5-cleft, sometimes with the recesses reflexed. *Corolla* campanulate, 5-cleft. *Stamens* 5, with the filaments broadest at the base. *Stigma* 4- or 5-parted. *Capsule* 3- or 5-celled, opening by perforations towards the base.

1. C. *rotundifolia* Linn. E. B. 13. 866.
Radical leaves heart or kidney-shaped, serrated; stem-leaves linear, entire. *Smith.*
On heaths, walls, banks, and about the borders of fields, common. — Perennial. *July, August.*

2. C. *patula* Linn. E. B. 1. 42.
Radical leaves obovate, or elliptic-lanceolate; the rest linear-lanceolate; all even, crenate, and roughish. Stem with several fringed angles. Panicle spreading. Calyx minutely toothed. *Smith.*
In pastures and hedges. — Biennial. *July, August.*

3. C. *Rapunculus* Linn. Rampion. E. B. 4. 283.
Leaves wavy, crenate, roughish; radical ones elliptic-lanceolate. Stem angular; hairy below. Panicle compact. Calyx entire. *Smith.*
On banks, and about the borders of fields. — Biennial. *July, August.*

4. C. *persicifolia* Linn.
Leaves smooth, slightly serrated; radical ones obovate; those of the stem linear-lanceolate, sessile, remote. Stem round, very smooth, with few flowers. *Smith.*
Near Cullen, in Scotland. — Perennial. *July.*

5. C. *latifolia* Linn. E. B. 5. 302.
Leaves roughish, ovate-lanceolate. Stem unbranched, round. Stalks single-flowered. Fruit drooping. *Smith.*
In moist woods and thickets. — Perennial. *July, August.*

6. C. *rapunculoides* Linn. E. B. 20. 1369.
Leaves roughish; radical ones heart-shaped, crenate, stalked; uppermost sessile, lanceolate. Flowers drooping, unilateral, in a terminal, bracteated, upright cluster. Calyx reflexed. *Smith.*
In woods and fields. — Perennial. *July, August.*

7. C. *Trachelium* Linn. E. B. 1. 12.
Stem angular. Leaves lanceolate, partly heart-shaped, sharply serrated, bristly as well as the calyx. Stalks axillary, with few flowers. *Smith.*
In groves, thickets, and hedges. — Perennial. *July.*

8. C. *glomerata* Linn. E. B. 2. 90.
Stem angular, simple. Flowers sessile, most of them in a terminal head. Leaves ovate, crenate. *Smith.*
In dry open chalky pastures. — Perennial. *July, August.*

9. C. *hederacea* Linn. E. B. 2. 73.
Stem flaccid, much branched, procumbent. Leaves stalked, smooth heart-shaped, with angular lobes. *Smith.*
In watery shady places. — Perennial. *June—August.*

Order 44. LOBELIACEÆ *Juss.*

Calyx superior, 5-lobed, or entire.
Corolla monopetalous, irregular, inserted in the calyx, 5-lobed, or deeply 5-cleft.
Stamens 5, inserted into the calyx alternately with the lobes of the corolla; *anthers* cohering; *pollen* oval.
Ovarium inferior, with from 1 to 3 cells; *ovula* very numerous, attached either to the axis or the lining; *style* simple; *stigma* surrounded by a cup or fringe.
Fruit capsular, 1- or more-celled, many seeded, dehiscing at the apex.
Seeds attached either to the lining or the axis of the pericarpium; *embryo* straight in the axis of fleshy albumen; *radicle* pointing to the hilum.
Herbaceous plants or *shrubs*. *Leaves* alternate, without stipulæ. *Flowers* axillary or terminal.

1. JASIONE *Linn.*

Calyx 5-cleft. *Corolla* rotate, with a very short tube, and 5 long linear segments. *Stamens* 5. *Stigma* 2-fid. *Capsule* 2-celled. — Flowers *collected within a many-leaved* involucrum.

1. J. *montana* Linn. Common Sheep's-bit. Sheep's Scabious. E. B. 13. 882.

In dry sandy fields, and heathy ground, plentiful. — Annual. *June, July.*

2. LOBELIA *Linn.*

Calyx 5-toothed. *Corolla* 2-lipped. *Stigma* blunt, usually 2-lobed. *Capsule* 2- or 3-celled. — Flowers *axillary, or in terminal* spikes.

1. L. *Dortmanna* Linn. E. B. 2. 140.
Leaves linear, entire, of 2 longitudinal cells. Stem nearly naked. *Smith.*

In the lakes of Wales, Scotland, Ireland, and the north of England. — Perennial. *July.*

2. L. *urens* Linn. E. B. 14. 953.
Stem nearly upright. Lower leaves obovate, slightly toothed; upper lanceolate, serrated. Clusters terminal. *Smith.*

On bushy heaths in Devonshire. — Perennial. *August, September.*

Order 45. VALERIANEÆ *Dec.*

Calyx superior; the limb either membranous or resembling pappus.
Corolla monopetalous, tubular, inserted into the top of the corolla, with

from 3 to 5 lobes, either regular or irregular; sometimes calcarate at the base.

Stamens from 1 to 5, inserted into the tube of the corolla, and alternate with its lobes.

Ovarium inferior, with 1 cell, and sometimes 2 other abortive ones; *ovulum* solitary, pendulous; *style* simple; *stigmas* from 1 to 3.

Fruit dry, indehiscent, with 1 fertile cell, and 2 empty ones.

Seed solitary, pendulous; *embryo* straight, destitute of albumen; *radicle* superior.

Herbs. *Leaves* opposite, without stipulæ. *Flowers* corymbose, panicled, or in heads.

1. VALERIANELLA *Tournef.*

Corolla regular, 5-lobed, without a spur. *Stamens* 3. *Fruit* membranous, with 3 cells, crowned with the erect, not involute limb of the calyx. *Dec.*

1. V. *olitoria* Mönch. Corn-salad or Lamb's Lettuce. E. B. 12. 811.

Stem weak. Leaves lanceolate, entire. Fruit naked, roundish, compressed. *Dec.*

Valeriana olitoria *Linn.*

Fedia olitoria *Vahl.*

In corn-fields. — Annual. *April, June.*

2. V. *dentata* Dec. E. B. 20. 1370.

Stem smooth. Lower leaves lanceolate, entire; upper irregularly toothed at the base. Bracteæ smooth. Fruit smooth, ovate, crowned with from 3 to 5 unequal teeth, of which 1 is longer than the rest. *Dec.*

Fedia dentata *Vahl.*

Valeriana dentata *Willd.*

In corn-fields. — Annual. *June, July.*

2. VALERIANA *Linn.*

Corolla regular, 5-lobed, without a spur. *Stamens* 3, otherwise as Centranthus. *Dec.*

1. V. *dioica* Linn. E. B. 9. 628.

Flowers diœcious. Stem-leaves pinnatifid; radical ones ovate.

In moist boggy meadows. — Perennial. *June.*

2. V. *officinalis* Linn. E. B. 10. 698.

Leaves all pinnate; leaflets lanceolate, nearly uniform.

In marshes, and about the banks of rivers. — Perennial. *June.*

3. V. *pyrenaica* Linn. E. B. 23. 1591.

Stem-leaves heart-shaped, serrated, stalked; the uppermost pinnate.

In woods in Scotland. — Perennial. *July.*

3. CENTRANTHUS *Dec.*

Corolla 5-lobed, regular, with a spur. *Stamen* 1. *Fruit* 1-celled, crowned with the limb of the calyx, which changes into a feathery pappus. *Dec.*

1. C. *latifolius* Dufresne. Red Valerian. E. B. 22. 1531.
Leaves ovate-lanceolate.
Valeriana rubra *Linn.*
On chalk cliffs, and old walls. — Perennial. *June—September.*

Order 46. Dipsaceæ *Juss.*

Calyx superior, membranous, resembling pappus; surrounded by a scarious involucellum.

Corolla monopetalous, tubular, inserted on the calyx; limb oblique, 4 or 5-lobed, with an imbricated æstivation.

Stamens usually 4 or 5, alternate with the lobes of the corolla; *anthers* distinct.

Ovarium inferior, 1-celled, with a single, pendulous ovulum; *style* 1; *stigma* simple.

Fruit dry, various, indehiscent, 1-celled, crowned by the pappus-like calyx; *embryo* straight in the axis of fleshy albumen; *radicle* superior.

Herbaceous plants or *under-shrubs*. *Leaves* opposite or whorled. *Flowers* collected upon a common receptacle, and surrounded by a many-leaved *involucrum*.

1. DIPSACUS. *Linn.*

Involucellum with 4 sides, and 8 little excavations. *Calyx* with a somewhat cyathiform limb. *Stigma* longitudinal. Leaflets of the *involucrum* longer than the bracteæ. *Receptacle* with spiny paleæ.

1. D. *fullonum* Linn. Fuller's Teasel. E B. 29. 2080.
Leaves combined, serrated. Scales of the receptacle hooked backwards. Involucrum reflexed.
About hedges, but scarcely wild. — Biennial. *July.*

2. D. *sylvestris* Linn. E. B. 15. 1032.
Leaves opposite, serrated. Scales of the receptacle straight. Involucrum inflexed, longer than the head.
About moist hedges, and by road sides. — Biennial. *July.*

3. D. *pilosus* Linn. Shepherd's Staff. E. B. 13. 877.
Leaves stalked, with lateral leaflets. Involucrum deflexed, about the length of the head.
In moist shady places. — Biennial. *August, September.*

2. SCABIOSA *Linn.*

Involucellum nearly cylindrical, with 8 little excavations. *Calyx* with a limb consisting of 5 setæ, occasionally partially abortive. *Coulter.*

1. S. *succisa* Linn. Devil's-bit. E. B. 13. 878.
Corolla in 4 equal segments. Heads nearly globular. Stem-leaves distantly toothed. *Smith.*
In grassy, rather moist pastures. — Perennial. *August—October.*

2. S. *columbaria* Linn. E. B. 19. 1311.
Corol in 5 unequal segments. Radical leaves ovate, or lyrate, notched; the rest pinnatifid, linear. *Smith.*
In pastures and waste ground. — Perennial. *June—August.*

3. KNAUTIA *Linn.*

Involucellum compressed, with 4 little excavations, closely surrounding the fruit, placed on a short stalk. *Calyx* with a somewhat cup-shaped limb.

1. K. *arvensis* Coulter. E. B. 10. 659.
Heads many-flowered. Teeth of the crown very small. Ciliæ of the calyx 8 or 16, somewhat awned. *Coulter.*
Scabiosa arvensis *Linn.*
In corn-fields and pastures. — Perennial. *July.*

Order 47. Compositæ *Juss.*

Calyx superior, closely adhering to the ovarium, and undistinguishable from it; its limb either wanting, or membranous, divided into bristles, paleæ, hairs or feathers, and called *pappus.*

Corolla monopetalous, superior, usually deciduous, either ligulate or funnel-shaped; in the latter case, 4 or 5-toothed, with a valvate æstivation.

Stamens equal in number to the teeth of the corolla, and alternate with them; the anthers cohering into a cylinder.

Ovarium inferior, 1-celled, with a single erect ovulum; *style* simple; *stigmas* 2, either distinct or united.

Fruit a small, indehiscent, dry pericarpium, crowned with the limb of the calyx.

Seed solitary, erect; *embryo* with a taper, inferior radicle; *albumen* none.

Herbaceous plants or *shrubs. Leaves* alternate or opposite, without stipulæ; usually simple. *Flowers* (called *florets*) collected in dense *heads* upon a common *receptacle,* surrounded by an *involucrum. Bracteæ* either present or absent; when present, stationed at the base of the florets, and called *paleæ of the receptacle.*

ANALYSIS OF THE TRIBES AND GENERA.

TRIBES.

Florets floscular in the disk. Stigma not articulated with the style - - - - - - - - - i. Corymbiferæ.
Florets floscular in the disk. Stigma articulated with the style - - - - - - - - - ii. Cynarocephalæ.
Florets ligulate in the disk, all hermaphrodite - - iii. Cichoraceæ.

GENERA.

i. Corymbiferæ

- Pappus hairy, abundant
  - Involucrum imbricated
    - Scales herbaceous
      - Flowers flosculous
        - Heads cylindrical - - - - - 1. Eupatorium.
        - Heads hemispherical or roundish
          - Florets all hermaphrodite - - 2. Chrysocoma.
          - Florets of the centre hermaphrodite; of the circumference female, imperfect - - - - - 3. Conyza.
      - Flowers radiant
        - Anthers with 2 bristles at the base
          - Pappus simple
            - Scales of involucrum leafy - 4. Inula.
            - Scales of involucrum simple - 5. Limbarda.
          - Pappus double - - - - 6. Publicaria.
        - Anthers naked at the base
          - Rays white or purple
            - oblong or lanceolate - - 7. Aster.
            - linear, in 2 rows - - - 8. Erigeron.
          - Rays yellow, about 5 - - - 9. Solidago.
    - Scales scarious, dry
      - Flowers diœcious - - - - - - 10. Antennaria
      - Flowers hermaphrodite
        - Scales of the involucrum unequal - - 11. Gnaphalium.
        - Scales of the involucrum equal - - 12. Filago.
  - Involucrum with external bracteolæ - - - 13. Senecio.
  - nvolucrum double - - - - - - - 14. Doronicum.
  - nvolucrum simple
    - Scales membranous at the margin - - - 15. Tussilago.
    - Scales not membranous at the margin
      - Receptacle flat - - - - - - - 16. Cineraria.
      - Receptacle conical - - - - - - 17. Bellis.
- Pappus wanting, or a membranous margin
  - Sexes in the same head
    - Receptacle naked or hairy
      - Flowers radiant
        - Scales of involucrum scarious at margin 18. Chrysanthemum.
        - Scales of involucrum not scarious - - 19. Matricaria.
      - Flowers flosculous
        - Pappus wholly wanting - - - 20. Artemisia.
        - Pappus membranous, entire - - 21. Tanacetum.
    - Receptacle paleaceous
      - Corolla with 2 sacs at the base - - - 22. Diotis.
      - Corolla not saccate at the base
        - Involucrum hemispherical, many-flowered
          - Pappus none - - - - - - 23. Maruta.
          - Pappus a membrane - - - - 24. Anthemis.
        - Involucrum ovate, few-flowered - - 25. Achillea.
  - Sexes in different heads - - - - - - - 26. Xanthium.
- Pappus aristate - - - - - - - - - 27. Bidens.

ii. Cynarocephalæ

- Receptacle excavated like a honey-comb - - - 28. Onopordum.
- Receptacle paleaceous
  - Pappus double - - - - - - - - 29. Saussurea.
  - Pappus single
    - feathery
      - Scales of the involucrum all spiny - - 30. Cnicus.
      - Inner scales of the involucrum scarious - 31. Carlina.
    - hairy
      - rigid, unequal - - - - - - - 32. Lappa.
      - soft, equal
        - between paleaceous and bristly - - 33. Silybum.
        - bristly
          - Scales of involucrum acute - - 34. Serratula.
          - Scales of involucrum spiny or torr
            - Paleæ jagged - - - - - 35. Centaurea.
            - Paleæ bristly - - - - - 36. Carduus.

iii. Cichoraceæ
Fruit compressed or 4-cornered, short
Involucrum imbricated
Pappus sessile - - - - - - - - - 37. Sonchus.
Pappus stalked - - - - - - - 38. Lactuca.
Involucrum with external bracteolæ
Pappus stalked - - - - - - - - - 39. Chondrilla.
Pappus sessile - - - - - - - - - 40. Prenanthes.
Fruit tapering to a point, long
Pappus wanting - - - - - - - - - 41. Lapsana.
Pappus hairy
Involucrum of the ripe fruit smooth - - 42. Leontodon.
Involucrum of the ripe fruit ribbed
Pappus stipitate - - - - - - - 43. Barkhausia.
Pappus sessile - - - - - - - - 44. Crepis.
Pappus feathery
stalked - - - - - - - - - - 45. Helminthia.
sessile - - - - - - - - - - 46. Picris.
Fruit taper, truncate at the base, short. Pappus hairy - 47. Hieracium.
Fruit cylindrical. Pappus feathery or scaly
Receptacle paleaceous - - - - - - - - 48. Hypochæris.
Receptacle hairy or naked
Involucrum simple - - - - - - - - 49. Tragopogon.
Involucrum imbricated
Pappus of the ray wanting - - - 50. Thrincia.
Pappus all feathery - - - - - 51. Apargia.
Involucrum with external bracteolæ - - 52. Cichorium.

Tribe 1. *Corymbiferæ* Juss.

Flowers flosculous, or radiant. Receptacle membranous, or not fleshy. Stigmata not articulated with the style.

1. EUPATORIUM *Linn.*

Involucrum cylindrical; scales imbricated, oval-oblong. *Florets* few, all tubular, hermaphrodite. *Receptacle* naked. *Pappus* pilose. *Dec.*

1. E. *cannabinum* Linn. Hemp-agrimony. E. B. 6. 428.
Leaves in 3, or 5, deep, lanceolate segments; the middle one longest. *Smith.*

In watery boggy places. — Perennial. *July, August.*

2. CHRYSOCOMA *Linn.*

Involucrum imbricated, hemispherical, or ovate; scales linear. *Florets* all hermaphrodite, tubular. *Pappus* hairy, ciliated. *Receptacle* excavated. *Dec.*

1. Ch. *Linosyris* Linn. Goldylocks. E. B. 35. 2505.
Herbaceous. Leaves linear, smooth. Involucrum loosely spreading.

On rocky cliffs, on the southern sea-coast. — Perennial. *August, September.*

3. CONYZA. *Linn.*

Involucrum roundish, imbricated. *Florets* all tubular; in the centre 5-toothed, hermaphrodite, in the ray barren, slender, 3-toothed. *Receptacle* naked. *Pappus* hairy, ciliated. *Dec.*

1. C. *squarrosa* Linn. Plowman's Spikenard. E. B. 17. 1195.
Leaves ovate-lanceolate, downy, crenate. Stem herbaceous, corymbose. Scales of involucrum leafy, recurved.

In chalky or limestone countries. — Biennial. *July, August.*

4. INULA *Linn.*

Involucrum imbricated; outer scales terminated by a leafy appendage. *Flowers* radiant, with the ray yellow. *Receptacle* naked. *Pappus* hairy, simple.

1. I. *Helenium* Linn. Elecampane. E. B. 22. 1546.
Leaves ovate, rugged, clasping the stem; downy beneath.
Corvisartia Helenium *Mérat.*
In moist meadows and pastures. — Perennial. *July, August.*

5. LIMBARDA *Cassini.*

Involucrum imbricated; scales without any appendage. *Flowers* radiant, with the ray yellow. *Receptacle* naked. *Pappus* hairy, simple.

1. L. *tricuspis* Cassini. Golden Samphire. E. B. 1. 68.
Leaves linear, fleshy, partly 3-pointed. Involucrum smooth.
Inula crithmoides *Linn.*
I. crithmifolia *Linn.*
On the sea-coast in the south. — Perennial. *August.*

6. PULICARIA *Cassini.*

Involucrum imbricated. *Flowers* radiant, with the ray yellow. *Receptacle* naked. *Pappus* double; the outer membranous.

1. P. *dysenterica* Cassini. E. B. 16. 1115.
Leaves oblong, downy, clasping the stem with their heart-shaped base. Stem woolly, panicled. Scales of involucrum bristle-shaped, hairy.
Inula dysenterica *Linn.*
In ditches. — Perennial. *August.*

2. P. *vulgaris* Cassini. E. B. 17. 1196.
Leaves clasping the stem, wavy. Stem much branched, hairy. Heads hemispherical; radius scarcely longer than the disk.
Inula pulicaria *Linn.*
I. uliginosa *Sibth.*
I. cylindrica *Withering.*
On damp commons. — Annual. *September*

7. ASTER *Linn.*

Involucrum imbricated; scales linear, acute. *Flowers* radiant; of the ray female, in a single row, oblong, not yellow. *Pappus* hairy. *Receptacle* naked.

1. A. *Tripolium* Linn. Sea Star-wort. E. B. 2. 87.
Herbaceous, corymbose. Leaves lanceolate, entire, fleshy, smooth, obscurely 3-ribbed. Scales of involucrum obtuse, somewhat membranous.
On the muddy sea-coast. — Perennial. *August, September.*

8. ERIGERON *Linn.*

Involucrum oblong, imbricated; scales linear, acute, very numerous. *Flowers* radiant; of the ray female, very narrow and numerous, in a double row, not yellow. *Pappus* hairy. *Receptacle* naked.

1. E. *canadense* Linn. E. B. 29. 2019.
Stem hairy, panicled, many-headed. Leaves lanceolate; lower ones toothed. *Smith.*
In cultivated, as well as waste ground. — Annual. *August, September.*

2. E. *acre* Linn. E. B. 17. 1158.
Stem racemose. Stalks mostly single-headed. Leaves lanceolate or tongue-shaped, sessile. Radius erect, scarcely taller than the pappus.
In dry gravelly or chalky pastures. — Biennial. *July, August;* sometimes early in the spring.

3. E. *alpinum* Linn. E. B. 7. 464.
Stem mostly single-headed. Involucrum hairy. Radius spreading, twice the length of the pappus.

β. *uniflorum.* E. B. 34. 2416.
Stem single-headed. Involucrum woolly. Radius erect, twice the length of the pappus.
E. uniflorum *Linn.*
On the mountains of Scotland. — Perennial. *July.*

9. SOLIDAGO *Linn.*

Involucrum imbricated. *Flowers* radiant; of the ray yellow, about 5. *Receptacle* naked. *Pappus* hairy. *Dec.*

1. S. *Virgaurea* Linn. Golden-rod. E. B. 5. 301.
Stem slightly zigzag, angular. Clusters downy, panicled, crowded, erect. Leaves partly serrated. *Smith*
In groves and mountains. — Perennial. *July—September.*

2. S. *cambrica* Hudson.
Stem simple, downy. Leaves wedge-shaped, lanceolate, serrated, slightly hairy. Racemes erect. Rays long. *Hort. Kew.*
On mountains. — Perennial. *July—September.*

10. ANTENNARIA *Gærtn.*

Involucrum imbricated, hemispherical; *scales* scarious, coloured. *Flowers* diœcious. *Florets* all tubular. *Males: Anthers* with 2 bristles at the base. *Stigmata* truncate. *Pappus* filiform or clavate. *Females: Florets* filiform, with a minute limb. *Pappus* capillary.

1. A. *margaritacea* Gærtn. E. B. 29. 2018.
Leaves linear-lanceolate, acute, loosely cottony on the upper side; densely underneath. Stem branched in the upper part. Panicles corymbose, level-topped.
Gnaphalium margaritaceum *Linn.*
In moist meadows. — Perennial. *August.*

2. A. *dioica* Gærtn. E. B. 4. 267.
Shoots procumbent. Stem unbranched. Corymb simple, terminal.
Gnaphalium dioicum *Linn.*
In dry mountainous pastures. — Perennial. *June, July.*

11. GNAPHALIUM *Linn.*

Involucrum imbricated; *scales* scarious, unequal, blunt, coloured, connivent at top, and of equal height with the truncate head. *Florets* filiform, tubular; female in the circumference, hermaphrodite in the disk. *Receptacle* flat, naked. *Pappus* capillary, deciduous.

1. G. *luteo-album* Linn. E. B. 14. 1002.
Leaves half stem-clasping, linear, oblong, wavy, woolly on both sides; lower ones blunt. Heads densely tufted.
In dry sandy ground in the south. — Annual. *July, August.*

2. G. *sylvaticum* Linn. E. B. 13. 913.
Stem simple, erect. Spike leafy, somewhat compound. Leaves lanceolate, tapering at the base, cottony on both sides. *Smith.*
On the Highland mountains of Scotland. — Perennial. *August.*

3. G. *rectum* Smith. E. B. 2. 124.
Stem erect, panicled, many-headed, leafy. Leaves linear-lanceolate, naked on the upper side, silky beneath.
In groves, thickets, and pastures. — Perennial. *August.*

4. G. *supinum* Linn. E. B. 17. 1193.
Stem recumbent, quite simple, with a simple cluster of very few heads. Leaves linear-lanceolate, somewhat cottony on both sides.
G. fuscum *Scopoli.*
G. alpinum *Lightfoot.*
On the summits of Highland mountains. — Perennial. *July*

5. G. *uliginosum* Linn. E. B. 17. 1194.
Stem much branched, spreading. Leaves linear-lanceolate, cottony on both sides. Heads in dense terminal tufts.
In sandy watery places. — Annual. *August.*

12. FILAGO *Linn.*

Involucrum imbricated; scales equal, acuminate, scarious, discoloured, longer than the few-flowered head. *Florets* filiform, tubular; female in the circumference, hermaphrodite in the disk. *Receptacle* conical, toothed, tuberculated, or paleaceous.

1. F. *gallica* Linn. E. B. 33. 2369.
Stem erect, branched. Leaves linear, revolute, acute. Heads awl-shaped, axillary, tufted.
Gnaphalium gallicum *Hudson.*
Logfia subulata *Cassini.*
In gravelly corn-fields, but very rare. Annual. *July, August.*

2. F. *minima* E. B. 17. 1157.
Stem erect, branched. Leaves lanceolate, acute, flat. Heads conical, in lateral and terminal tufts.
Gnaphalium minimum *Smith.*
G. montanum *Hudson.*
Logfia brevifolia *Cassini.*
In barren, sandy, and gravelly ground. — Annual. *July*

3. F. *germanica* Linn. Cudweed. E. B. 14. 946.
Stem erect, proliferous. Leaves lanceolate. Heads globose, many-flowered, lateral as well as terminal. Scales of involucrum bristle-pointed.

Gnaphalium germanicum *Hudson.*
Gifola vulgaris *Cassini.*
In fields and waste ground. — Annual. *July, August.*

13. SENECIO *Linn.*

Involucrum with bracteolæ at the base; the scales scorched at the apex. *Flowers* either flosculous or radiant. *Receptacle* naked. *Pappus* soft, hairy. *Dec.*

1. S. *vulgaris* Linn. Groundsel, or Simpson. E. B. 11. 747.
Heads dispersed, without rays. Leaves pinnatifid, toothed, obtuse, smoothish; clasping at the base.
Everywhere. — Annual. *At all seasons.*

2. S. *viscosus* Linn. E. B. 1. 32.
Rays revolute. Leaves pinnatifid, viscid. Bracteolæ lax, almost as long as the involucrum. Stem with many spreading branches.
In waste ground, on a chalky or sandy soil. — Annual. *July—October.*

3. S. *lividus* Linn. E. B. 35. 2515.
Rays revolute. Leaves clasping the stem, lanceolate, pinnatifid and toothed. Bracteolæ short, with sharp, not discoloured points.
On barren heaths, and newly-enclosed moorland. — Annual. *September, October.*

4. S. *sylvaticus* Linn. E. B. 11. 748.
Rays revolute. Leaves sessile, pinnatifid, lobed and toothed. Bracteolæ short, with bluntish discoloured tips.
In bushy heathy places. — Annual. *July.*

5. S. *squalidus* Linn. E. B. 9. 600.
Rays spreading, elliptical, entire. Leaves smooth, pinnatifid, with distant, and somewhat linear, segments. *Smith*
S. chrysanthemifolius *Bivona.*
On walls at Oxford; a doubtful native. — Annual. *June—October.*

6. S. *tenuifolius* Jacq. E. B. 8. 574.
Rays spreading, oblong. Leaves pinnatifid, somewhat revolute; paler and shaggy beneath. Stem erect, loosely cottony. *Smith.*
S. erucifolius *Hudson.*
In woods, hedges, and by road sides. — Perennial. *July, August.*

7. S. *Jacobæa* Linn. Ragwort. E. B. 16. 1130.
Rays spreading, oblong, toothed. Leaves doubly pinnatifid, somewhat lyrate, with spreading, toothed, smooth segments. Stem erect. Fruit of the disk silky.
By road sides, very common. — Perennial. *July, August.*

8. S. *aquaticus* Hudson. E. B. 16. 1131.
Rays spreading, elliptic-oblong. Leaves lyrate, serrated; the lowermost obovate and undivided. Fruit all smooth.
In marshes and watery places. — Perennial. *July, August.*

9. S. *paludosus* Linn. Bird's-tongue. E. B. 10. 650.
Rays spreading, toothed. Heads corymbose. Leaves lanceolate,

tapering, sharply serrated, somewhat cottony beneath. Stem perfectly straight, hollow.

In the ditches and fens of the east part of England. — Perennial. *June, July.*

10. S. *saracenicus* Linn. E. B. 31. 2211.
Rays spreading, nearly entire. Heads corymbose. Leaves lanceolate, serrated, minutely downy. Stem solid.

In moist meadows and pastures, rare. — Perennial. *July, August.*

14. DORONICUM *Linn.*

Involucrum equal, with the scales in a double row. *Flowers* radiant; those of the ray female. *Pappus* of the flosculous florets simple, of the radiant florets none. *Receptacle* naked. *Dec.*

1. D. *Pardalianches* Linn. Leopard's-bane. E. B. 9. 630.
Leaves heart-shaped, toothed; radical ones stalked; the rest clasping the stem. *Smith.*

n mountainous pastures or meadows. — Perennial. *May.*

15. TUSSILAGO *Linn.*

Involucrum simple; the scales membranous at the margin. *Florets* either flosculous or radiant; either all hermaphrodite, or hermaphrodite in the centre, female in the ray. *Receptacle* naked. *Pappus* hairy. *Dec.*

1. T. *Farfara* Linn. Colt's-foot. E. B. 6. 429.
Stalks single-headed, clothed with scaly bracteas. Flowers radiant. Leaves heart-shaped, angular and toothed.

In moist shady situations. — Perennial. *March, April.*

2. T. *Petasites* Linn. Butter-bur. E. B. 6. 431.
Panicle dense, ovate-oblong. Flowers flosculous. Leaves heart-shaped, unequally toothed, 3-ribbed at the base. *Smith.*

β. *fœmina* Willd. E. B. 6. 430.
Stamens abortive.

Tussilago hybrida *Linn.*

In moist boggy meadows. — Perennial. *April.*

16. CINERARIA *Linn.*

Involucrum simple, many-leaved, equal. *Flowers* radiant. *Receptacle* naked. *Pappus* hairy. *Dec.*

1. C. *palustris* Linn. Marsh Flea-wort. E. B. 3. 151.
Heads corymbose. Leaves broadly lanceolate, toothed or sinuated. Stem shaggy, leafy, hollow.
Othonna palustris *Linn.*

In ditches and pools. — Perennial. *June, July.*

2. C. *campestris* Retz. E. B. 3. 152.
Heads simply and imperfectly umbellate, with several lanceolate bracteas. Radical leaves elliptical, obscurely toothed; the rest lanceolate; all shaggy. Stem unbranched.
C. integrifolia *Jacq.*

C. alpina *Hudson.*

On chalky downs. — Perennial. *May, June.*

17. BELLIS *Linn.*

Involucrum hemispherical, many-leaved, simple; scales lanceolate. *Flowers* radiant. *Receptacle* naked, conical. *Pappus* 0.

1. B. *perennis* Linn. Daisy. E. B. 6. 424.
Root creeping. Scapes radical, naked.
In pastures and meadows everywhere. — Perennial. *March—November.*

18. CHRYSANTHEMUM *Linn.*

Involucrum hemispherical, imbricated; scales scarious at the margin. *Flowers* radiant. *Receptacle* naked. *Pappus* none, or a short membrane.

* *Rays white.*

1. Ch. *Leucanthemum* Linn. Great White Ox-eye. Moon Daisy. E. B. 9. 601.
Leaves clasping the stem, oblong, obtuse, cut; pinnatifid at the base; radical ones obovate, stalked. *Smith.*
In fields and by way sides. — Perennial. *June, July.*

2. C. *Parthenium* Smith. Feverfew. E. B. 18. 1231.
Leaves stalked, compound, flat; leaflets ovate, cut; the uppermost confluent. Flower-stalks corymbose. Stem erect. Rays shorter than the diameter of the disk. *Smith.*
Pyrethrum Parthenium *Smith.*
Matricaria Parthenium *Linn.*
In waste ground. — Biennial. *June, July.*

3. C. *inodorum* Linn. Mayweed. E. B. 10. 676.
Leaves sessile, pinnate, in numerous, capillary, pointed segments. Stem branched, spreading. Pappus entire.
Pyrethrum inodorum *Smith.*
Matricaria inodora *Linn.*
In fields and by way sides. — Annual. *August, September.*

4. C. *maritimum* Smith. E. B. 14. 979.
Leaves sessile, doubly pinnate, fleshy, pointless; convex above; keeled beneath. Pappus lobed. Stems diffuse.
Pyrethrum maritimum *Smith.*
On the sea-coast. — Perennial. *July, August.*

** *Rays yellow.*

5. Ch. *segetum Linn.* Yellow Ox-eye. Corn Marigold. E. B. 8. 540.
Leaves clasping the stem, glaucous; jagged upwards; toothed at the base. *Smith.*
In fields. — Annual. *June—August.*

19. MATRICARIA *Linn.*

Involucrum hemispherical, imbricated; scales obtuse. *Flowers* radiant. *Receptacle* naked, conical. *Pappus* 0.

1. M. *Chamomilla* Linn. E. B. 18. 1232.
Leaves smooth, pinnate; leaflets linear, simple, or divided. Rays spreading. Scales of involucrum dilated, bluntish.
In cultivated and waste ground. — Annual. *May—July.*

20. ARTEMISIA *Linn.*

Involucrum ovate or round, imbricated. *Florets* all tubular; of the disk hermaphrodite, 5-toothed; of the ray slender, less numerous, entire, female. *Receptacle* naked or hairy. *Pappus* 0.

1. A. *campestris* Linn. E. B. 5. 338.
Leaves in many linear segments. Stems procumbent before flowering, wand-like. *Smith.*
Oligosporus campestris *Cassini.*
On dry open sandy heaths, rare. — Perennial. *August.*

2. A. *maritima* Linn.
Leaves downy, pinnatifid; uppermost undivided. Heads drooping, oblong, downy, sessile. Receptacle naked.
On the sea-shore. — Perennial. *August.*

3. A. *gallica* Willd. E. B. 24. 1706. and 14. 1001.
Leaves downy, pinnatifid; radical ones capillary; uppermost undivided. Heads erect, oblong, downy, partly stalked, of few florets. Receptacle naked.
About the banks of rivers. — Perennial. *August.*

4. A. *Absinthium* Linn. Wormwood. E. B. 18. 1230.
Leaves in many deep segments, clothed with close silky down. Heads drooping, hemispherical. Receptacle hairy.
In waste ground. — Perennial. *August.*

5. A. *vulgaris* Linn. Mugwort. E. B. 14. 978.
Leaves pinnatifid, flat, cut; downy beneath. Clusters simple. Heads ovate. Receptacle naked.
About hedges, and in the borders of fields. — Perennial. *August.*

6. A. *cærulescens* Linn. E. B. 34. 2426.
Leaves hoary; most of them lanceolate, undivided, tapering at the base; lower ones variously divided. Heads erect, cylindrical. Receptacle naked.
On the sea-coast; a very doubtful native. — Perennial. *August, September*

21. TANACETUM *Linn.*

Involucrum hemispherical, imbricated. *Florets* all tubular; of the disk hermaphrodite, 5-lobed; of the ray female, 3-lobed. *Receptacle* naked. *Pappus* membranous, entire.

1. T. *vulgare* Linn. Tansy. E. B. 18. 1229.
Leaves doubly pinnatifid, deeply serrated, naked. *Smith.*
In hedges, and by road sides. — Perennial. *July, August.*

22. DIOTIS *Desfontaines.*

Involucrum hemispherical, imbricated. *Florets* all tubular, hermaphrodite, 5-toothed, contracted in the middle, auricled or saccate on each side at the base. *Receptacle* convex, paleaceous. *Pappus* none; in its place the fruit is crowned by the persistent lower 2-eared half of the corolla.

1. D. *maritima* Desf. Cotton-weed. E. B. 2. 141.
Santolina maritima *Linn.*
Athanasia maritima *Linn.*
On the sandy sea-coast. — Perennial. *August, September.*

23. MARUTA *Cassini.*

Involucrum hemispherical; scales nearly equal, scarious at the margin, imbricated. *Flowers* radiant; of the disk neuter; of the ray female. *Receptacle* paleaceous in the disk, naked towards the edges. *Pappus* none.

1. M. *fœtida* Cassini. E. B. 25. 1772.
Leaves doubly pinnatifid, slightly hairy, fetid, covered with resinous dots.
Anthemis cotula *Linn.*
In waste ground. — Annual. *June, July*

24. ANTHEMIS *Linn.*

Involucrum hemispherical; scales nearly equal, scarious at the margin, imbricated. *Flowers* radiant; of the disk hermaphrodite; of the ray lanceolate, female. *Receptacle* conical, paleaceous. *Pappus* a membrane.

* *Rays white.*

1. A. *maritima* Linn. E. B. 33. 2370.
Leaves doubly pinnatifid, acute, fleshy, dotted, somewhat hairy. Stem prostrate. Paleæ prominent, sharp-pointed.
On the sea-coast. — Annual. *July.*

2. A. *nobilis* Linn. E. B. 14. 980.
Leaves doubly pinnate, semicylindrical, acute, a little downy. Stem procumbent. Paleæ membranous, obtuse, shorter than the florets.
On open pastures or commons. — Perennial. *August, September.*

3. A. *arvensis* Linn. E. B. 9. 602.
Receptacle conical; paleæ lanceolate, acute, keeled, prominent. Leaves doubly pinnatifid, hairy; segments parallel. Pappus a quadrangular border.
In cultivated fields. — Annual or biennial. *June, July.*

** *Rays yellow like the disk.*

4. A. *tinctoria* Linn. E. B. 21. 1472.
Leaves doubly pinnatifid, serrated; downy beneath. Stem corymbose, erect. Pappus membranous, undivided.
In stony mountainous places. — Biennial? *July, August.*

25. ACHILLEA *Linn.*

Involucrum ovate, imbricated. *Flowers* radiant; of the disk hermaphrodite, of the ray short, female, and few. *Receptacle* narrow, flat, paleaceous. *Pappus* 0.

* *Flowers white.*

1. A. *Ptarmica* Linn. Sneeze-wort. Goose-tongue. E. B. 11. 757.
Leaves linear, pointed, equally and sharply serrated, smooth. *Smith.*
In wet hedges and thickets, or about the banks of rivers. — Perennial. *July, August.*

2. A. *serrata* Retz. E. B. 36. 2531.
Leaves linear-lanceolate, sessile, downy, deeply serrated; laciniated at the base. Heads almost simply corymbose.
In mountainous limestone countries. — Perennial. *August.*

3. A. *Millefolium* Linn. Yarrow, or Milfoil. E. B. 11. 758.
Leaves doubly pinnatifid, hairy; segments linear, toothed, pointed. Stem furrowed. *Smith.*
In meadows and pastures. — Perennial. *June—August.*

** *Flowers yellow.*

4. A. *tomentosa* Linn. E. B. 36. 2532.
Leaves doubly pinnatifid, woolly; segments crowded, linear, acute. Corymbs repeatedly compound. *Smith.*
In dry hilly pastures in Scotland and Ireland. — Perennial. *July, August.*

26. XANTHIUM *Linn.*

Monœcious. — Male. *Involucrum* many-leaved, many-flowered. *Florets* all tubular. *Receptacle* paleaceous. — Female. *Involucrum* 1-leaved, prickly, containing 2 florets. *Fruit* enclosed in the hard indurated involucrum.

1. X. *strumarium* Linn. Bur-weed. E. B. 36. 2544.
Prickles none. Leaves heart-shaped; 3-ribbed at the base. *Smith.*
In rich moist ground in the south of England. — Annual. *August, September.*

27. BIDENS *Linn.*

Involucrum with bracteolæ at the base; outer scales longer than the rest, and spreading. *Flowers* mostly floscular; florets all hermaphrodite, or if ligulate in the ray, then female, or hermaphrodite. *Receptacle* flat, paleaceous. *Pappus* from 2 to 5 persistent awns.

1. B. *tripartita* Linn. Bur-marigold. E. B. 16. 1113.
Leaves in 3 segments. Bracteas unequal. Bristles of the pappus 2 or 3, erect.
In watery places. — Annual. *August, September*

2. B. *cernua* Linn. E. B. 16. 1114.
Leaves lanceolate, serrated. Heads drooping. Bracteas nearly equal, entire. Bristles of the pappus about 4, erect.
β. radiata
Flowers radiant.
Coreopsis Bidens *Linn.*
γ. minima
Bidens minima *Linn.*
In ditches and ponds. *γ.* in dry places. — Annual. *September.*

Tribe ii. *Cynarocephalæ* Juss.

Florets all tubular. Receptacle paleaceous. Stigma articulated with the apex of the style. — Leaves usually spiny.

28. ONOPORDUM *Linn.*

Involucrum imbricated; scales pungent. *Receptacle* excavated like honey-comb. *Fruit* compressed, 4-cornered, furrowed transversely. *Pappus* hairy, deciduous; its hairs connected in a ring at the base.

1. O. *Acanthium* Linn. Cotton-thistle. E. B. 14. 977.
Scales of involucrum awl-shaped, spreading in every direction. Leaves ovate-oblong, sinuated, woolly on both sides.
On banks, and by road sides. — Biennial. *July, August.*

29. SAUSSUREA *Dec.*

Involucrum imbricated; scales unarmed. *Florets* all hermaphrodite. *Receptacle* paleaceous. *Pappus* in 2 rows, hairy; the outer hairs short, with minute notches; the inner long and feathery. *Dec.*

1. S. *alpina* Dec. E. B. 9. 599.
Leaves undivided, distantly toothed; cottony beneath. Involucrum ovate, finely downy.
Serratula alpina *Linn.*
In the fissures of alpine rocks. — Perennial. *July, August.*

30. CNICUS *Linn.*

Involucrum ovate, imbricated; scales spiny at the points. *Florets* all hermaphrodite. *Style* equal, simple. *Receptacle* paleaceous; paleæ split into bristle-shaped segments. *Pappus* feathery; hairs collected into a ring at the base.

* *Leaves decurrent. Stem winged.*

1. C. *lanceolatus* Willd. E. B. 2. 107.
Leaves decurrent, pinnatifid, hispid, with variously-spreading spinous lobes. Involucrum ovate, shaggy. Stem furrowed, hairy.
Cirsium lanceolatum *Scopoli.*
Eriolepis lanceolata *Cassini.*
Carduus lanceolatus *Linn.*
In waste ground. — Biennial. *June—September.*

2. C. *palustris* Willd. E. B. 14. 974.
Leaves decurrent, pinnatifid, toothed, spinous, rough. Heads aggregate. Involucrum ovate, minutely spinous, nearly smooth.
Cirsium palustre *Scopoli.*
Onotrophe palustris *Cassini.*
Carduus palustris *Linn.*
In meadows. — Biennial. *July, August.*

** *Leaves sessile, or partially decurrent. Stem not winged.*

3. C. *arvensis* Smith. E. B. 14. 975.
Leaves sessile, pinnatifid, spinous, nearly smooth. Stem panicled, solid. Involucrum ovate; outer scales spinous. Root creeping, tuberous.
Cirsium arvense *Lam.*
Carduus arvensis *Curtis.*
In fields and by way sides. — Perennial. *July.*

4. C. *Forsteri* Smith.
Leaves slightly decurrent, pinnatifid, spinous; downy beneath. Stem panicled, hollow. Involucrum ovate, rather cottony; outer scales spinous.
Near Frant, Sussex, two miles from Tonbridge Wells. — Perennial. *July, August.*

5. C. *eriophorus* Willd. E. B. 6. 386.
Leaves sessile, pinnatifid, with 2-ranked spinous segments; bristly above. Involucrum globular, densely woolly, leafy at the base.
Cirsium eriophorum *Scopoli.*
Eriolepis lanigera *Cassini.*
Carduus eriophorus *Linn.*
In waste mountainous ground. — Biennial. *August.*

6. C. *tuberosus* Willd. E. B. 36. 2562.
Leaves deeply pinnatifid, lobed, fringed with prickles; lower ones on long stalks. Stem almost single-headed, without wings or prickles. Scales of involucrum minutely spinous, nearly smooth. Root creeping, tuberous.
Cirsium bulbosum *Dec.*
Carduus tuberosus *Linn.*
In thickets on the downs of Wiltshire. — Perennial. *August.*

7. C. *heterophyllus* Willd. E. B. 10. 675.
Leaves clasping the stem, fringed; undivided or pinnatifid; very smooth above; densely cottony beneath. Stem downy, almost single-headed.
Cirsium heterophyllum *Dec.*
Carduus heterophyllus *Linn.*
C. helenioides *Hudson.*
In moist mountain pastures. — Perennial. *July, August.*

8. C. *pratensis* Willd. E. B. 3. 177.
Leaves lanceolate, wavy or lobed, fringed with prickles; loosely cottony beneath. Stem downy, slightly leafy, single-headed. Involucrum cottony, with tapering sharp scales.
Cirsium pratense *Dec.*
Carduus pratensis *Hudson.*
C. dissectus *Villars.*
In wet pastures and meadows. — Perennial. *June.*

9. C. *acaulis* Willd. E. B. 3. 161.
Stalks radical, single-headed, shorter than the smooth involucrum.
Cirsium acaule *Allioni.*
Onotrophe acaulis *Cassini.*

Carduus acaulis *Linn.*

In pastures and meadows. — Perennial. *July, August.*

31. CARLINA *Linn.*

Involucrum imbricated; the outer scales sinuated, spiny, spreading at the points; the inner generally simple and acute; the innermost much longer, scarious, ligulate, coloured, and resembling radiant florets. *Receptacle* paleaceous; paleæ split into bristle-like segments. *Pappus* feathery. *Dec.*

1. C. *vulgaris* Linn.
Stem corymbose, many-headed. Heads terminal. Outer scales of involucrum pinnatifid; inner whitish.
In dry sandy heathy pastures. — Biennial. *June.*

32. LAPPA *Tournef.*

Involucrum imbricated; scales ending in a soft spine hooked at the end. *Receptacle* paleaceous. *Pappus* short, persistent, with rigid, unequal hairs. *Dec.*

1. L. *glabra* Lam. E. B. 18. 1228.
Involucrum in fruit smooth. Leaves wavy.
Arctium Lappa *Linn.*
In waste ground. — Biennial. *July, August.*

2. L. *tomentosa* Allioni. E. B. 35. 2478.
Involucrum in fruit cottony. Leaves even.
Arctium Bardana *Willd.*
In waste ground. — Biennial. *July, August.*

33. SILYBUM *Vaillant.*

Involucrum imbricated; scales leafy at the base, closely pressing upon each other, spreading and spiny at the points. *Receptacle* paleaceous. *Pappus* between hairy and paleaceous, deciduous; the hairs connected in a ring at the base.

1. S. *marianum* Gærtn. Milk Thistle. E. B. 14. 976.
Leaves wavy, spinous, clasping the stem; radical ones pinnatifid. Scales of involucrum leafy, recurved, channelled.
Carduus marianus *Linn.*
On ditches, common. — Annual. *June, July.*

34. SERRATULA *Linn.*

Diœcious. *Involucrum* imbricated; scales unarmed, acute. *Receptacle* paleaceous; the paleæ split into numerous linear bristles. *Pappus* persistent, with rigid unequal hairs.

1. S. *tinctoria* Linn. Saw-wort. E. B. 1. 38.
Leaves with copious bristly serratures, pinnatifid, somewhat lyrate; terminal lobe largest.
In groves and pastures. — Perennial. *July, August.*

35. CENTAUREA *Linn.*

Involucrum imbricated; scales leafy, scarious, or spiny in various ways. *Florets* of the disk hermaphrodite; of the ray neuter and larger than the others. *Receptacle* paleaceous; paleæ jagged. *Fruit* inserted obliquely at the base. *Pappus* hairy.

* *Scales of involucrum jagged or fringed.* (Cyanus).

1. C. *Jacea* Linn. E. B. 24. 1678.
Scales of involucrum membranous, torn; lower ones pinnatifid. Leaves linear-lanceolate; radical ones elliptic-lanceolate, toothed. Flowers radiant.
In meadows and groves. — Perennial. *August, September.*

2. C. *nigra* Linn. Knapweed. E. B. 4. 278.
Scales of involucrum oval, fringed with upright capillary teeth. Lower leaves somewhat lyrate, with angular lobes; upper ones ovate. Flowers discoid.
In pastures, and by road sides. — Perennial. *June—August.*

3. C. *Cyanus* Linn. Blue-bottle. E. B. 4. 277.
Scales of involucrum serrated. Leaves linear-lanceolate, entire; lower ones toothed towards their base.
In corn-fields. — Annual. *July, August.*

4. C. *Scabiosa* Linn. E. B. 1. 56.
Scales of involucrum ovate, fringed, somewhat downy. Leaves pinnatifid; segments lanceolate, roughish, partly toothed.
In corn-fields, and by way sides. — Perennial. *July, August.*

** *Scales of involucrum palmate or pinnate, and spinous.* (Calcitrapa).

5. C. *Isnardi* Linn. E. B. 32. 2256.
Scales of involucrum with palmate spines. Leaves toothed, somewhat lyrate, roughish, slightly clasping the stem. Heads terminal, solitary; leafy at the base.
In pastures in the isle of Jersey. — Perennial. *July, August.*

6. C. *Calcitrapa* Linn. Star-thistle. E. B. 2. 125.
Heads lateral, sessile. Scales of involucrum doubly spinous. Leaves pinnatifid, toothed. Stem hairy, widely spreading.
In waste ground near the sea. — Annual. *July, August.*

7. C. *solstitialis* Linn. St. Barnaby's Thistle. E. B. 4. 243.
Heads terminal, solitary. Scales of involucrum doubly spinous. Stem winged, from the decurrent, lanceolate, unarmed leaves; radical leaves lyrate.
In cultivated fields, and about hedges. — Annual. *July—September.*

36. CARDUUS *Linn.*

Involucrum imbricated; scales simple, spiny at the points. *Receptacle* paleaceous; the paleæ split into numerous linear bristles. *Pappus* deciduous, hairy; the hairs collected into a ring at the base. *Dec.*

1. C. *nutans* Linn. Musk Thistle. E. B. 16. 1112.
Leaves interruptedly decurrent, spinous. Heads solitary, drooping. Scales of involucrum lanceolate; their upper part spreading.
In waste ground and dry pastures. — Annual. *July, August.*

2. C. *acanthoides* Linn. E. B. 14. 973.
Leaves decurrent, sinuated, very spinous. Heads aggregate, somewhat stalked. Involucrum globose; scales linear, partly recurved.

Carduus crispus *Hudson.*
C. polyacanthos *Curtis.*
About hedges and in waste ground. — Annual. *June, July.*

3. C. *tenuiflorus* Curtis. E. B. 6. 412.
Leaves decurrent, sinuated, spinous. Heads aggregate, sessile. Involucrum nearly cylindrical; scales ovate at the base, somewhat recurved at the point.
Carduus acanthoides *Hudson.*
In dry sandy ground. — Annual. *June, July.*

Tribe iii. *Cichoraceæ* Juss.

Florets all ligulate and hermaphrodite. Receptacle scarcely fleshy. — Sap generally milky.

37. SONCHUS *Linn.*

Involucrum oblong, imbricated, ovate at the base. *Receptacle* naked. *Fruit* striated longitudinally. *Pappus* short, sessile, hairy. *Dec.*

1. S. *alpinus* Willd. E. B. 34. 2425.
Peduncles and involucrum bristly, racemose. Leaves somewhat lyrate; their terminal lobe triangular and very large.
S. cæruleus *Smith.*
S. canadensis *Linn.*
On Loch-na-gore, Aberdeenshire, and on the Clova mountains. — Perennial. *July, August.*

2. S. *palustris* Linn. E. B. 13. 935.
Peduncles and involucrum bristly, somewhat umbellate. Leaves runcinate, rough-edged; arrow-shaped at the base.
In marshes near the banks of rivers. — Perennial. *July, August.*

3. S. *arvensis* Linn. E. B. 10. 674.
Peduncles and involucrum bristly, somewhat umbellate. Leaves runcinate, finely toothed, heart-shaped at the base. Root creeping.
In corn-fields and hedges, on a clayey soil. — Perennial. *August.*

4. S. *oleraceus* Linn. Sow-thistle. E. B. 12. 843.
Peduncles cottony. Involucrum smooth. Leaves runcinate, toothed.
Everywhere. — Annual. *July—September.*

38. LACTUCA *Linn.*

Involucrum oblong, imbricated; scales membranous at the margin. *Receptacle* naked. *Pappus* stipitate, hairy, soft, fugacious. *Dec.*

1. L. *virosa* Linn. E. B. 28. 1957.
Leaves horizontal, finely toothed; the keel prickly. *Smith.*
About hedges, old walls, and the borders of fields. — Biennial. *August, September.*

2. L. *Scariola* Linn. E. B. 4. 268.
Leaves perpendicular, sinuated, finely toothed; the keel prickly. *Smith.*
L. sylvestris *Lam.*
In waste ground, and dry stony borders of fields. — Biennial. *August.*

3. L. *saligna* Linn. E. B. 10. 707.
Leaves linear, hastate or pinnatifid, entire, sessile; the keel prickly. *Smith.*

In chalky waste ground, or about salt marshes. — Biennial. *August.*

39. CHONDRILLA *Gærtn.*

Involucrum with external bracteolæ; scales erect, linear, bracteolæ very minute or obsolete. *Receptacle* naked. *Pappus* stipitate, hairy. *Dec.*

1. C. *muralis* Lam. E. B. 7. 457.
Leaves lyrate, pinnatifid, toothed; their terminal lobe with 5 angles. Scales 5 or 6. *Stipes* of the pappus shorter than the seed. *Dec.*
Prenanthes muralis *Linn.*
Mycelis angulosa *Cassini.*

On old walls. — Perennial. *July.*

40. PRENANTHES. *Linn.*

Involucrum with external, unequal, ovate bracteolæ; scales imbricated at their margin. *Florets* few. *Receptacle* naked. *Pappus* sessile, hairy, in a single row. *Dec.*

1. P. *hieracifolia* Willd. E. B. 33. 2325.
Leaves downy, toothed; radical ones obovate; the rest somewhat arrow-shaped and clasping. Panicle corymbose, spreading. Involucrum pyramidal.
Lapsana chondrilloides *Linn.*
Crepis pulchra *Linn.*
Prenanthes pulchra *Dec.*
Phæcasium lampsanoides *Cassini.*

Amongst crumbling rocks on the hill of Turin, near Forfar. — Annual. *June— September.*

41. LAPSANA *Linn.*

Involucrum with external bracteolæ; scales linear-lanceolate. *Receptacle* naked. *Fruit* quickly deciduous, not enveloped in the scales of the involucrum. *Pappus* 0.

1. L. *communis* Linn. Nipple-wort. E. B. 12. 844.
Involucrum of the fruit angular. Stem branched, panicled, leafy. Leaves ovate, stalked, toothed. Peduncles cylindrical, eve

In waste ground. — Annual. *June, July.*

2. L. *pusilla* Willd. Swine's succory. E. B. 2. 95.
Peduncles radical, leafless, subdivided; swelling and tubular at the summit. Leaves obovate, rough-edged, toothed.
L. minima *Lam.*
Arnoseris pusilla *Gærtn.*
Hyoseris minima *Linn.*

In gravelly fields. — Annual. *June.*

42. LEONTODON *Linn.*

Involucrum with short external spreading bracteolæ; not ribbed when in fruit. *Receptacle* naked. *Pappus* stipitate, hairy. — Scapes *always single-flowered.*

1. L. *Taraxacum* Linn. Dandelion. E. B. 8. 510.
Outer scales of the involucrum reflexed. Leaves runcinate, toothed, smooth.
Taraxacum Dens Leonis *Desf.*
T. officinale *Sibth.*
Leontodon officinalis *Withering.*
In meadows, pastures, waste and cultivated ground, everywhere. — Perennial. *April—July.*

2. L. *palustre* Smith. E. B. 8. 553.
Outer scales of the involucrum shorter, imbricated, ovate. Leaves sinuated and toothed, not quite smooth.
Taraxacum palustre *Dec.*
Hedypnois paludosa *Scopoli.*
L. taraxacum var. *Bentham.*
In low boggy meadows. — Perennial. *June, July.*

43. BARKHAUSIA *Mœnch.*

Involucrum with external bracteolæ; when in fruit ribbed and furrowed; outer scales lax. *Receptacle* naked. *Pappus* hairy, stipitate. *Dec.*

1. B. *fœtida* Dec. E. B. 6. 406.
Leaves hairy, pinnatifid, with reversed teeth; on toothed footstalks. Stem hairy. Involucrum downy.
Crepis fœtida *Linn.*
On dry chalky ground. — Biennial. *June, July.*

44. CREPIS. *Linn.*

Involucrum lax, with external bracteolæ; when in fruit usually ribbed. *Receptacle* naked. *Pappus* hairy, sessile, snowy white. *Dec.*

1. C. *tectorum* Linn. E. B. 16. 1111.
Radical leaves runcinate; the rest clasping, lanceolate and toothed. Stem smooth. Involucrum rough. Pappus sessile.
Hedypnois tectorum *Hudson.*
In dry pastures and waste ground, on cottage roofs, old walls, and banks. — Annual. *June—September.*

2. C. *biennis* Linn. E. B. 3. 149.
Leaves pinnatifid, runcinate, rough; their lobes toothed in front. Involucrum somewhat bristly and downy.
Hedypnois biennis *Hudson.*
In chalky pastures. — Biennial. *June, July.*

45. HELMINTHIA *Juss.*

Involucrum of 8 equal scales, surrounded by 5 leafy lax bracteæ. *Fruit* transversely striated. *Pappus* feathery, stipitate.

1. H. *echioides* Gærtn. E. B. 14. 972.
Bracteolæ broad, ovate-cordate, somewhat spiny.
Picris echioides *Linn.*
About hedges and the borders of fields. — Annual. *June, July.*

46. PICRIS *Linn.*

Involucrum with small, linear-lanceolate, external bracteolæ. *Fruit* transversely striated. *Pappus* feathery, sessile, or nearly so.

1. P. *hieracioides* Linn. E. B. S. 196.
Leaves lanceolate, wavy; radical ones toothed. Stem rough. Hedypnois hieracioides *Hudson.*
On dry banks, or in the borders of fields. — Biennial. *July, August.*

47. HIERACIUM *Linn.*

Involucrum imbricated. *Receptacle* naked, or with a few short hairs. *Pappus* hairy, sessile, generally dirty brown. *Dec.*

* *Stalk radical, naked, single-headed.*

1. H. *alpinum* Linn. E. B. 16. 1110.
Leaves oblong, undivided, somewhat toothed. Stalk almost leafless, single-headed. Involucrum shaggy.
On dry rocky mountains, in Wales and Scotland. — Perennial. *July.*

2. H. *Pilosella* Linn. Mouse-ear Hawkweed. E. B. 16. 1093.
Leaves elliptical, entire; cottony beneath. Runners creeping. Stalks single-headed, naked.
In dry open pastures, and on banks, park walls, cottage roofs, &c., common. — Perennial. *May—July.*

** *Stalk radical, naked, many-headed.*

3. H. *dubium* Linn. E. B. 33. 2332.
Leaves elliptic-lanceolate, obtuse, nearly entire, besprinkled with coarse hairs; rather glaucous beneath. Runners creeping, elongated. Stalk nearly smooth, loosely corymbose. Involucrum bristly.
On mountains, in rather moist situations. — Perennial. *July.*

4. H. *Auricula* Linn. E. B. 33. 2368.
Leaves lanceolate, acute, nearly entire, coarsely hairy; green on both sides. Runners scarcely so long as the leaves. Stalk downy and hairy, corymbose. Involucrum shaggy.
On Dalehead, not far from Grass-mere, Westmoreland, but sparingly. *Hudson.* — Perennial. *July.*

5. H. *aurantiacum* Linn. E. B. 21. 1469.
Leaves elliptical, acute, entire. Stalk almost leafless, hairy, densely corymbose, many-headed. Involucrum shaggy.
In rather mountainous woods. — Perennial. *June, July.*

*** *Stem leafy.*

6. H. *murorum* Linn. E. B. 29. 2082.
Stem corymbose, with a solitary leaf. Leaves ovate-heart-shaped, wavy, with radiating teeth chiefly at the base. *Smith.*
On rocks and old walls. — Perennial. *June.*

7. H. *maculatum* Smith. E. B. 20. 2121.
Stem cymose, many-leaved, tubular. Leaves ovate-lanceolate, strongly toothed; teeth pointing forward. *Smith.*
On the mountains of Wales, Westmoreland, and Scotland. — Perennial. *June—September.*

8. H. *sylvaticum* Smith. E. B. 29. 203 .
Stem simply racemose, many-leaved, solid. Leaves ovate-lanceolate, toothed chiefly about the base; teeth pointing forward. *Smith.*

In dry chalky woods, and on dry banks, or especially old park walls, frequent. — Perennial. *June, July.*

9. H. *pulmonarium* Smith. E. B. 33. 2307.
Stem somewhat corymbose, solid, slightly leafy. Leaves lanceolate, deeply and unequally toothed throughout; teeth pointing forward. *Smith.*
H. pulmonarioides *Villars.*

On rocks about rivers in Scotland. — Perennial. *July.*

10. H. *Lawsoni* Villars. E. B. 29. 2083.
Stem remotely and simply branched, solid, slightly leafy. Radical leaves stalked, elliptic-lanceolate, decurrent, glaucous, fringed, nearly entire. *Smith.*

On the mountains of Westmoreland, Craven, and Scotland. — Perennial. *July.*

11. H. *paludosum* Linn. E. B. 16. 1094.
Stem angular, tubular, leafy, smooth, corymbose. Leaves smooth, toothed, clasping the stem with their heart-shaped base. Calyx hairy.

In watery shady places in Wales, the north of England, and lowlands of Scotland. — Perennial. *July.*

12. H. *molle* Jacq. E. B. 31. 2210.
Stem angular, tubular, leafy, downy, corymbose. Leaves lanceolate, slightly toothed, hairy, clasping the stem; lower ones stalked, elliptical and obtuse. *Smith.*

In woods in the south of Scotland. — Perennial. *July, August.*

13. H. *cerinthoides* Linn. E. B. 34. 2378.
Stem solid, leafy, corymbose, somewhat angular. Leaves hairy, slightly toothed; the uppermost ovate, pointed, clasping; radical ones elliptic-oblong, with shaggy fringed footstalks. *Smith.*

On rocks in the Highlands of Scotland. — Perennial. *August.*

14. H. *villosum* Linn. E. B. 34. 2379.
Stem tubular, leafy, shaggy, with very few heads. Leaves oblong, wavy, unequally toothed, shaggy as well as the calyx. Fruit angular.

On moist alpine rocks. — Perennial. *August.*

15. H. *Halleri* Villars.
Stem erect, with 1 or 2 heads, slightly leafy. Leaves hairy; lower ones obovate-oblong, stalked, toothed; upper lanceolate, much diminished. Involucrum shaggy.
H. pumilum *Willd.*

On the Clova mountains. — Perennial. *July, August.*

16. H. *sabaudum* Linn. E. B. 6. 349.
Stem erect, copiously leafy, many-headed. Leaves ovate-lanceolate, sharply toothed, rough-edged, somewhat clasping; hairy beneath.

In coppices, groves, and thickets, frequent. — Perennial. *August, September.*

17. H. *denticulatum* Smith. E. B. 30. 2122.
Stem erect, leafy, solid, many-headed, cymose, with downy glandular stalks. Leaves sessile, elliptic-lanceolate, finely toothed, smoothish; glaucous beneath.
H. prenanthoides *Fl. Brit.*
In woods in the south of Scotland. — Perennial. *July, August.*

18. H. *prenanthoides* Villars. E. B. 32. 2235.
Stem erect, leafy, solid, many-headed; corymbose, with downy glandular stalks. Leaves somewhat toothed, clasping, rough near the edge; glaucous beneath; upper ones heart-shaped.
H. spicatum *Allioni.*
In woods and thickets in the south of Scotland. *J. Mackay.* — Perennial. *August.*

19. H. *umbellatum* Linn. E. B. 25. 1771.
Stem erect, leafy, almost solid, imperfectly umbellate. Leaves scattered, linear, slightly toothed, nearly smooth as well as the involucrum.
In groves, gravelly thickets, and shady rocky situations. — Perennial. *August, September.*

48. HYPOCHŒRIS *Linn.* Cat's-ea

Involucrum oblong, imbricated. *Receptacle* paleaceous. *Pappus* feathery, stipitate, or sessile in the disk. *Dec.*

1. H. *maculata* Linn. E. B. 4. 225.
Stem solitary, nearly naked, mostly simple. Leaves ovate-oblong, undivided, toothed. *Smith.*
In open high chalky pastures. — Perennial. *July.*

2. H. *glabra* Linn. E. B. 8. 575.
Nearly smooth. Involucrum oblong, regularly imbricated. Stems branched, somewhat leafy. Leaves toothed or sinuated. Pappus of the marginal seeds sessile.
On heaths and gravelly ground. — Annual. *June—August.*

3. H. *radicata* Linn. E. B. 12. 831.
Leaves runcinate, bluntish, rough. Stems branched, naked, smooth. Peduncles scaly. Pappus of all the seeds stalked.
In pastures and waste ground, everywhere. — Perennial. *July, August.*

49. TRAGOPOGON *Linn.* Goat's-beard.

Involucrum simple, of 8 or 10 connected scales. *Receptacle* naked. *Fruit* striated longitudinally. *Pappus* feathery, with a slender stipes. *Dec.*

1. T. *pratensis* Linn. E. B. 7. 434.
Involucrum about equal to the florets. Leaves keeled, tapering; dilated and somewhat undulated at the base. Peduncle cylindrical.
In grassy pastures and meadows. — Biennial. *June.*

2. T. *porrifolius* Linn. E. B. 9. 638.
Involucrum half as long again as the florets. Leaves tapering, straight. Peduncle swelling upwards.
In moist meadows. — Biennial. *May, June.*

50. THRINCIA *Roth.*

Involucrum imbricated. *Receptacle* excavated like honeycomb. *Pappus* of the disk sessile, feathery, and unequal; of the ray almost abortive.

1. T. *hirta* Roth. E. B. 8. 554.
Leaves toothed, rough. Involucrum nearly smooth.
Hedypnois hirta *Fl. Brit.*
Apargia hirta *Hoffm.*
Leontodon hirtum *Linn.*
On gravelly heaths and commons. — Perennial. *July, August.*

51. APARGIA *Schreb.*

Involucrum imbricated. *Receptacle* excavated like honeycomb. *Pappus* feathery, sessile; some of the hairs scaly, others silky. *Dec.*

1. A. *hispida* Willd. E. B. 8. 554.
Stalks naked, single-headed. Leaves runcinate, rough. Florets hairy at their orifice; glandular at the tip. Fruit scarcely beaked.
Leontodon hispidum *Linn.*
Hedypnois hispida *Smith.*
In pastures. — Perennial. *July.*

2. A. *Taraxaci* Willd. E. B. 16. 1109.
Stalks mostly single-headed; tumid and hairy at the summit. Leaves smooth, runcinate. Involucrum shaggy.
Leontodon montanum *Lam.*
Hieracium Taraxaci *Linn.*
Hedypnois Taraxaci *Vill.*
Picris Taraxaci *Allioni.*
On the Highland mountains of Scotland and Wales. — Perennial. *August.*

3. A. *autumnalis* Willd. E. B. 12. 830.
Common stalk branched; partial ones scaly. Leaves lanceolate, toothed or pinnatifid, nearly smooth. *Smith.*
Leontodon autumnale *Linn.*
Hedypnois autumnalis *Hudson.*
In meadows and pastures. — Perennial. *August.*

52. CICHORIUM *Linn.*

Involucrum of 8 scales, united at the base, and surrounded by 5 external bracteæ. *Receptacle* naked, or rather hairy. *Pappus* sessile, scaly, shorter than the fruit. *Dec.*

1. C. *Intybus* Linn. Wild Succory. E. B. 8. 539.
Heads in pairs, each nearly sessile. Leaves runcinate.
About the borders and ridges of fields. — Perennial. *July, August.*

Order 48. Boragineæ *Juss.*

Calyx persistent, with 4 or 5 divisions.
Corolla hypogynous, monopetalous, generally regular, 5-cleft, sometimes 4-cleft; with an imbricate æstivation.
Stamens inserted upon the petals, equal to the number of lobes of the corolla, and alternate with them, seldom in greater number.
Ovarium 4-parted, 4-seeded; *ovula* attached to the lowest point of the cavity; *style* simple; *stigma* simple or bifid.
Nuts 4, distinct.
Seed separable from the pericarpium, destitute of albumen. *Embryo* with a superior *radicle*, and flat *cotyledons* parallel with the axis.
Herbaceous plants or *shrubs*. *Stems* round. *Leaves* alternate, covered with asperities, consisting of hairs proceeding from an indurated enlarged base. *Flowers* in 1-sided spikes or racemes.

ANALYSIS OF THE GENERA.

Corolla naked at the orifice
  irregular - - - - - - - - - - - 1. Echium.
  regular
    Calyx prismatic, 5-cleft at the margin - - - 2. Pulmonaria.
    Calyx 5-parted - - - - - - - - 3. Lithospermum.
Corolla with scales or valves at the orifice
  Scales subulate
    Limb campanulate - - - - - - - 4. Symphytum.
    Limb rotate - - - - - - - - 5. Borago.
  Scales obtuse, concave
    Corolla funnel-shaped
      Tube doubly bent - - - - - - - 6. Lycopsis.
      Tube straight - - - - - - - - 7. Anchusa.
    Corolla hypocrateriform
      Calyx of the fruit compressed - - - 8. Asperugo.
      Calyx of the fruit taper
        Nuts smooth, oblong - - - - - 9. Myosotis.
        Nuts depressed, muricated - - - 10. Cynoglossum.

---

1. ECHIUM *Linn.*

Calyx 5-parted. *Corolla* with a short tube; limb large, campanulate, obliquely 5-lobed: segments unequal; the 2 upper largest, the lowest small, acute, and reflexed. *Nuts* covered with little tubercles. *Dec.*

1. E. *vulgare* Linn. Viper's-bugloss. E. B. 3. 181.
Stem bristly and warty. Stem-leaves lanceolate, bristly, single-ribbed. Spikes lateral, deflexed, hairy. *Smith.*
In fields and waste ground.—Biennial. *June, July.*

2. PULMONARIA *Linn.*

Calyx campanulate, 5-cleft, 5-cornered. *Corolla* funnel-shaped, somewhat 5-lobed; the limb nearly erect. *Stigma* obtuse, emarginate. *Dec.*

1. P. *officinalis* Linn. Lungwort. E. B. 2. 118.
Leaves ovate. *Smith.*

In woods and thickets. — Perennial. *May.*

2. P. *angustifolia* Linn. E. B. 23. 1628.
Leaves lanceolate. *Smith.*

In woods and thickets, rare. — Perennial. *May, June.*

3. LITHOSPERMUM *Linn.*

Calyx 5-parted. *Corolla* small, 5-lobed, funnel-shaped. *Anthers* oblong, included in the throat of the corolla. *Stigma* obtuse, bifid. *Nuts* shining, even, or wrinkled. *Dec.*

1. L. *officinale* Linn. Common Gromwell. Grey Mill. E. B. 2. 134.
Nuts even. Corolla not much longer than the calyx. Leaves lanceolate, rather acute, with lateral transverse ribs.

In dry situations, amongst rubbish. — Perennial. *May.*

2. L. *arvense* Linn. Bastard Alkanet. E. B. 2. 123.
Nuts wrinkled. Corolla not much longer than the calyx. Leaves obtuse, without lateral ribs.

In corn-fields and waste ground. — Annual. *May, June.*

3. L. *purpuro-cœruleum* Linn. E. B. 2. 117.
Nuts even. Corolla much longer than the calyx. Leaves lanceolate, acute, without lateral ribs. Barren stems prostrate.

In dry situations, amongst rubbish. — Perennial. *May.*

4. L. *maritimum* Lehm. E. B. 6. 368.
Nuts keeled, even. Leaves ovate, glaucous, besprinkled with callous points. Stems all procumbent.
Pulmonaria maritima *Linn.*

On the sea-shore. — Perennial. *July, August.*

4. SYMPHYTUM *Linn.*

Calyx 5-cleft. *Corolla* cylindrical, campanulate; tube very short; limb ventricose, with 5 short lobes. *Scales* of the orifice subulate, converging.

1. S. *officinale* Linn. Comfrey. E. B. 12. 817.
Leaves ovate-lanceolate, decurrent, finely hairy. *Smith.*

β. *patens.* Flowers purple.
S. patens *Sibth.*

In watery weadows. — Perennial. *May, June.*

2. S. *tuberosum* Linn. E. B. 21. 1502.
Leaves ovate, slightly decurrent, rather harsh; upper ones opposite. *Smith.*

In moist shady places in the north. — Perennial. *July.*

5. BORAGO *Linn.*

Calyx 5-parted. *Corolla* rotate, 5-cleft, usually spreading. *Scales* of the orifice obtuse, emarginate. *Nuts* wrinkled.

1. B. *officinalis* Linn. Borage. E. B. 1. 36.
Limb of the corolla flat, much longer than the tube; mouth with a

double row of valves; the innermost awl-shaped, bearing the stamens. *Smith.*

In waste ground by road sides. — Biennial. *June, July.*

6. LYCOPSIS *Linn.*

Calyx 5-cleft. *Corolla* funnel-shaped, 5-lobed; limb nearly erect; tube incurved. *Scales* of the orifice ovate, prominent, converging. *Stigma* emarginate. *Nuts* sculptured at the base. *Dec.*

1. L. *arvensis* Linn. Bugloss. E. B. 14. 938.
Leaves lanceolate, wavy, somewhat toothed, very bristly. Stalks of the flowers and fruit erect. Limb of the corolla slightly unequal. *Smith.*
Anchusa arvensis *Lehm.*
In fields, waste ground, and on dry banks. — Annual. *June.*

ANCHUSA *Linn.*

Calyx 5-cleft. *Corolla* funnel-shaped, 5-lobed; the limb erect; the lobes entire. The other characters of Lycopsis. *Dec.*

1. A. *officinalis* Linn. Alkanet. E. B. 10. 662.
Spikes imbricated, unilateral. Bracteas ovate, as long as the calyx. Leaves lanceolate. *Smith.*
In waste ground near the sea. — Perennial. *June, July.*

2. A. *sempervirens* Linn. E. B. 1. 45.
Flower-stalks axillary, each bearing 2 dense spikes, with an intermediate flower, and 2 principal ovate bracteas. Leaves ovate. *Smith.*
In waste ground. — Perennial. *May, June.*

8. ASPERUGO *Linn.*

Calyx 5-cleft, unequal, with intermediate teeth. *Corolla* with a short tube, and a 5-lobed limb. *Scales* of the orifice convex, converging. *Nuts* covered by the doubled, compressed calyx.

1. A. *procumbens* Linn. German Madwort E. B. 80. 661.
In rich waste ground. — Annual. *June, July.*

9. MYOSOTIS *Linn.*

Calyx 5-cleft, or 5-toothed. *Corolla* hypocrateriform, with a short tube; limb flat, with 5 emarginate lobes. *Scales* of the orifice convex, converging. *Nuts* smooth.

1. M. *palustris* Roth. E. B. 28. 1973.
Nuts smooth. Leaves and calyx roughish, with close bristles. Clusters leafless. Calyx funnel-shaped, with short broad spreading teeth. Limb of the corolla horizontal, longer than the tube. Root creeping. *Smith.*
M. scorpioides *Willd.*
In rivulets and ditches. — Perennial. *June—August.*

2. M. *cæspitosa* Schulz.
Nuts smooth. Leaves and calyx besprinkled with erect bristles. Clusters leafy at the base. Calyx funnel-shaped, with broad

spreading teeth. Limb of the corolla the length of the tube. Root fibrous. *Smith.*

In watery places.— Perennial, or biennial. *May, June.*

3. M. *intermedia* Link.
Nuts smooth. Leaves hairy. Clusters leafless. Tube of the calyx clothed with hooked bristles; segments with straight upright hairs. Root creeping. Stems decumbent. *Smith.*

In dry shady places. — Perennial. *April, May.*

4. M. *sylvatica* Lehm.
Nuts smooth. Leaves hairy. Clusters with a leaf at the base. Tube of the calyx clothed with hooked bristles; segments with straight upright hairs. Root fibrous. Stems erect. *Smith.*

In woods and dry shady places. — Perennial. *June, July.*

5. M. *alpestris* Lehm. E. B. 36. 2559.
Nuts smooth. Leaves hairy, radical ones but half the length of their footstalks. Clusters forked at the base, leafless. Calyx deeply 5-cleft, clothed with upright hairs; the lowermost incurved. Root fibrous, tufted.
M. alpina *Don.*
M. rupicola *Smith.*

On the mountains of Scotland. — Perennial. *July, August.*

6. M. *arvensis* Roth. E. B. 36, 2558.
Nuts smooth. Leaves hairy. Clusters stalked, with 1 remote axillary flower. Calyx half-5-cleft, clothed with spreading hairs; those of the tube hooked. Root fibrous. *Smith.*
M. annua *Mœnch.*

In dry shady places. — Annual. *June—August.*

7. M. *versicolor* Lehm. E. B. 36. 2558.
Seeds smooth. Leaves hairy. Clusters on long, naked stalks. Calyx longer than the partial stalks; hairs of its tube hooked. Root fibrous. *Smith.*

In dry sandy fields, as well as in moist meadows. — Annual. *April—June.*

10. CYNOGLOSSUM *Linn.*

Calyx 5-parted. *Corolla* short, funnel-shaped, 5-lobed. *Scales* of the orifice convex, converging. *Stigma* emarginate. *Nuts* depressed.

1. C. *officinale* Linn. Hound's-tongue. E. B. 13. 921.
Stamens shorter than the corolla. Stem-leaves broadly lanceolate, downy, sessile. Flowers without bracteas. *Smith.*

In waste ground, and by road sides. — Biennial. *June, July.*

2. C. *sylvaticum* Hænke. E. B. 23. 1642.
Stamens shorter than the corolla. Leaves lanceolate, somewhat spatulate; the upper ones clasping the stem; all smooth and shining above; hairy and warty beneath. Flowers without bracteas. *Smith.*

In shady situations, rare. — Biennial. *June.*

Order 49. CONVOLVULACEÆ *Juss.*

Calyx persistent, in 5 divisions.

Corolla monopetalous, hypogynous, regular, deciduous; the limb 5-lobed, generally plaited.

Stamens 5, inserted into the base of the corolla, and alternate with its segments.

Ovarium simple, with 2 or 4 cells, seldom with 1; sometimes in 2 or 4 divisions, few-seeded; the ovules definite and erect; *style* 1, usually divided at the top, sometimes down to the base; *stigmas* obtuse or acute.

Disk annular, hypogynous.

Capsule with from 1 to 4 cells; the valves fitting to the angles of a loose dissepiment, bearing the seeds at its base; sometimes valveless or dehiscing transversely.

Seeds with a small quantity of mucilaginous albumen; *embryo* curved; *cotyledons* shrivelled; *radicle* inferior.

Herbaceous plants or *shrubs*, usually twining and milky, smooth, with a simple pubescence. *Leaves* alternate, undivided, or lobed, seldom pinnatifid, with no stipulæ. *Inflorescence* axillary or terminal; peduncles 1- or many-flowered; the partial ones generally with 2 bracteæ. *R. Br.*

1. CONVOLVULUS *Linn.* BINDWEED.

Calyx 5-parted, naked, or with 2 small bracteæ at the base. *Corolla* campanulate, with 5 plaits. *Stamens* shorter than the limb. *Ovarium* 2-celled, rarely 3-celled; cells 2-seeded. *Style* undivided. *Stigmas* 2, filiform. *Capsule* valvular. *R. Br.*

1. C. *arvensis* Linn. E. B. 5. 312.

Leaves arrow-shaped, acute at each end. Stalks mostly single-flowered.

In hedges, fields, and gardens, very common; an almost unconquerable weed, especially on a gravelly soil. — Perennial. *June, July.*

.2. CALYSTEGIA *R. Br.*

Calyx 5-parted, inclosed in 2 foliaceous bracteæ. *Corolla* campanulate, with 5 plaits. *Stamens* nearly equal, shorter than the limb. *Ovarium* half 2-celled, 4-seeded. *Style* undivided. *Stigmata* 2, obtuse (taper or round). *Capsule* 1-celled. *R. Br.*

1. C. *sepium* E. B. 5. 313.

Leaves arrow-shaped, abrupt at the posterior lobes. Stalks square, single-flowered.

Convolvulus sepium *Linn.*

In moist hedges. — Perennial. *July, August.*

2. C. *Soldanella.* E. B. 5. 314.

Leaves kidney-shaped, somewhat angular. Stalks single-flowered, heir angles membranous. Stems procumbent. *Smith.*

Convolvulus Soldanella *Linn.*

On the sandy sea-shore. — Perennial. *June, July.*

3. CUSCUTA *Linn.* Dodder.

Calyx 4- or 5-cleft. *Corolla* roundish, urceolate; limb 4- or 5-cleft, with as many scales at the base. *Stamens* 4 or 5. *Ovarium* 2 cells, each with 2 seeds. *Stigmata* 2. *Capsule* dehiscing transversely. *Embryo* without cotyledons. — Parasitical *twining leafless plants, with thread-like* stems.

1. C. *europæa* Linn. E. B. 6. 378.
Flowers nearly sessile. Corolla with reflexed segments.
C. major *Dec.*
On thistles, and other annual or biennial herbs. — Annual. *August, September.*

2. C. *Epithymum* Linn. E. B. 6. 378. *b.*
Flowers sessile. Corolla with erect segments.
C. minor *Dec.*
On common, and other small shrubby plants. — Perennial? *August.*

Order 50. Polemoniaceæ *Juss.*

Calyx inferior, monosepalous, 5-parted, persistent, sometimes irregular.

Corolla regular, 5-lobed.

Stamens 5, inserted into the middle of the tube of the corolla, and alternate with its segments.

Ovarium superior, 3-celled, with a few ovula; *style* simple; *stigma* trifid.

Capsule 3-celled, 3-valved, few-seeded, with a loculicidal dehiscence; the valves separating from the axis.

Seeds angular or oval, often enveloped in mucus; *embryo* straight in the axis of horny albumen; *radicle* inferior; *cotyledons* elliptical, foliaceous.

Herbaceous plants, with opposite, or occasionally alternate, compound, or simple *leaves.*

1. POLEMONIUM *Linn.*

Calyx 5-cleft. *Corolla* rotate, with a short tube; limb 5-lobed. *Filaments* broadest at the base; *anthers* incumbent. *Dec.*

1. P. *cæruleum* Linn. Jacob's Ladder. Greek Valerian.
E. B. 1. 14.
Leaves pinnate. Flowers erect. Root fibrous. *Smith.*
In bushy places, in the north of England, and south of Scotland. — Perennial. *June.*

Order 51. Plantagineæ *Juss.*

Flowers usually hermaphrodite, seldom unisexual.
Calyx 4-parted, persistent.
Corolla monopetalous, hypogynous, scarious, with a 4-parted limb.
Stamens 4, inserted into the corolla, alternately with its segments; *filaments* filiform, doubled inwards in æstivation; *anthers* versatile, 2-celled.
Ovarium sessile, 2-, very seldom 4-celled; *ovula* peltate or erect, solitary, twin, or indefinite; *style* simple, subulate; *stigma* subulate, simple.
Capsule membranous, dehiscing transversely.
Seeds sessile, peltate, or erect, solitary, twin, or indefinite; *testa* mucilaginous; *embryo* in the axis of fleshy albumen; *radicle* inferior; *plumula* inconspicuous.
Herbaceous plants, usually stemless, occasionally with a stem; *hair* simple, articulated. *Leaves* flat and ribbed, or taper and fleshy. *Flowers* in spikes, rarely solitary.

1. PLANTAGO *Linn.*

Flowers hermaphrodite. *Capsule* with 2 or 4 cells, and 2 or several seeds.

1. P. *major* Linn. Greater Plantain. E. B. 22. 1558.
Leaves ovate, smoothish, somewhat toothed, on longish foot-stalks. Flower-stalks round. Spike tapering. Seeds numerous. *Smith.*
In meadows and cultivated ground. — Perennial. *All summer.*

2. P. *media* Linn. E. B. 22. 1559.
Leaves ovate, downy, with very short foot-stalks. Flower-stalks round. Spike cylindrical. Seeds solitary. *Smith.*
In dry pastures. — Perennial. *June—August.*

3. P. *lanceolata* Linn. E. B. 8. 507.
Leaves lanceolate, entire, tapering at each end, woolly at the base. Flower-stalks angular. Spike ovate. *Smith.*
In meadows and pastures. — Perennial. *June, July.*

4. P. *maritima* Linn. E. B. 3. 175.
Leaves linear, channelled, nearly entire. Flower-stalks round, longer than the leaves. Spike cylindrical. *Smith.*
In muddy salt marshes. — Perennial. *August, September.*

5. P. *Coronopus* Linn. Star of the Earth. E. B. 13. 892.
Leaves in many pinnate linear segments. Flower-stalks round. *Smith.*
In dry, sandy, or gravelly ground. — Annual. *June—August.*

2. LITTORELLA *Linn.*

Monœcious. Males. *Flowers* 4-parted. *Stamens* hypogynous. Female. *Flowers* sessile. *Calyx* 3-parted. *Corolla* urceolate, contracted at the orifice, with obsolete toothings. *Ovarium* 1-seeded.

1. L. *lacustris* Linn. Plantain Shore-weed. E. B. 7. 468.

In watery sandy places, especially about the margins of lakes and pools. — Perennial. *June.*

Order 52. PLUMBAGINEÆ *Juss.*

Calyx tubular, plaited, persistent.
Corolla monopetalous or 5-petalous, regular.
Stamens definite; in the monopetalous species hypogynous! in the polypetalous inserted on the petals!
Ovarium single, 1-seeded; *ovulum* inverted, pendulous from the point of an umbilical cord arising from the bottom of the cavity; *styles* 5! seldom 3 or 4; *stigmas* the same number.
Pericarpium a nearly indehiscent utriculus.
Seed inverted; *testa* simple; *embryo* straight; *radicle* superior.
Herbaceous plants or *under-shrubs*, variable in appearance. *Leaves* alternate or clustered, undivided, somewhat sheathing at the base. *Flowers* in spikes or heads. *R. Br.*

1. ARMERIA *Dec.*

Calyx scarious, plaited, entire. *Corolla* monopetalous or pentapetalous. *Stamens* 5, inserted on the lobes of the corolla. *Styles* 5. *Fruit* indehiscent. *Flowers* capitate, in solitary heads, surrounded by a common imbricated scarious involucrum. *Leaves* radical, tufted.

1. A. *maritima* Willd. Common Thrift. Sea Gilliflower. E. B. 4. 226.

Leaves linear, flat, obtuse. Scape twice or 4 times as long as the leaves. Scales of the involucrum scarious, obovate, very obtuse, shorter than the flowers. Calyx hairy at the base, with 5 sharp teeth, shorter than the corolla.

Statice Armeria *Linn.*
S. linearifolia *Laterr.*

On the sea-coast. — Perennial. *July, August.*

2. STATICE *Linn.*

Flowers in loose panicles, arranged on one side of the branches in long rows, surrounded by scarious scales. Otherwise the same as Armeria.

1. S. *Limonium* Linn. Sea Lavender. E. B. 2. 102.

Stalks round. Spikes level-topped. Leaves elliptic-oblong, single-ribbed, smooth, with a nearly terminal bristle. *Smith.*

On muddy sea-shores. — Perennial. *July, August.*

2. S. *reticulata* Linn. E. B. 5. 328.
Stalks prostrate, zigzag, dotted; lower branches barren. Leaves wedge-shaped, rather acute, without points. *Smith.*
On muddy sea-shores. — Perennial. *July, August.*

Order 53. Oleineæ *Hoffmannsegg & Link.*

Flowers hermaphrodite, sometimes diœcious.
Calyx monophyllous, divided, persistent, inferior.
Corolla hypogynous, monopetalous, 4-cleft, occasionally of 4 petals, connected in pairs by the intervention of the filaments, sometimes without petals; *æstivation* somewhat valvate.
Stamens 2, alternate with the segments of the corolla or the petals; *anthers* 2-celled, opening longitudinally.
Ovarium simple, without any hypogynous disk, 2-celled; the cells 2-seeded; the *ovules* pendulous and collateral. *Style* 1 or 0. *Stigma* bifid or undivided.
Fruit drupaceous, berried, or capsular; often by abortion 1-seeded.
Seeds with dense, fleshy, abundant albumen; *embryo* straight; *cotyledons* foliaceous, partly asunder; *radicle* superior; *plumula* inconspicuous.
Trees or *shrubs*. *Leaves* opposite, simple, seldom pinnated. *Flowers* in terminal or axillary racemes or panicles; the *pedicels* opposite, with single bracteæ. *R. Br.*

1. LIGUSTRUM *Linn.*

Calyx very small, 4-toothed. *Corolla* with a short tube; *limb* 4-cleft spreading. *Berry* 1-celled, with from 2 to 4 seeds. *Dec.*

1. L. *vulgare* Linn. Common Privet. E. B. 11. 764.
Leaves elliptic-lanceolate, obtuse, with a small point. *Smith.*
In thickets and hedges. — Shrub. *May, June.*

2. FRAXINUS *Linn.*

Calyx and *corolla* none. *Samara* pendulous, with 1 seed.

1. F. *excelsior* Linn. Common Ash. E. B. 24. 1692.
Leaflets serrated. Flowers without calyx or corolla. *Smith.*
In woods and hedges. — Tree. *April, May.*

2. F. *heterophylla* Vahl. E. B. 35. 2476.
Leaves both simple and compound, with tooth-like serratures. *Smith.*
F. simplicifolia *Willd.*
In woods, rare. — Tree. *April, May.*

Order 54. Ericeæ Juss.

Calyx monophyllous, 4- or 5-cleft, nearly equal, inferior, persistent.

Corolla hypogynous, monopetalous, 4- or 5-cleft, occasionally separable into 4 or 5 petals, regular, often withering, with an imbricated æstivation.

Stamens definite, equal in number to the segments of the corolla, or twice as many, hypogynous, or inserted into the base of the corolla; *anthers* 2-celled; the cells hard and dry, separate either at the apex or base, where they are furnished with some kind of appendage, and dehiscing by a pore or a cleft.

Ovarium surrounded at the base by a disk or secreting scales, many-celled, many-seeded. *Style* 1, straight. *Stigma* 1, undivided, or toothed.

Fruit capsular, many-celled, with central placentæ; dehiscence various.

Seeds indefinite, minute; *testa* firmly adhering to the nucleus; *embryo* cylindrical, in the axis of fleshy albumen; *radicle* opposite the hilum.

Shrubs or *undershrubs*. *Leaves* evergreen, rigid, entire, whorled, or opposite, without stipulæ. *Inflorescence* variable; the pedicels generally bracteate.

ANALYSIS OF THE GENERA.

Fruit dry
- Divisions of the calyx 5
  - Calyx coloured - - - - - - - - - 1. Azalea.
  - Calyx green
    - Corolla 5-parted - - - - - - - - 2. Ledum.
    - Corolla ventricose - - - - - - - 3. Andromeda.
- Divisions of the calyx 4
  - Capsule dehiscing through the dissepiments
    - Calyx coloured - - - - - - - 4. Calluna.
    - Calyx green - - - - - - - 5. Menziesia.
  - Capsule dehiscing through the cells - - - 6. Erica.

Fruit fleshy
- Berry granular - - - - - - - - - 7. Arbutus.
- Berry smooth - - - - - - - - - 8. Arctostaphylos.

---

1. AZALEA *Linn.*

Calyx 5-cleft, coloured. *Corolla* campanulate, 5-cleft. *Stamens* 5, hypogynous. *Capsule* 5-celled, many-seeded, with 5 valves dehiscing through the dissepiments.

1. A. *procumbens* Linn. E. B. 13. 865.

B anches spreading and reclining. Leaves opposite, revolute very smooth. *Smith.*

On alpine moors. — Shrub. *July.*

2. LEDUM *Linn.*

Calyx minute, 5-toothed. *Corolla* 5-parted. *Stamens* 5 or 10, inserted into the base of the calyx. *Capsule* 5-celled, many-seeded, with 5 valves dehiscing th ough the dissepiments.

1. L. *palustre* Linn.
Leaves linear, revolute at the margin, downy beneath. *Dec.*
On the north-west coast of Ireland. — Shrub. *July, August.*

3. ANDROMEDA *Linn.*

Calyx small, 5-parted. *Corolla* ovate, with a 5-cleft reflexed limb. *Stamens* 10, erect. *Capsule* 5-celled, 5-valved. *Seeds* numerous, minute.

1. A. *polifolia* Linn. Wild Rosemary. E. B. 10. 713.
Clusters terminal. Leaves alternate, lanceolate, revolute; glaucous beneath. *Smith.*
On mossy peat bogs, in the mountainous parts of England and Ireland, and the lowlands of Scotland. — Shrub. *June.*

4. CALLUNA *Salisb.*

Calyx of 4 coloured sepals, surrounded by 4 coloured bracteæ. *Corolla* campanulate, 4-cleft. *Stamens* 8. *Capsule* 4-celled, the dissepiments adhering to the axis, and with 4 valves dehiscing through the dissepiments.

1. C. *vulgaris* Salisb. Common Ling. E. B. 15. 1013.
Erica vulgaris *Linn.*
Common on dry heaths. — Shrub. *June, July.*

5. MENZIESIA *Smith.*

Calyx 4-parted. *Corolla* ventricose, with a spreading 4- or 5-toothed limb. *Stamens* 8 or 10, inserted into the base of the corolla. *Capsule* 4-celled, many-seeded, with 4 valves dehiscing through the dissepiments.

1. M. *cærulea* Swartz. E. B. 25. 2469.
Leaves linear, obtuse, with cartilaginous teeth. Flower-stalks terminal, aggregate, simple. Flowers 5-cleft, decandrous. *Smith.*
Andromeda cærulea *Linn.*
A. taxifolia *Pall.*
Erica cærulea *Willd.*
On dry heathy moors in Scotland. — Shrub. *June, July.*

2. M. *polifolia Smith.* E. B. 1. 35.
Leaves ovate, revolute; downy and white beneath. Flowers cleft, octandrous, in terminal leafy clusters. *Smith.*
Andromeda Daboecia *Linn.*
Vaccinium cantabricum *Huds.*
Erica Dabeoci *Linn.*
Menziesia Dabeoci *Smith.*
On mountains in the west of Ireland. — Shrub. *June, July.*

6. ERICA *Linn.* HEATH.

Calyx 4-parted. *Corolla* campanulate, often ventricose, 5-toothed. *Stamens* 8. *Capsule* with from 4 to 8 cells, and the same number of valves.

1. E. *Tetralix Linn.* E. B. 15. 1014.
Anthers horned. Style nearly concealed. Corolla ovate. Leaves fringed, 4 in a whorl. Flowers in round tufts. *Smith.*
On heathy boggy ground. — Shrub. *July, August.*

2. E. *cinerea* Linn. E. B. 15. 1015.
Anthers crested. Style a little prominent. Stigma capitate. Corolla ovate. Leaves 3 in a whorl. *Smith.*
On dry turfy heaths. — Shrub. *July—October.*

3. E. *vagans* Linn. E. B. 1. 3.
Anthers simple, deeply cloven, prominent as well as the style. Corolla bell-shaped. Leaves 4 in a whorl. Flowers on simple, crowded, axillary stalks. *Smith.*
E. multiflora *Hudson.*
E. didyma *Withering.*
On heaths in Cornwall. — Shrub. *July, August.*

4. E. *ciliaris* Linn.
Leaves 3 or 4 in a whorl, somewhat ovate, ciliated. Racemes 1-sided. Segments of the calyx lanceolate. Corolla ovate, inflated. Anthers simple, included. Style protruded. *Dec.*
Sent from a bog near Truro, by the Rev. I. S. Tozer, to *Dr. Greville*, 1828. — Shrub. *July, August.*

7. ARBUTUS *Linn.*

Calyx small, 5-parted. *Corolla* ovate, with a small 5-cleft revolute limb. *Stamens* 10, villous at base; *anthers* with 2 pores at the tip. *Berry* granular, 5-celled, many-seeded.

1. A. *Unedo* Linn. Common Strawberry-tree. E. B. 34. 2377.
Stem arboreous. Leaves smooth, bluntly serrated. Panicle terminal. Berry with many seeds. *Smith.*
About the lake of Killarney. — Tree. *September.*

8. ARCTOSTAPHYLOS *Kunth.*

Calyx small, 5-parted. *Corolla* ovate, with a small 5-cleft revolute limb. *Stamens* 10, smooth; *anthers* without pores. *Berry* smooth; seeds solitary.

1. A. *alpina* Sprengel. E. B. 29. 2030.
Stems procumbent. Leaves rugged, serrated. Clusters terminal. *Smith.*
Arbutus alpina. *Linn.*
On stony mountainous heaths. — Shrub. *May.*

2. A. *Uva ursi* Sprengel. E. B. 10. 714.
Stems procumbent. Leaves obovate, entire. Clusters terminal. *Smith.*
Arbutus Uva ursi *Linn.*
On dry stony alpine heaths. — Shrub. *June.*

Order 55. Pyroleæ.

Calyx 5-leaved, persistent, inferior.
Corolla monopetalous, hypogynous, regular, deciduous, 4 or 5-toothed, with an imbricated æstivation.
Stamens hypogynous, twice as numerous as the divisions of the corolla; *anthers* 2-celled, opening longitudinally, and furnished with appendages at the base.
Ovarium superior, 4- or 5-celled, many-seeded, with an hypogynous disk; *style* 1, straight or declinate; *stigma* simple.
Fruit capsular, 4- or 5-celled, dehiscent, with central placentæ.
Seeds indefinite, minute, winged; *embryo* minute, at one extremity of a fleshy albumen.
Herbaceous plants, rarely *undershrubs*, sometimes parasitical and leafless. *Stems* round, covered with scales; in the frutescent species leafy. *Leaves* either wanting or simple, entire or toothed. *Flowers* in terminal racemes, rarely solitary.

1. PYROLA *Linn.* Winter-green.

Calyx 5-cleft or 5-parted. *Petals* 5, cohering very slightly at the base. *Style* longer than the stamens. *Stamens* 10, subulate. *Capsule* 5-celled, dehiscing at the angles of the base. — Herbaceous plants, *with evergreen entire leaves, rarely leafless.* Scapes *simple.* Flowers *white or pink, unilateral.*

1. P. *rotundifolia* Linn. E. B. 3. 213.
Stamens ascending. Style twice as long, declining and recurved. Cluster many-flowered. Calyx as long as the stamens. *Smith.*
In bushy places, and in dry heathy woods. — Perennial. *July, August.*

2. P. *media* Swartz. E. B. 28. 1945.
Stamens regularly inflexed. Style twice as long, deflexed. Cluster of many pendulous flowers. Calyx shorter than the stamens. *Smith.*
In woods in the north. — Perennial. *July, August.*

3. P. *minor* Linn. E. B. 36. 2543.
Stamens regularly inflexed. Style the same length, straight. Stigma 5-lobed, pointless, without a ring. Cluster of many drooping flowers. *Smith.*
P. rosea *E. Bot.*
In mossy woods and thickets, in mountainous situations. — Perennial. *July.*

4. P. *secunda* Linn. E. B. 8. 517.
Leaves ovate, acute, serrated. Flowers drooping, unilateral. Pores of the anthers dilated. Style straight. Stigma 5-lobed. *Smith.*
In mossy alpine woods. — Perennial. *July.*

5. P. *uniflora* Linn. E. B. 3. 146.
Stalk single-flowered. Pores of the anthers contracted, tubular stigma with 5 pointed rays. *Smith.*
In alpine woods. — Perennial *July.*

2. MONOTROPA *Linn.*

Sepals 4 or 5, coloured. *Petals* 4 or 5, slightly cohering by the base. *Stamens* 8 or 10. *Style* simple, cylindrical. *Capsule* with 4 or 5 cells, and 4 or 5 valves, many-seeded. —Herbaceous *parasitical plants, with leafless scaly* stems.

1. M. *Hypopitys* Linn. Yellow Bird's-nest. E. B. 1. 69.
Flowers in a drooping cluster; lateral ones with 8 stamens; terminal one with 10. *Smith.*
About the roots of beeches and firs, in woods. — Perennial. *June.*

Order 56. APOCYNEÆ *Juss.*

Calyx divided in 5, persistent, inferior.

Corolla monopetalous, hypogynous, regular, 5-lobed, deciduous, with an imbricated æstivation.

Stamens 5, inserted upon the corolla, alternate with the segments of the limb. *Filaments* distinct. *Anthers* 2-celled, opening longitudinally. *Pollen* granular, applied immediately to the stigma.

Ovaries 2, or 1 with 2 cells, usually many-seeded; *styles* 2 or 1; *stigma* 1.

Fruit follicular, capsular, drupaceous, or berried, double or single.

Seeds usually containing *albumen; embryo* leafy; *plumula* inconspicuous.

Trees or *shrubs*, generally flowing with milk. *Leaves* opposite, sometimes whorled, sometimes scattered, quite entire, generally hairy; fringes or glands between the petioles. *Inflorescence* somewhat corymbose. *R. Br.*

1. VINCA *Linn.* PERIWINKLE.

Calyx 5-parted. *Corolla* hypocrateriform; the tube long, the orifice prominent, 5-angled, the limb flat, bluntly 5-lobed. *Anthers* converging. *Style* 1. *Stigma* capitate, with a ring at the base. *Seeds* naked. *Dec.*

1. V. *minor* Linn. E. B. 13. 917.
Stems procumbent. Leaves elliptic-lanceolate, smooth-edged. Flowers stalked. Segments of the calyx lanceolate. *Smith.*
In bushy places, and about hedges and banks. — Perennial. *May.*

2. V. *major* Linn. E. B. 8. 514.
Stems ascending. Leaves ovate, fringed. Flowers stalked. Segments of the calyx bristle-shaped, elongated. *Smith.*
In thickets and groves. — Perennial. *May.*

Order 57. Gentianeæ *Juss.*

Calyx monophyllous, divided, inferior, persistent.

Corolla monopetalous, hypogynous, usually regular, withering or deciduous; the limb divided, equal, its lobes of the same number as those of the calyx, generally 5, sometimes 4 or 8, with an imbricated æstivation.

Stamens inserted upon the corolla, equal in number to the segments, and alternate with them; some of them occasionally abortive.

Ovarium single, 1- or 2-celled, many-seeded. *Styles* 1 or 2, either partially or wholly cohering. *Stigmas* 1 or 2.

Capsule, or *Berry*, many-seeded, with 1 or 2 cells, generally 2-valved; the margins of the valves turned inwards, and in the genera with 1 cell, bearing the seeds; in the 2-celled genera inserted into a central placenta.

Seeds small; *embryo* straight in the axis of soft fleshy albumen; *radicle* next the hilum.

Herbaceous plants, seldom *shrubs*, generally smooth. *Leaves* opposite, entire, without stipulæ. *Flowers* terminal or axillary. *R. Br.*

ANALYSIS OF THE GENERA.

Leaves opposite
  Stamens 4 . . . . 1. Exacum.
  Stamens 5
    Style simple . . . . 2. Erythræa.
    Style 2-lobed
      Corolla funnel-shaped . . . . . . 3. Gentiana.
      Corolla rotate . . . . 4. Swertia.
  Stamens 8 . . . . 5. Chlora.
Leaves alternate
  Margin of the corolla flat . . . . . . 6. Menyanthes.
  Margin of the corolla inflexed . . . . . 7. Villarsia.

1. EXACUM *Linn.*

Calyx 4-parted. *Corolla* 4-lobed, with a globose tube. *Stamens* 4. *Anthers* when burst remaining straight. *Style* 1. *Stigma* 2-fid. *Capsule* 2-celled. *Dec.*

1. E. *filiforme* Smith. E. B. 4. 235.
Leaves sessile. Stem thread-shaped forked. Flowers on long stalks. *Smith.*
Gentiana filiformis *Linn.*
On sandy or turfy bogs.—Annual. *July.*

2. ERYTHRÆA *Rencalm.*

Calyx 5-cleft. *Corolla* funnel-shaped, withering, with a short limb. *Stamens* 5. *Anthers* when burst becoming spiral. *Style* erect. *Stigmas* 2, roundish. *Cavsule* linear. *R. Br.*

1. E. *Centaurium* Pers. E. B. 6. 417.
Stem nearly simple. Panicle forked, corymbose. Leaves ovate lanceolate. Calyx half the length of the tube; its segments partly combined by a membrane. *Smith.*
Chironia Centaurium *Curtis.*
Gentiana Centaurium *Linn.*
In dry gravelly pastures. — Annual. *July, August.*

2. E. *littoralis* Hooker.
Stem simple, straight. Leaves linear-obovate; obscurely 3-ribbed. Flowers densely corymbose, nearly sessile. Calyx as long as the tube; its segments combined below. *Smith.*
Chironia littoralis *Turner.*
Ch. pulchella *Don.*
On the sandy sea-coast. — Annual. *June, July.*

3. E. *latifolia* Smith.
Stem 3-cleft at the top. Flowers in dense forked tufts. Calyx as long as the tube. Segments of the corolla lanceolate. Lower leaves broadly elliptical, with 5 or 7 ribs. *Smith.*
In sandy ground near the sea, to the north of Liverpool. — Annual. *July.*

4. E. *pulchella* Hooker. E. B. 7. 458.
Stem forked, variously branched, or simple, winged. Flowers solitary, stalked. Calyx above half as long as the tube. Segments of the corolla lanceolate. Leaves ovate. *Smith.*
Chironia pulchella *Willd.*
Ch. ramosissima *Ehr.*
Gentiana pulchella *Swartz.*
In sandy ground, chiefly near the sea. — Annual. *August, September.*

3. GENTIANA *Linn.*

Calyx 4 or 5-cleft. *Corolla* funnel-shaped, or hypocrateriform, 4 or 5-cleft, with the orifice naked. *Stamens* 5. *Stigma* 2-lobed. *Seed* not bordered. *R. Br.*

1. G. *Pneumonanthe* Linn. Calathian Violet. E. B. 1. 20.
Corolla bell-shaped, 5-cleft. Flowers stalked. Leaves linear. *Smith.*
On moist turfy bogs. — Perennial. *August, September.*

2. G. *acaulis* Linn. E. B. 23. 1594.
Flower solitary, 5-cleft, bell-shaped, about as long as the quadrangular stem. *Smith.*
On mountains; a very doubtful native. — Perennial. *June, July.*

3. G. *verna* Linn. E. B. 7. 493.
Corolla 5-cleft, salver-shaped, crenate; segments auricled at the base. Leaves crowded, ovate. *Smith.*
In barren mountainous situations. — Perennial. *April.*

4. G. *nivalis* Linn. E. B. 13. 896.
Corolla funnel-shaped, 5-cleft, with notched intermediate segments. Angles of the calyx even, acute. Stem bearing several flowers. *Smith.*
On the top of Ben Lawers. — Annual. *August.*

5. G. *Amarella* Linn. E. B. 4. 236.
Corolla salver-shaped, 5-cleft; bearded in the throat. Segments of the calyx nearly equal. Stem flowering from top to bottom, with short axillary branches. *Smith.*

In limestone and chalky pastures. — Annual. *August, September.*

6. G. *campestris* Linn. E. B. 4. 237.
Corolla salver-shaped, 4-cleft; bearded in the throat. Two outer segments of the calyx ovate, very large. *Smith.*

In elevated pastures towards the sea-coast. — Annual. *September, October.*

4. SWERTIA *Linn.*

Calyx nearly 5-parted. *Corolla* rotate, with a very short tube; the limb flat, 5-parted, each segment lanceolate, and having at the base, in the inside, a double fringed gland. *Stamens* 5, shorter than the corolla. *Capsule* 1-celled, 2-valved, with numerous seeds adhering to the margins of the valves.

1. S. *perennis* Linn. Marsh Felwort. E. B. 21. 1441.
Flowers 5-cleft. Radical leaves ovate. *Smith.*

Found in Wales by Dr. Richardson. *Huds.* Some mistake is to be suspected, as no person has confirmed this discovery, and so fine a plant could scarcely be overlooked. *Smith.* — Perennial. *August.*

5. CHLORA *Linn.*

Calyx 8-parted. *Corolla* hypocrateriform; the tube short, the limb 8-parted. *Stamens* 8, very short, inserted in the orifice. *Style* 1. *Stigma* 4-cleft. *Capsule* 1-celled.

1. C. *perfoliata* Linn. E. B. 1. 60.
Leaves perfoliate. Panicle forked, many-flowered. *Smith.*

On chalky hills or banks. — Annual. *July, August.*

§ *Menyantheæ.*

Leaves alternate, usually toothed or divided.

MENYANTHES *Linn.*

Calyx 5-parted. *Corolla* funnel-shaped; the limb spreading, 5-parted, bearded internally, with a simple margin. *Stamens* 5. *Style* 1. *Stigma* capitate, with from 2 to 5 furrows. *Glands* 5, hypogynous, alternate with the stamens. *Capsule* 1-celled, 2-valved; the *valves* bearing the seed in their axis. — Leaves *ternate.*

1. M. *trifoliata* Linn. Common Buckbean, or Bog-bean. Marsh Trefoil. E. B. 7. 495.
Leaves ternate. Disk of the corolla densely shaggy. *Smith.*

In watery meadows. — Perennial. *June, July.*

7. VILLARSIA *Vent.*

Calyx 5-parted. *Corolla* somewhat rotate, the limb spreading, 5-parted, smooth in the disk, bearded or scaly at the base, with an inflexed margin. *Stamens* 5, alternate with the segments. *Style* 1. *Stigma*

2-lobed; the lobes toothed. *Glands* 5, hypogynous, alternate with the stamens. *Capsule* 1-celled, 2-valved, many-seeded, (in the floating species indehiscent); the valves bearing the seeds in their axis. — Leaves *simple*. R. Br.

1. V. *nymphæoides* Vent. E. B. 4. 217.
Leaves heart-shaped, wavy at the edges, floating. Corolla fringed. *Smith.*

In ponds, and the bays of rivers. — Perennial. *July, August.*

Order 58. Solaneæ *Juss.*

Calyx 5-parted, seldom 4-parted, persistent, inferior.

Corolla monopetalous, hypogynous; the limb 5-cleft, seldom 4-cleft, regular, or somewhat unequal, deciduous; the æstivation, in the genuine genera of the order, plaited; in the spurious genera, imbricated.

Stamens inserted upon the corolla, as many as the segments of the limb, with which they are alternate; 1 sometimes being abortive.

Pericarpium with 2 or 4 cells, either a capsule with a double dissepiment parallel with the valves, or a berry, with the placentæ adhering to the dissepiment.

Seeds numerous, sessile. *Embryo* more or less curved, often out of the centre, lying in fleshy *albumen*; *radicle* next the hilum.

Herbaceous plants or *shrubs*. *Leaves* alternate, undivided, or lobed; the floral ones sometimes double, and placed near each other. *Inflorescence* variable, often out of the axillæ; the pedicels without bracteæ. *R. Br.*

ANALYSIS OF THE GENERA.

Fruit dry
- Corolla funnel-shaped
  - Capsule with 4 valves - - - - - - - - 1. Datura.
  - Capsule dehiscing transversely - - - - - 2. Hyoscyamus.
- Corolla rotate - - - - - - - - - - - 3. Verbascum.

Fruit fleshy
- Corolla rotate - - - - - - - - - - - - 4. Solanum.
- Corolla campanulate - - - - - - - - - 5. Atropa.

---

1. DATURA *Linn.*

Calyx tubular, ventricose, with 5 angles, 5-toothed, deciduous, leaving behind a broad orbicular base. *Corolla* funnel-shaped; the tube long; the limb with 5 angles, 5 plaits, and 5 points. *Stamens* 5. *Stigma* of 2 plates. *Capsule* echinate or smooth, 2-celled; the cells divided occasionally by spurious dissepiments.

1. D. *Stramonium* Linn. Common Thorn-apple. E. B. 18. 1288.
Fruit spinous, ovate, erect. Leaves ovate, smooth, sinuated. *Smith.*
In waste ground, and on dunghills. — Annual. *July.*

2. HYOSCYAMUS *Linn.*

Calyx tubular, 5-cleft. *Corolla* funnel-shaped; the limb spreading, obliquely 5-lobed, unequal. *Stamens* 5. *Stigma* capitate. *Capsule* compressed, furrowed on each side, opening at the apex by a transverse aperture.

1. H. *niger* Linn. Common Henbane. E. B. 9. 591.
Leaves sinuated, clasping the stem. Flowers sessile. *Smith.*
On waste ground. — Annual. *July.*

3. VERBASCUM *Linn.*

Calyx 5-parted. *Corolla* rotate, 5-lobed, unequal. *Stamens* 5, unequal; filaments declinate, almost always villous at the base. *Capsule* with 2 valves, ovate, or globose. *Dec.*

1. V. *Thapsus* Linn. High Taper. E. B. 8. 549.
Leaves decurrent, crenate, woolly on both sides. Stem simple. Cluster dense. Flowers almost sessile. *Smith.*
On banks and waste ground. — Biennial. *July, August.*

2. V. *Lychnitis* Linn. White Mullein. E. B. 1. 58.
Leaves wedge-shaped-oblong; stripped of down on their upper side. Stem angular, panicled. *Smith.*
By road sides, and other waste places. — Biennial. *July, August.*

3. V. *thapsiforme* Schrad.
Stem simple. Leaves lanceolate-ovate, decurrent, crenulate, downy; the upper acuminate. Raceme spiked, dense. Bracteæ longer than the woolly calyx. Segments of the corolla obovate, rounded. Two anthers oblong. *Dec.*
V. thapsoides *Willd.*
By road sides in Kent. — Biennial. *July, August.*

4. V. *pulverulentum* Villars. E. B. 7. 487.
Leaves ovate-oblong, obscurely serrated, clothed on both sides with mealy deciduous wool. Stem round, panicled. *Smith.*
By road sides, and in the borders of fields, chiefly in Norfolk and Suffolk. — Biennial. *July.*

5. V. *nigrum* Linn. E. B. 1. 59.
Leaves oblong-heart-shaped, stalked, waved, and crenate, slightly downy. Cluster mostly solitary. *Smith.*
On banks in shady lanes. — Perennial. *July, August.*

6. V. *virgatum* Withering. E. B. 8. 550.
Leaves ovate-lanceolate, toothed, sessile; radical ones downy, somewhat lyrate. Stem branched. Flowers aggregate, partly sessile. *Smith.*
In fields, and by way sides, rare. — Biennial. *August.*

7. V. *Blattaria* Linn. E. B. 6. 393.
Leaves clasping the stem, oblong, smooth, serrated; radical ones

sinuated. Clusters panicled, simple. Flower-stalks longer than the bracteas. *Smith.*

On banks, in a gravelly soil. — Annual. *July.*

4. SOLANUM *Linn.*

Calyx persistent, with from 5 to 10 divisions. *Corolla* monopetalous, rotate; the tube very short; the limb spreading, with 4, 5, or 6 divisions. *Stamens* 4, 5, or 6. *Anthers* oblong, opening by 2 pores at the apex. *Berry* roundish, with 2, 3, 4, or 6 cells. *Embryo* spiral. *Dec.*

1. S. *Dulcamara* Linn. E. B. 8. 365.
Stem shrubby, zigzag, without thorns. Upper leaves hastate. Clusters cymose. *Smith.*

In hedges and thickets. — Shrub. *June, July.*

2 S. *nigrum* Linn. Nightshade. E. B. 8. 566.
Stem herbaceous, without thorns. Leaves ovate, bluntly toothed, or wavy. Umbels lateral, drooping. *Smith.*

Common in waste ground, and on dunghills. — Annual; occasionally perennial. *June—September.*

5. ATROPA *Linn.*

Calyx campanulate, 5-cleft. *Corolla* campanulate, twice as long as the calyx, 5-lobed, equal. *Filaments* 5, filiform. *Berry* globose, seated in the calyx. *Dec.*

1. A. *Belladonna* Linn. Common Dwale. Deadly Nightshade. E. B. 9. 592.
Stem herbaceous. Leaves ovate, undivided. Flowers solitary. *Smith.*

In hedges and waste ground. — Perennial. *June.*

Order 59. PRIMULACEÆ *Vent.*

Calyx divided, 5-cleft, seldom 4-cleft, inferior, regular, persistent.
Corolla monopetalous, hypogynous, regular; the limb 5-cleft, seldom 4-cleft.
Stamens inserted upon the corolla, equal in number to its segments, and opposite them.
Ovarium 1-celled. *Style* 1. *Stigma* capitate.
Capsule opening with valves. *Placenta* central, distinct.
Seeds numerous, peltate. *Embryo* included within fleshy albumen, and lying across the hilum; *radicle* with no determinate direction.
Herbaceous plants. *Leaves* usually opposite, either whorled or scattered. *R. Br.*

ANALYSIS OF THE GENERA.

Ovarium wholly superior
  Stamens 4 - - - - - - - - - - - 1. CENTUNCULUS.
  Stamens 5
    Fruit rather fleshy, indehiscent - - - 2 CYCLAMEN.
    Fruit dry, dehiscent
      Dehiscing by valves
        Corolla wanting - - - - - - - - 3. GLAUX.
        Corolla present
          Calyx 5-toothed - - - - - - 4. PRIMULA.
          Calyx 5-parted
            Corolla with no tube - - - 5. LYSIMACHIA.
            Corolla with a tube - - - 6. HOTTONIA.
      Dehiscing by a lid - - - - - - - - 7. ANAGALLIS.
  Stamens 7 - - - - - - - - - - - 8. TRIENTALIS.
Ovarium half inferior - - - - - - - - - - 9. SAMOLUS.

---

1. CENTUNCULUS *Linn.*

Calyx 4- or 5-parted. *Corolla* somewhat urceolate, 4- or 5-cleft, withering. *Stamens* 4 or 5; *filaments* beardless. *Capsule* globose, dehiscing by a transverse incision. *R. Br.*

1. C. *minimus* Linn. Chaff-weed. Bastard Pimpernel. E. B. 8. 531.

Flowers sessile. Corolla without glands at the base. *Smith.*

On sandy watery heaths. — Annual. *June, July.*

2. CYCLAMEN *Linn.*

Calyx campanulate, 5-cleft. *Corolla* with the tube ovate; the limb 5-parted and reflexed. *Stamens* 5, inserted in the base of the tube. *Anthers* sessile. *Fruit* globose, coriaceous, or rather fleshy, many-seeded.

1. C. *hederifolium* Willd. E. B. 8. 548.

Leaves heart-shaped, angular, finely toothed; their ribs and foot stalks roughish. *Smith.*

C. europæum *E. Bot.*

On a bank at Bramfield, Suffolk. — Perennial. *April.*

3. GLAUX *Linn.*

Calyx campanulate, 5-cleft, coloured. *Corolla* 0. *Stamens* 5, hypogynous. *Stigma* capitate. *Capsule* with 1 cell, and 5 valves. *Seeds* attached to a central globose placenta. *Embryo* straight, lying across the hilum. *Dec.*

1. G. *maritima* Linn. Sea-milkwort. Black Saltwort. E. B. 1. 13.

In muddy salt-marshes. — Perennial. *June, July.*

4. PRIMULA *Linn.*

Calyx 5-toothed. *Corolla* hypocrateriform; the limb 5-lobed, usually emarginate; the orifice dilated; the tube taper, as long as the calyx or longer. *Anthers* usually tapering to the point. *Capsule* ovate, dehiscing at the apex, with 5 or 10 teeth. *Seeds* minute, very numerous.

1. P. *vulgaris* Hudson. Common Primrose. E. B. 1. 4.
Leaves obovate-oblong, toothed, wrinkled. Stalks single-flowered. Limb of the corolla flat. *Smith.*
P. acaulis *Jacq.*
In groves and grassy places. — Perennial. *April, May.*

2. P. *elatior* Withering. Oxlip. E. B. 8. 513.
Leaves toothed, wrinkled, contracted towards the middle. Stalk many-flowered. Limb of the corolla flat. *Smith.*
In woods or pastures. — Perennial. *April.*

3. P. *veris* Linn. Common Cowslip. Paigle. E. B. 1. 5.
Leaves toothed, wrinkled, contracted towards the middle. Stalk many-flowered. Limb of the corolla concave. *Smith.*
P. officinalis *Jacq.*
In meadows and pastures. — Perennial. *April, May.*

4. P. *farinosa* Linn. Bird's-eye Primrose. E. B. 1. 6.
Leaves toothed, even; powdery beneath. Limb of the corolla flat; mouth with a notched border. Stigma undivided. *Smith.*
In wet pastures, and by rivulets, on mountains. — Perennial. *June, July.*

5. P. *scotica* Hooker.
Leaves finely toothed, even; powdery on both sides. Limb of the corolla flat; mouth with a notched border. Stigma 5-cleft. *Smith.*
Found by Mr. Gibb, of Inverness, on Holborn Head, near Thurso, in Caithness, abundantly; also, in the way from Thurso to Dunbeath. *Hooker.* — Perennial. *July.*

5. LYSIMACHIA *Linn.*

Calyx 5-parted. *Corolla* rotate, 5-cleft. *Stamens* 5. *Capsule* globose, with 5 or 10 valves. *R. Br.*

* *Stalks many-flowered.*

1. L. *vulgaris* Linn. Yellow Loosestrife. E. B. 11. 761.
Clusters panicled, terminal. Leaves ovate-lanceolate, acute. *Smith.*
In watery shady places. — Perennial. *July.*

2. L. *thyrsiflora* Linn. E. B. 3. 176.
Clusters lateral, axillary, stalked. *Smith.*
In watery places, very rare. — Perennial. *July.*

** *Stalks single-flowered.*

3. L. *nemorum* Linn. Yellow Pimpernel. E. B. 8. 527.
Leaves ovate, acute. Flowers solitary. Stem procumbent. Stamens smooth. *Smith.*
In moist woods and shady places. — Perennial. *May—September.*

4. L. *Nummularia* Linn. Moneywort, or Herb Twopence. E. B. 8. 528.
Leaves somewhat heart-shaped. Flowers solitary. Stem prostrate, creeping. Stamens glandular. *Smith.*
In wet meadows, and the borders of rivulets. — Perennial. *June, July.*

6. HOTTONIA *Linn.*

Calyx 5-parted. *Corolla* with a short tube, and a flat 5-lobed limb. *Stamens* 5, nearly sessile. *Capsule* globose, crowned by the long persistent style. *Dec.*

1. H. *palustris* Linn. Water Featherfoil. Common Water-violet. E. B. 6. 364.
Stalks solitary, many flowered; partial stalks whorled. *Smith.*
In clear ditches and ponds. — Perennial. *June.*

7. ANAGALLIS *Linn.*

Calyx 5-parted. *Corolla* rotate, 5-lobed. *Capsule* globose, dehiscing by a transverse incision.

1. A. *arvensis* Linn. Common Pimpernel. E. B. 8. 529.
Leaves ovate, dotted beneath. Stem procumbent. Corolla minutely notched. *Smith.*
In fields and gardens. — Annual. *June—August.*

2. A. *cærulea* Schreb. E. B. 26. 1823.
Leaves ovate, or somewhat lanceolate, dotted beneath. Stem erect. Corolla strongly notched. *Smith.*
In corn-fields, rare. — Annual. *July.*

3. A. *tenella* Linn. E. B. 8. 530.
Leaves roundish, somewhat pointed, stalked. Stem creeping. Stigma acute. *Smith.*
Lysimachia tenella *Linn.*
On wet mossy bogs. — Perennial. *July, August.*

8. TRIENTALIS *Linn.*

Calyx 7-parted. *Corolla* rotate, 7-parted. *Stamens* 7. *Fruit* fleshy, dehiscing at the sutures.

1. T. *europæa* Linn. Chickweed Winter-green. E. B. 1. 15.
Leaves obovate-oblong; the lowermost very obtuse. *Smith.*
On turfy heaths, in mountainous countries. — Perennial. *May, June.*

*** *Related to Primulaceæ.*

9. SAMOLUS *Linn.*

Calyx half superior. *Corolla* somewhat campanulate, 5-lobed. *Stamen.* 5, bearing anthers, and opposite the segments of the limb; 5 sterile and alternate. *Capsule* half inferior, ovate, half 5-valved, 1-celled. *Placenta* central, loose. *Seeds* numerous, fixed by one end, albuminous. *Embryo* included; *radicle* next the umbilicus. *R. Br.*

1. S. *Valerandi* Linn. Common Brook-weed. E. B. 10. 703.
Leaves obovate, obtuse. Clusters corymbose, many-flowered. Bracteas solitary, in the middle of each partial stalk. *Smith.*
In clear watery places. — Perennial. *July.*

Order 60. LENTIBULARIÆ *Richard.*

Calyx divided, persistent, inferior.
Corolla monopetalous, hypogynous, irregular, bilabiate, with a spur.
Stamens 2, included within the corolla, and inserted into its base.
Anthers simple, sometimes contracted in the middle.
Ovarium 1-celled. *Style* 1, very short. *Stigma* bilabiate.
Capsule 1-celled, many-seeded, with a large central placenta.
Seeds minute. *Embryo* within fleshy albumen, sometimes undivided.
Herbaceous plants, living in water or marshes. *Leaves* radical, undivided, or compound, resembling roots, and bearing little vesicles. *Scapes* either with minute stipula-like scales, or naked; sometimes with whorled vesicles; generally undivided. *Flowers* single, or in spikes, or in many-flowered racemes. *Flowers* with a single bractea, rarely without bracteæ. *R. Br.*

1. PINGUICULA *Linn.*

Calyx campanulate, 5-cleft. *Corolla* 2-lipped; the upper 3-lobed, the lower 2-lobed, shorter and spurred. *Stigma* bilabiate.

1. P. *lusitanica* Linn. E. B. 3. 145.
Spur obtuse, shorter than the nearly regular petal. Flower-stalk hairy. Capsule globose.
P. villosa *Hudson.*
In bogs. — Perennial. *June, July.*

2. P. *vulgaris* Linn. Common Butterwort. E. B. 1. 70.
Spur cylindrical, acute, as long as the very irregular petal. Segments of the calyx oblong. Capsule ovate.
On bogs. — Perennial. *May, June.*

3. P. *grandiflora* Willd.
Spur cylindrical, acute, as long as the nearly regular, 5-cleft, veiny petal. Segments of the calyx ovate, obtuse. Capsule ovate.
In marshy ground, in the western part of the county of Cork. — Perennial. *May*

2. UTRICULARIA *Linn.*

Calyx 2-leaved; the lips equal and undivided. *Corolla* personate; the lower lip spurred at the base. *Stamens* 2, the filaments bearing the anthers on their inner face at the top. *Stigma* bilabiate. *R. Br.*

1. U. *vulgaris* Linn. Bladderwort, or Hooded Milfoil. E. B. 4. 253.
Spur conical. Stalk straight. Cluster somewhat corymbose. Upper lip of the corolla the length of the palate, reflexed at the sides. *Sm.*
In ditches and deep standing pools. — Perennial. *June, July.*

2. U. *intermedia* Hayne. E. B. 35. 2489.
Spur conical. Stalk 2- or 3-flowered. Upper lip of the corolla flat,

twice as long as the palate. Leaves with deep, forked, flat segments. Bladders separate from the leaves. *Smith.*

In ditches and bogs. — Perennial. *July.*

3. U. *minor* Linn. E. B. 4. 254.
Spur short, obtuse, keeled, deflexed. Cluster of few flowers. Corolla gaping; palate nearly flat; lips undivided. *Smith.*

In ditches, on spongy bogs, but rare. — Perennial. *July.*

Order 61. SCROPHULARINEÆ *Juss.*

Calyx divided, persistent, inferior.

Corolla monopetalous, hypogynous, usually irregular, deciduous, with an imbricated æstivation.

Stamens 2, or 4, didynamous, very seldom equal.

Ovarium superior, 2-celled, many-seeded. *Style* 1. *Stigma* 2-lobed.

Fruit capsular, very seldom succulent, with from 2 to 4 valves, which are either entire or bifid; the dissepiment either double, arising from the incurved margins of the valves; or simple, and in that case, either parallel with, or opposite to, the valves. *Placentæ* central, either adhering to the dissepiment or separating from it.

Seeds indefinite. *Embryo* included within fleshy albumen; *radicle* turned towards the hilum.

Herbaceous plants, seldom *shrubs,* with opposite leaves. *Inflorescence* very variable.

ANALYSIS OF THE GENERA.

Stamens 2	1. VERONICA.
Stamens 4	
Calyx ventricose	
4-cleft	2. RHINANTHUS.
5-cleft	3. PEDICULARIS.
Calyx not ventricose	
4-cleft	
Galea entire	4. BARTSIA.
Galea emarginate	5. EUPHRASIA.
5-cleft	
Corolla with a spur	6. LINARIA.
Corolla without a spur	
gibbous at the base in front	7. ANTIRRHINUM.
equal at the base in front	
campanulate	
5-lobed	8. LIMOSELLA.
4-lobed	9. DIGITALIS.
rotate	10. SIBTHORPIA.
globose	11. SCROPHULARIA.

1. VERONICA *Linn.*

Calyx 4- or 5-parted. *Corolla* rotate; the *limb* 4-parted, unequal, with entire lobes. *Stamens* 2. *Capsule* either separable in 2, or bearing the septa in the middle of the valves.

* *Clusters or spikes terminal. Root perennial.*

1. V. *spicata* Linn. E. B. 1. 2.
Spike terminal. Leaves bluntly serrated about the middle only; their base tapering into a footstalk: radical ones obovate. Stem ascending, quite simple. *Smith.*
In high dry chalky pastures. — Perennial. *July—September.*

2. V. *hybrida* Linn. E. B. 10. 673.
Spikes terminal. Leaves elliptical, obtuse, unequally and bluntly serrated: lowermost ovate, stalked. Stem nearly upright, not perfectly simple. *Smith.*
At the sides of mountains, very rare. — Perennial. *July, August.*

3. V. *fruticulosa* Linn. E. B. 15. 1028.
Cluster terminal, elongated, many-flowered. Leaves elliptic-lanceolate. Stems erect; shrubby below. Capsule ovate, with 4 lanceolate valves. *Smith.*
On the mountains of Scotland. — Perennial. *July.*

4. V. *saxatilis* Linn. E. B. 15. 1027.
Cluster terminal, corymbose, of few flowers. Leaves elliptical. Stems spreading; shrubby below. Capsule ovate, with 4 lanceolate valves. *Smith.*
On rocks and stony ground, on the mountains of Scotland. — Perennial. *July.*

5. V. *alpina* Linn. E. B. 7. 484.
Cluster terminal, dense, corymbose. Leaves ovate, smoothish, somewhat serrated. Calyx fringed. Stem ascending, simple. *Smith.*
V. pumila *All.*
On the margins of rivulets, on the highest mountains of Scotland. — Perennial *July, August.*

6. V. *serpyllifolia* Linn. Paul's Betony. E. B. 15. 1075.
Cluster terminal, somewhat spiked. Leaves ovate, slightly crenate, 3-ribbed, smooth. Capsule inversely heart-shaped, shorter than the style. *Smith.*
V. humifusa *Dicks.*
In meadows and pastures. — Perennial. *May, June.*

** *Clusters or spikes lateral. Root perennial.*

7. V. *Beccabunga* Linn. Brooklime. E. B. 10. 655.
Clusters lateral. Leaves elliptical, flat. Stem creeping. *Smith.*
In clear ditches and rivulets. — Perennial. *June, July.*

8. V. *Anagallis* Linn. E. B. 11. 781.
Clusters lateral, opposite. Leaves lanceolate, serrated. Stem erect. *Smith.*
In ditches, and muddy watery places. — Perennial. *July.*

9. V. *scutellata* Linn. E. B. 11. 782.
Clusters lateral, alternate; fruit-stalks reflexed. Leaves linear, slightly indented. *Smith.*

V. parmularia *Poit.*

In watery, spongy bogs. — Perennial. *July, August.*

10. V. *officinalis* Linn. Common Speedwell. E. B. 11. 765.

Clusters lateral; partial stalks shorter than their bracteas. Leaves elliptical, serrated, roughish. Stem procumbent. Stigma capitate. *Smith.*

β. V. Allionii *Hooker's Fl. Scot.*

Flowers densely spiked. Leaves ovate, subserrated, rigid, and as well as the procumbent stem perfectly glabrous. *Hooker.*

About dry sandy banks, woods, and pastures. β. On various mountains in Scotland and Ireland. — Perennial. *May, June.*

11. V. *hirsuta* Hopkirk.

Clusters lateral; partial stalks shorter than their bracteas. Leaves elliptic-lanceolate, somewhat serrated, slightly hairy. Stems ascending. Capsule abrupt, undivided. *Smith.*

V. setigera *D. Don.*

In dry heathy places in Carrick, Ayrshire. — Perennial. *June.*

12. V. *Chamædrys* Linn. Germander. E. B. 9. 623.

Clusters lateral. Leaves ovate, sessile, rugged, deeply serrated. Stem diffuse, with a hairy line at each side. Calyx 4-cleft, lanceolate. *Smith.*

In groves and hedges. — Perennial. *May, June.*

13. V. *montana* Linn. E. B. 11. 766.

Clusters lateral, lax, of few flowers. Leaves ovate, stalked, serrated. Stem diffuse, hairy all round. *Smith.*

In woods, chiefly on a moist calcareous soil. — Perennial. *May, June.*

*** *Flowers axillary, solitary. Root annual.*

14. V. *agrestis* Linn. E. B. 11. 783.

Flowers solitary. Leaves ovate, deeply serrated, shorter than the flower-stalks. Stems procumbent. Segments of the calyx ovate. Seeds cupped. *Smith.*

A common weed. — Annual. *April—September.*

15. V. *arvensis* Linn. E. B. 11. 734.

Flowers solitary, nearly sessile. Leaves ovate, deeply serrated; the floral ones lanceolate, entire. Stem erect. Seeds flat. *Smith.*

On walls, and dry gravelly ground. — Annual. *May.*

16. V. *hederifolia* Linn. E. B. 11. 784.

Flowers solitary. Leaves heart-shaped, flat, 5-lobed. Segments of the calyx heart-shaped, acute. Seeds cupped, wrinkled. *Smith.*

In cultivated and waste ground. — Annual. *April, May.*

17. V. *triphyllos* Linn. E. B. 1. 26.

Flowers solitary. Upper leaves in deep, finger-like, obtuse segments. Flower-stalks longer than the calyx. Seeds flat. *Smith.*

In sandy fields. — Annual. *April.*

18. V. *verna* Linn. E. B. 1. 25.

Flowers solitary. Leaves pinnatifid; uppermost lanceolate, undivided. Flower-stalks shorter than the calyx. Stem erect. *Smith.*

V. Bellardi *Willd.*
V. succulenta *Willd.*

In fields of the most barren sand. — Annual. *April.*

2. RHINANTHUS *Linn.*

Calyx 4-cleft, ventricose. *Corolla* tubular, 2-lipped; the upper lip compressed, the lower flat and 3-lobed. *Stamens* 4, didynamous. *Capsule* compressed, obtuse, covered by the calyx. *Seeds* compressed, bordered. *Dec.*

1. R. *Crista galli* Linn. E. B. 10. 657.
Stem slightly branched. Leaves lanceolate, serrated. Calyx smooth. Style concealed by the upper lip. Seeds with a dilated membranous border. *Smith.*

In meadows and pastures. — Annual. *June.*

2. R. *major* Ehr.
Stem much branched. Leaves linear-lanceolate, serrated. Bracteas taper-pointed. Calyx smooth. Style prominent. Seeds slightly bordered. *Smith.*

In corn-fields, in the north of England. — Annual. *July, 2 or 3 weeks later than the former.*

3. PEDICULARIS *Linn.*

Calyx ventricose, generally 5-cleft, or unequally 2-3-lobed. *Corolla* tubular, 2-lipped; the upper lip long, compressed, often emarginate; the lower flat, spreading, 3-lobed. *Stamens* 4, didynamous. *Capsule* compressed, acuminate, often oblique and longer than the calyx. *Dec.*

1. P. *palustris* Linn. E. B. 6. 399.
Stem solitary, branched. Calyx ovate, hairy, ribbed, in 2 unequally notched lobes. *Smith.*

In marshes and boggy meadows. — Perennial? *June, July.*

2. P. *sylvatica* Linn. Louse-wort. Red Rattle. E. B. 6. 400.
Stems several, spreading, simple. Calyx oblong, angular, smooth, in 5 unequal notched segments. *Smith.*

In heathy pastures. — Perennial. *June, July.*

4. BARTSIA *Linn.*

Calyx not ventricose, 4-cleft. *Corolla* tubular, 2-lipped; the upper lip galeate, erect, entire, the lower very small, reflexed, trifid. *Stamens* shorter than the galea; *anthers* downy. *Capsule* ovate, acuminate, compressed. *Dec.*

1. B. *alpina* Linn. E. B. 6. 361.
Leaves opposite, ovate, somewhat heart-shaped, bluntly serrated. Stem square. Root creeping. *Smith.*

In boggy alpine meadows. — Perennial. *July.*

2. B. *viscosa* Linn. E. B. 15. 1045.
Leaves serrated; upper ones alternate. Flowers lateral and distant. Stem round. Root fibrous. *Smith.*

In marshy ground, rare. — Annual. *July, August.*

3. B. *Odontites* Linn. E. B. 20. 1415.
Leaves lanceolate, serrated; the upper ones alternate. Flowers forming unilateral clusters. Stem square, branched. Root fibrous. *Smith.*

In meadows and pastures. — Annual. *July, August.*

5. EUPHRASIA *Linn.* EYE-BRIGHT.

Calyx 4-cleft. *Corolla* tubular, 2-lipped; the upper lip galeate, emarginate, the lower 3-lobed, equal. *Anthers* 2, or 4, acuminate at the base. *Capsule* ovate, compressed, obtuse, emarginate. *Dec.*

1. E. *officinalis* Linn. E. B. 20. 1416.
Leaves ovate, furrowed, sharply toothed. *Smith.*

On heaths, and in mountainous pastures. — Annual. *July—September.*

6. LINARIA *Desf.*

Calyx 5-parted; the 2 lower segments far apart from the rest. *Corolla* ringent, calcarate at the base; the tube inflated; the limb 2-lipped; the upper lip bifid, reflexed; the lower 3-lobed. *Capsule* ovate or globose, opening with several valves at the apex. *Seeds* bordered. *Dec.*

* *Leaves dilated. Stems flaccia.*

1. L. *Cymbalaria* Mill. E. B. 7. 502.
Leaves heart-shaped, 5-lobed, alternate, smooth. Stems procumbent. *Smith.*
Antirrhinum Cymbalaria *Linn.*

On old walls. — Perennial. *May—November.*

2. L. *spuria* Miller. E. B. 10. 691.
Leaves ovate, downy, chiefly alternate. Stems procumbent, hairy. *Smith.*
Antirrhinum spurium *Linn.*

In corn-fields. — Annual. *July—September.*

3. L. *Elatine* Desf. E. B. 10. 692.
Leaves chiefly halberd-shaped, alternate; lowermost ovate, opposite. Stems procumbent, hairy. *Smith.*
Antirrhinum Elatine *Linn.*

In corn-fields, after harvest. — Annual. *July—September.*

** *Leaves narrower. Stems upright.*

4. L. *repens* H. Kew. E. B. 18. 1253.
Leaves linear, glaucous, scattered; partly whorled. Stem panicled. Calyx smooth, the length of the spur. *Smith.*
Antirrhinum repens *Linn.*
Antirrhinum monspessulanum *Linn.*

On chalky banks. — Perennial. *July—September.*

5. L. *vulgaris* Mœnch. E. B. 10. 658.
Leaves linear-lanceolate, crowded. Stem erect. Spikes terminal. Flowers imbricated. Calyx smooth, shorter than the spur. *Smith.*
Antirrhinum Linaria *Linn.*

In hedges and fields. — Perennial. *June, July.*

6. L. *minor* Desf. E. B. 28. 2014.
Leaves lanceolate, obtuse, downy, mostly alternate. Stem mucl branched, spreading. Calyx longer than the spur. *Smith.*
Antirrhinum minus *Linn.*

In sandy fields. — Annual. *June—August.*

7. ANTIRRHINUM *Linn.*

Calyx 5-parted. *Corolla* without a spur, gibbous at the base; the tube inflated; the limb 2-lipped; the upper lip bifid and reflexed; the lower 3-lobed, with a projecting palate. *Capsule* oblique at the base, dehiscing by 3 small holes at the apex. *Dec.*

1. A. *majus* Linn. Snapdragon. E. B. 2. 129.
Flowers in a dense cluster. Leaves lanceolate. Segments of the calyx ovate, obtuse.

n old walls and cliffs. — Perennial. *July, August.*

2. A. *Orontium* Linn. E. B. 17. 1155.
Corolla scarcely tumid at the base. Flowers loosely spiked. Calyx finger-shaped, longer than the corolla. *Smith.*

In dry fields. — Annual. *July, August.*

8. LIMOSELLA *Linn.*

Calyx 5-cleft, irregular. *Corolla* campanulate, very small, 5-lobed, nearly equal. *Stamens* 4, or by abortion 2. *Stigma* globose. *Ovarium* 2-celled at the base, 1-celled at the apex. *Capsule* ovate. *Dec.*

1. L. *aquatica* Linn. Common Mudwort. E. B. 5. 357.
Leaves lanceolate, somewhat spatulate. Foot-stalks twice as long as the flower-stalks. *Smith.*

In muddy spots, where water has stagnated during winter. — Annual. *July, August.*

9. DIGITALIS *Linn.*

Calyx 5-parted, unequal. *Corolla* campanulate; the limb unequal, obliquely 4-lobed. *Stigma* simple or bilabiate. *Capsule* ovate, acuminate. *Dec.*

1. D. *purpurea* Linn. E. B. 19. 1297.
Segments of the calyx ovate, acute. Corolla obtuse; its upper lobe scarcely cloven. Leaves downy. *Smith.*

In pastures, and about hedges. — Biennial. *June, July.*

10. SIBTHORPIA *Linn.*

Calyx 5-parted. *Corolla* somewhat rotate; the tube short; the limb regular, 5-lobed. *Stamens* 4, didynamous. *Stigma* capitate. *Capsule* orbicular, compressed, dehiscing at the apex. *Dec.*

1. S. *europæa* Linn. Cornish Moneywort. E. B. 10. 649.

In moist shady places. — Perennial. *July, August.*

11. SCROPHULARIA *Linn.*

Calyx 5-lobed. *Corolla* globose; the limb contracted, 2-lipped; the upper lip 2-lobed, with an occasional intermediate scale; the lower shorter, and 3-lobed. *Stigma* simple. *Capsule* roundish, acuminate; *valves* entire, turned inwards at the base. *Dec.*

1. S. *nodosa* Linn. E. B. 22. 1544.
Leaves heart-shaped, acute, 3-ribbed at the base. Stem sharp-edged. Root tuberous. *Smith.*
In hedges, woods, and thickets. — Perennial. *July.*

2. S. *aquatica* Linn. Figwort. Water Betony. E. B. 12. 854.
Leaves heart-shaped, bluntish, on decurrent foot-stalks. Stem winged. Root fibrous. *Smith.*
In watery places. — Perennial. *July.*

3. S. *Scorodonia* Linn. E. B. 31. 2209.
Leaves heart-shaped, doubly serrated, downy beneath. Cluster leafy. *Smith.*
On the banks of rivulets. — Perennial. *July, August.*

4. S. *vernalis* Linn. E. B. 8. 567.
Leaves heart-shaped, doubly serrated, downy. Flower-stalks axillary, solitary, forked, leafy. Corolla without an interior lobe. *Smith.*
In thickets, and under hedges. — Biennial. *April, May.*

Order 62. Orobancheæ *Ventenat.*

Calyx divided, persistent, inferior.
Corolla monopetalous, hypogynous, irregular, persistent, with an imbricated æstivation.
Stamens 4, didynamous.
Ovarium superior, 1-celled, seated in a fleshy disk, with 2 or 4 parietal polyspermous placentæ; *style* 1; *stigma* 2-lobed.
Fruit capsular, enclosed within the withered corolla, 1-celled, 2-valved, each valve bearing 1 or 2 placentæ in the middle.
Seeds indefinite, very minute; *embryo* minute, at the base of a fleshy albumen.
Herbaceous leafless plants, growing parasitically upon the roots of other species. *Stems* covered with brown or colourless scales.

1. OROBANCHE *Linn.*

Calyx 1- or 2-parted, with from 1 to 3 bracteæ. *Corolla* tubular, ringent, 4- or 5-cleft. *Stamens* 4, didynamous. *Ovarium* surrounded by a disk at the base. *Style* 1. *Stigma* capitate, emarginate, 2-lobed. *Dec.*

* *Bracteas solitary.*

1. O. *major* Linn. Broom-rape. E. B. 6. 421.
Stem simple. Corolla inflated; upper lip slightly notched; lower

with acute, nearly equal segments. Stamens quite smooth below Style downy. *Smith.*

In bushy places, on the roots of broom or furze. — Perennial. *June, July.*

2. O. *elatior* Sutton. E. B. 8. 568.
Stem simple. Corolla funnel-shaped; lower lip with acute, nearly equal segments. Stamens downy. Style smooth. *Smith.*

In clover fields. — Perennial. *July, August.*

3. O. *minor* Smith. E. B. 6. 422.
Stem simple. Corolla nearly cylindrical; lower lip with curled segments, the middle one largest and lobed. Stamens fringed. Style smooth. *Smith.*

In clover fields. — Annual? *July, August.*

4. O. *rubra* Smith. E. B. 25. 1786.
Stem simple. Corolla somewhat tumid; upper lip cloven; lower in 3 nearly equal segments. Stamens fringed at the base. Style partially hairy. Calyx-leaves lanceolate, undivided. *Smith.*

On basaltic rocks in Ireland and Scotland. — Perennial. *July.*

** *Bracteas 3 to each flower.*

5. O. *cærulea* Villars. E. B. 6. 423.
Stem simple. Bracteas 3. Upper lip of the corolla cloven and notched; lower in 3 equal entire segments. Style downy. *Smith.*

O. purpurea *Jacq.*

In grassy pastures, near the sea. — Perennial. *July.*

6. O. *ramosa* Linn. E. B. 3. 184.
Stem branched. Bracteas 3. Upper lip of the corolla deeply cloven; lower equally 3-lobed; segments all rounded and entire. Style smoothish. *Smith.*

In low, moist, rich fields, attached to the roots of hemp. — Annual. *August, September.*

2. LATHRÆA *Linn.*

Calyx campanulate, 4-cleft. *Corolla* tubular, 2-lipped; the upper lip galeate. *Stamens* 4, didynamous. *Ovarium* surrounded by a disk at the base. *Style* 1. *Stigma* 2-lobed.

1. L. *Squamaria* Linn. Tooth-wort. E. B. 1. 50.
Flowering branches erect, simple. Flowers axillary, unilateral, pendulous; lower lip in 3 lobes; upper cloven. *Smith.*

In dry shady places, mostly at the roots of hazels or elms. — Perennial. *April.*

Order 63. MELAMPYRACEÆ *Richard.*

Calyx divided, persistent, unequal, inferior.
Corolla monopetalous, hypogynous, deciduous, personate.

Stamens 4, didynamous; *anthers* with acuminate lobes.
Ovarium superior, 2-celled, 2-seeded; *style* 1; *stigma* obtuse.
Fruit capsular, 2-celled, 2-valved, covered by the calyx.
Seeds in pairs, erect; *embryo* minute, inveried in the apex of fleshy *albumen*; *radicle* superior.
Herbaceous plants. *Leaves* opposite, without stipulæ. *Flowers* axillary, with coloured floral leaves.

1. MELAMPYRUM *Linn.*

Calyx tubular, 4-cleft. *Corolla* tubular, 2-lipped, compressed; the upper lip galeate, folded back at the margin, the lower sulcate, trifid. *Stamens* 4, didynamous. *Capsule* oblong, obliquely acuminate, compressed; cells monospermous. *Seeds* large, somewhat ovate.

1. M. *cristatum* Linn. E. B. 1. 41.
Spikes quadrangular. Bracteas heart-shaped, closely imbricated, finely toothed. *Smith.*
In woods and thickets. — Annual. *July.*

2. M. *arvense* Linn. E. B. 1. 53.
Spikes conical. Bracteas lax, lanceolate, pinnatifid. Calyx-teeth longer than the tube. Corolla closed. *Smith.*
In corn-fields. — Annual. *July.*

3. M. *pratense* Linn. Cow-wheat. E. B. 2. 113.
Flowers axillary, in partly distant pairs, turned to one side. Corolla closed; lip direct. Upper floral leaves toothed at the base. *Smith.*
Woods and bushy places. — Annual. *July, August.*

4. M. *sylvaticum* Linn. E. B. 12. 804.
Flowers axillary, in distant pairs, turned to one side. Corolla gaping; lip deflexed. Leaves nearly all entire. *Smith.*
In alpine woods. — Annual. *July, August.*

Order 64. VERBENACEÆ *Juss.*

Calyx tubular, persistent, inferior.
Corolla hypogynous, monopetalous, tubular, deciduous, generally with an irregular limb.
Stamens usually 4, didynamous, seldom equal, occasionally 2.
Ovarium 2- or 4-celled; *ovules* erect, solitary or twin; *style* 1; *stigma* bifid or undivided.
Fruit drupaceous, or baccate.
Seeds erect; *albumen* none, or in very small quantity; *embryo* erect.
Trees or *shrubs*, sometimes *herbaceous* plants. *Leaves* generally opposite, simple or compound, without stipulæ. *Flowers* in opposite corymbs, or spiked alternately; sometimes in dense heads; very seldom axillary and solitary. *R. Br.*

1. VERBENA *Linn.*

Calyx 5-cleft. *Corolla* with a somewhat 2-lipped, 5-lobed, unequal limb. *Stamens* 4, didynamous. *Seeds* inclosed in a vesicular tissue. *Dec.*

1. V. *officinalis* Linn. Vervain. E. B. 11. 767.
Stamens 4. Spikes slender, panicled. Leaves deeply cut. Stem mostly solitary. *Smith.*
In waste ground or pastures. — Perennial. *July.*

Order 65. LABIATÆ *Juss.*

Calyx tubular, 5- or 10-toothed, inferior, persistent; the odd tooth being next the axis; regular or irregular.
Corolla monopetalous, hypogynous, bilabiate; the upper lip undivided or bifid, overlapping the lower, which is larger and 3-lobed.
Stamens 4, didynamous, inserted upon the corolla, alternately with the lobes of the lower lip; the 2 upper sometimes wanting; *anthers* 2-lobed; the lobes sometimes so far apart at the base that the 2 cells are confluent at the apex; sometimes 1 cell altogether obsolete.
Ovarium deeply 4-lobed, inserted in a fleshy hypogynous disk; the lobes each containing 1 erect ovulum; *style* 1; *stigma* bifid, usually acute.
Fruit 4 small nuts, enclosed within the persistent calyx.
Seeds erect, with little or no albumen; *embryo* erect; *cotyledons* flat.
Herbaceous plants or *under-shrubs.* *Stem* 4-cornered, with opposite ramifications. *Leaves* opposite, divided or undivided, without stipulæ, replete with receptacles of aromatic oil. *Flowers* in opposite, nearly sessile cymes, resembling whorls; sometimes as if capitate.

ANALYSIS OF THE GENERA.

Stamens 2
  Corolla ringent - - - - - - - - - - 1. SALVIA.
  Corolla nearly equal - - - - - - - - - 2. LYCOPUS.
Stamens 4
  Calyx regular
    Upper lip of the corolla minute - - - - - 3. AJUGA.
    Upper lip of the corolla deeply cleft - - - 4. TEUCRIUM.
    Upper lip of the corolla arched
      Anthers covered with granulations - - - 5. LEONURUS.
      Anthers free from granulations
        cohering by pairs - - - - - - 6. GLECHOMA.
        distinct
          Corolla nearly equal - - - - - 7. MENTHA.
          Corolla ringent
            Calyx with 10 furrows
              Upper lip of corolla vaulted - 8. BALLOTA.
              Upper lip of corolla linear - 9. MARRUBIUM.

Calyx with 5 furrows or angles
Lower lip of corolla reflexed at the edges - - - - 10. STACHYS.
Lower lip of corolla not reflexed
Stamens included in the throat - - - - - 11. BETONICA.
Stamens longer than the throat
Lower lip of corolla in 3 acute segments - 12. GALEOBDOLON.
Lower lip of corolla in 3 blunt segments - 13. LAMIUM.
Lower lip of corolla notched - - - 14. NEPETA.
Lower lip of corolla with 2 hollow projections at the base - 15. GALEOPSIS.
Calyx irregular
with an appendage at the top - - - - - 16. SCUTELLARIA.
with no appendage
with 5 ribs
closed at the orifice by hairs - - - 17. THYMUS.
without hairs at the orifice
Lobes of the anthers contiguous - - 18. MELITTIS.
Lobes of the anthers distant - - 19. PRUNELLA.
with many ribs - - - - - - - - 20. CLINOPODIUM.
with no ribs - - - - - - - - - 21. ORIGANUM.

---

1. SALVIA *Linn.*

Calyx somewhat campanulate, 2-lipped; the upper lip 3-toothed, the lower bifid; the orifice naked. *Corolla* ringent; the upper lip fornicate and emarginate. *Stamens* 2. *Anthers* with 2 cells, the one fertile, the other abortive, separated by a long linear connectivum.

1. S. *pratensis* Linn. Meadow Clary. E. B. 3. 153.
Leaves oblong, crenate; heart-shaped at the base: uppermost clasping the stem. Bracteas very small. Summit of the corolla glutinous. *Smith.*

In dry meadows, and about hedges. — Perennial. *July.*

2. S. *verbenaca* Linn. E. B. 3. 154.
Leaves serrated, sinuated, smoothish. Corolla much more contracted than the calyx. *Smith.*

In meadows, pastures, and waste ground. — Perennial. *June—October.*

2. LYCOPUS *Linn.*

Calyx tubular, 5-cleft, with a naked orifice. *Corolla* tubular, 4-lobed, nearly equal; the upper lip broader and emarginate. *Stamens* 2.

1. L. *europæus* Linn. Common Gipsy-wort. Water Horehound. E. B. 16. 1105.
Leaves deeply serrated. *Smith.*

On the banks of ditches and rivers. — Perennial. *July, August.*

3. AJUGA *Linn.*

Calyx 5-cleft, nearly equal. *Corolla* tubular, labiate; the upper lip very small, and with 2 teeth; the lower 3-lobed, with a large intermediate obcordate lobe. *Nuts* reticulated. *Dec.*

* *Bugula* Tourn. *Flowers whorled, blue.*

1. A. *reptans* Linn. Common Bugle. E. B. 7. 489.
Almost smooth, with a solitary stem, and creeping runners. Lower lip of the corolla 4-cleft.
In woods and moist pastures. — Perennial. *May.*

2. A. *alpina* Linn. E. B. 7. 477.
Leaves almost smooth, irregularly toothed; uppermost entire. Runners none. Whorls not crowded, of many flowers.
A. pyramidalis *Hudson.*
A. genevensis *With.*
On mountains. — Perennial. *July.*

3. A. *pyramidalis* Linn. E. B. 18. 1270.
Hairy. Whorls crowded into a pyramidal form, many-flowered. Radical leaves very large, obovate, crenate, obtuse. Upper lip of the corolla deeply cloven. *Smith.*
In the Highlands of Scotland. — Perennial. *June.*

** *Chamæpitys* Tourn. *Flowers solitary, yellow.*

4. A. *Chamæpitys* Smith. Ground Pine. Yellow Bugle. E. B. 2. 77.
Stem diffuse, branched. Leaves in 3 deep, linear, entire segments. Flowers axillary, solitary, shorter than the leaves. *Smith.*
Teucrium Chamæpitys *Linn.*
In sandy or gravelly fields. — Annual. *April, May.*

4. TEUCRIUM *Linn.*

Calyx tubular, seldom campanulate, 5-cleft. *Corolla* with a short tube, labiate; the upper lip 2-parted; the segments reflexed at the sides; the lower 3-lobed, the intermediate lobe largest. *Stamens* protruded between the fissure of the upper lip. *Dec.*

1. T. *Scorodonia* Linn. Wood Sage. E. B. 22. 1543.
Leaves heart-shaped, hairy, serrated, stalked. Clusters aggregate, unilateral. Stem erect. *Smith.*
In woods, and heathy bushy places. — Perennial. *July.*

2. T. *Scordium* Linn. E. B. 12. 828.
Leaves oblong, sessile, downy, with tooth-like serratures. Flowers axillary, stalked, in pairs. Stem procumbent. *Smith.*
In low wet meadows. — Perennial. *July, August.*

3. T. *Chamædrys* Linn. E. B. 10. 680.
Leaves somewhat ovate, stalked, deeply crenate or cut. Flowers axillary, 3 together, stalked. Stem roundish, hairy. *Smith.*
On old ruined buildings. — Perennial. *July.*

5. LEONURUS *Linn.*

Calyx cylindrical, 5-cornered, 5-toothed, with a naked orifice. *Corolla* scarcely longer than the calyx, 2-lipped; the upper lip villous, entire, concave, the lower reflexed, 3-parted, nearly equal. *Anthers* covered with shining dots. *Dec.*

1. L. *Cardiaca* Linn. Motherwort. E. B. 4. 286.
Upper leaves lanceolate, either 3-lobed or undivided. *Smith.*
About hedges, on a gravelly or calcareous soil. — Perennial. *July, August.*

6. GLECHOMA *Linn.*

Calyx striated, cylindrical, when in fruit naked. *Corolla* twice as long as the calyx, bilabiate; the upper lip bifid, the lower 3-fid, with the intermediate segment larger and emarginate. *Anthers* cohering by pairs in a cross-like manner. *Nuts* nearly smooth, cylindrical, ovate. *Dec.*

1. G. *hederacea* Linn.. Ground-ivy, Gill, or Alehoof. E. B. 12. 853.
Leaves kidney-heart-shaped, crenate. *Smith.*
By road sides, and about hedge banks. — Perennial. *April, May.*

7. MENTHA *Linn.*

Corolla little longer than the calyx, 4-lobed, nearly equal; the upper lobe broadest and often emarginate. *Stamens* distant. *Dec.*

1. M. *sylvestris* Smith. Horse Mint. E. B. 10. 686.
Spikes shaggy, scarcely interrupted. Leaves acute, with deep-toothed serratures; chiefly downy beneath. Bracteas awl-shaped. Calyx all over hairy. *Smith.*
M. gratissima *of Willd.; according to Decandolle.*

α. longifolia.
Leaves lanceolate, acute. *Smith.*
M. longifolia *Hudson.*

β. villosa.
Leaves ovate, acute. *Smith.*
M. villosa *Hudson.*

γ. candicans.
Leaves shorter. Spikes more obtuse. *Smith.*

δ. nemorosa.
Leaves elliptical, broad and obtuse. *Smith.*
M. rotundifolia *Sole.*
M. nemorosa *Willd.*
M. alopecuroides *Hull.*
In waste ground, especially in watery places. — Perennial. *August, September.*

2. M. *rotundifolia* Linn. E. B. 7. 446.
Spikes interrupted, somewhat hairy. Leaves elliptical, obtuse, wrinkled, sharply crenate; shaggy beneath. Bracteas lanceolate. *Smith.*
M. crispa *Linn.*
M. sylvestris *Sole.*
In wet places, amongst rubbish. — Perennial. *August, September.*

3. M. *viridis* Linn. Spear Mint. E. B. 34. 2424.
Spikes interrupted. Leaves sessile, lanceolate, acute, naked. Bracteas bristle-shaped, somewhat hairy as well as the teeth of the calyx. Flower-stalks very smooth. *Smith.*
In marshy places. — Perennial. *August.*

4. M. *piperita* Smith. Pepper Mint. E. B. 10. 687.
Spikes blunt, interrupted below. Leaves stalked, somewhat ovate, smoothish. Calyx very smooth at the base. *Smith.*

α. *officinalis.*
Leaves ovate-lanceolate. Spikes elongated. *Smith.*
M. officinalis Hull

β. *ovata.*
Leaves ovate. Spikes shorter and blunter, almost capitate. *Smith.*

γ. *hircina.*
Leaves ovate, slightly heart-shaped. Spikes more acute. *Smith.*
M. hircina *Hull.*

In watery places. — Perennial. *August, September.*

5. M. *citrata* Ehr. Bergamot Mint. E. B. 15. 1025.
Spikes capitate, very blunt. Leaves stalked, heart-shaped, naked on both sides. Calyx and flower-stalks perfectly smooth. *Smith.*
M. odorata *Sole.*

In watery places. — Perennial. *August, September.*

6. M. *hirsuta* Linn. E. B. 7. 447.
Flowers capitate or whorled. Leaves stalked, ovate. Calyx clothed with erect hairs. Flower-stalks with recurved ones. *Smith.*

* Flowers capitate.

α. M. aquatica *Huds.*
M. aquatica major *Sole.*

β. M. palustris *Sole.*

γ. M. paludosa *Sole*

** Flowers whorled.

δ. M. sativa *Linn.*
M. verticillata *Linn.*
M. rivalis β, γ, δ. *Sole.*

In watery places, everywhere. — Perennial. *August, September.*

7. M. *acutifolia* Smith. E. B. 34. 2415.
Flowers whorled. Leaves ovate-lanceolate, tapering at each end. Calyx hairy all over. Hairs of the flower-stalks spreading. *Smith.*

About the banks of rivers. — Perennial. *September?*

8. M. *rubra* Smith. E. B. 20. 1413.
Flowers whorled. Leaves ovate. Stem upright, zigzag. Flower-stalks, and lower part of the calyx, very smooth; teeth hairy. *Smith.*
M. sativa *Sole.*

About wet hedges and thickets. — Perennial. *September.*

9. M. *gentilis* Linn. E. B. 30. 2118.
Flowers whorled. Leaves ovate. Stem much branched, spreading. Flower-stalks, and base of the bell-shaped calyx, nearly smooth. *Smith.*

M. rubra *Sole.*

In watery waste places. — Perennial. *August.*

10. M. *gracilis* Smith. E. B. 7. 449.

Flowers whorled. Leaves lanceolate, nearly sessile. Stem upright, much branched. Flower-stalks, and base of the calyx, quite smooth. *Smith.*

M. gentilis *E. Bot.*

M. rubra *Hudson.*

In watery places, or moist meadows. — Perennial. *August, September.*

11. M. *arvensis* Linn. E. B. 30. 2119.

Flowers whorled. Leaves ovate. Stem much branched, diffuse. Calyx bell-shaped, covered all over with horizontal hairs. *Smith.*

M. præcox *Sole.*

In sandy corn-fields. — Perennial. *June—September.*

12. M. *agrestis* Sole. E. B. 30. 2120.

Flowers whorled. Leaves somewhat heart-shaped, strongly serrated, rugose. Stem erect. Calyx bell-shaped, covered all over with horizontal hairs. *Smith.*

In corn-fields, and neglected gardens. — Perennial. *August, September.*

13. M. *Pulegium* Linn. Penny-royal. E. B. 15. 1026.

Flowers whorled. Leaves ovate. Stem prostrate. Flower-stalks and calyx all over downy; teeth fringed. *Smith.*

On wet commons, and about the margins of small brooks. — Perennial. *September.*

8. BALLOTA *Linn.*

Calyx campanulate, 5-cornered, with 10 streaks and 5 teeth. *Corolla* 2-lipped; the upper lip concave, crenate; the lower 3-lobed; the middle lobe larger and emarginate. *Nuts* triangular.

1. B. *nigra* Linn. Black Horehound. E. B. 1. 46.

Leaves ovate, undivided, serrated. Calyx funnel-shaped, abrupt, with short spreading teeth. *Smith.*

About hedges and waste places, common. — Perennial. *July, August.*

9. MARRUBIUM *Linn.*

Calyx cylindrical, with 10 streaks, and 5 or 10 teeth. *Corolla* a little longer than the calyx, 2-lipped; the upper narrow and bifid; the lower trifid, with the middle segment broader and emarginate. *Dec.*

1. M. *vulgare* Linn. White Horehound. E. B. 6. 410.

Calyx-teeth 10, bristle-shaped, hooked backwards. Leaves roundish-ovate, unequally serrated. *Smith.*

In dry waste ground, by road sides. — Perennial. *July.*

10. STACHYS *Linn.*

Calyx angular, 5-cleft, or 5-toothed, acuminate. *Corolla* with a short tube; the upper lip vaulted; the lower 3-lobed, with the sides reflexed. *Stamens*, after the anthers are burst, bent back on each side. *Nuts* obsoletely 3-cornered, ovate, or roundish. *Dec.*

1. S. *sylvatica* Linn. E. B. 6. 416.
Six flowers in a whorl. Leaves heart-shaped, stalked. Stem solid. *Smith.*
Under hedges. — Perennial. *July, August.*

2. S. *ambigua* Smith. E. B. 30. 2089.
Six flowers in a whorl. Leaves oblong, stalked, heart-shaped at the base. Stem hollow. *Smith.*
In waste ground, chiefly in Scotland. — Perennial. *August, September.*

3. S. *palustris* Linn. E. B. 24. 1675.
Six to 10 flowers in a whorl. Leaves linear-lanceolate, half embracing the stem. Root tuberous. *Smith.*
In wet hedges and fields. — Perennial. *August.*

4. S. *germanica* Linn. E. B. 12. 829.
Many flowers in a whorl. Leaves crenate, densely silky; woolly beneath. Stem woolly, erect. *Smith.*
In fields, and by road sides. — Perennial. *September.*

5. S. *arvensis* Linn. E. B. 17. 1154.
Six flowers in a whorl. Stem weak. Leaves heart-shaped, obtuse, crenate, slightly hairy. *Smith.*
In sandy or chalky fields. — Annual. *July, August.*

11. BETONICA *Linn.*

Calyx cylindrical, 5-toothed, awned, with a naked orifice. *Corolla* with a slender tube; the limb bilabiate; the upper lip erect, roundish, entire, or emarginate; the lower trifid. *Dec.*

1. B. *officinalis* Linn. Betony. E. B. 16. 1142.
Spike interrupted. Middle segment of the lower lip notched. *Smith.*
In woods and thickets. — Perennial. *July, August.*

12. GALEOBDOLON *Hudson.*

Calyx campanulate, 5-toothed; the teeth unequal, acute. *Corolla* longer than the calyx, 2-lipped; the upper lip fornicate and entire; the lower trifid, with acute segments, the middle of which is the longest. *Dec.*

1. G. *luteum* Hudson. Yellow Archangel. E. B. 11. 787.
G. Galeopsis *Curtis.*
Galeopsis Galeobdolon *Linn.*
Leonurus Galeobdolon *Scop.*
In shady, rather moist situations, in groves or hedge bottoms, not uncommon. — Perennial. *May.*

13. LAMIUM *Linn.*

Calyx 5-toothed, awned, naked, spreading at the point. *Corolla* longer than the calyx; its orifice inflated; the upper, lip vaulted, entire; the lower with 2 small lateral lobes, and a large emarginate one in the middle. *Anthers* smooth. *Nuts* 3-cornered, smooth. *Dec.*

1. L. *album* Linn. White Dead-nettle. E. B. 11. 768.
Leaves heart-shaped, pointed, strongly serrated, hairy. Flowers about 20 in a whorl. Tube of the calyx shorter than its teeth. Upper lip of the corolla notched; lateral teeth solitary, lanceolate. *Smith.*
In waste ground. — Perennial. *May, June; also September.*

2. L. *maculatum* Linn. E. B. 36. 2550.
Leaves heart-shaped, pointed, strongly serrated, hairy. Flowers about 10 in a whorl. Tube of the calyx curved, as long as its teeth. Upper lip of the corolla notched; lateral teeth solitary, bristle-shaped. *Smith.*
On banks, in warm situations. — Perennial. *April.*

3. L. *purpureum* Linn. Red Dead-nettle. E. B. 11. 769.
Leaves heart-shaped, bluntish, unequally crenate, stalked; the upper ones crowded. Stem leafless in the middle. Calyx-teeth lanceolate. Tube of the corolla closed, near the bottom, with hairs. *Smith.*
In waste ground. — Annual. *May.*

4. L. *incisum* Willd. E. B. 27. 1933.
Leaves heart-shaped, dilated, stalked, irregularly cut; the upper ones crowded. Stem leafless in the middle. Tube of the corolla internally naked; marginal teeth dilated, combined. *Smith.*
L. dissectum *With.*
In waste ground. — Annual. *May.*

5. L. *amplexicaule* Linn. Great Henbit. E. B. 11. 770.
Floral leaves sessile, kidney-shaped, obtuse, deeply crenate, partly lobed, clasping the stem. Teeth of the calyx linear-awl-shaped, as long as its tube. *Smith.*
In sandy fields. — Annual. *February—June.*

14. NEPETA *Linn.*

Calyx cylindrical, with a naked orifice. *Corolla* with a long tube; the orifice gaping; the upper lip emarginate; the lower 3-lobed; the lateral lobes very short, reflexed; the intermediate one larger, crenate, and concave.

1. N. *cataria* Linn. Cat-mint, or Nep. E. B. 2. 137.
Whorls stalked, crowded into spikes. Leaves finely downy, heart-shaped, stalked, with tooth-like serratures. *Smith.*
On banks, and by road sides. — Perennial. *July.*

15. GALEOPSIS *Linn.*

Calyx campanulate, 5-toothed, spiny. *Corolla* with a short tube; the orifice dilated, with 2 teeth; the limb bilabiate; the upper lip vaulted and crenate; the lower with 3 unequal lobes. *Anthers* somewhat hairy internally. *Dec.*

1. G. *Ladanum* Linn. E. B. 13. 884.
Stem not swelled below the joints. Leaves lanceolate, somewhat serrated, hairy. Upper lip of the corolla slightly notched. *Smith.*
Galeopsis angustifolia *Ehr.*
In dry gravelly, or chalky fields. — Annual. *August, September.*

2. G. *villosa* Hudson. E. B. 33. 2353.
Stem not swelled below the joints. Leaves ovate-lanceolate, serrated, very soft and downy. Upper lip of the corolla deeply notched.
G. grandiflora *Wild.*
G. latifolia *Ehr.*
In sandy corn-fields. — Annual. *July, August.*

3. G. *Tetrahit* Linn. Common Hemp-nettle. E. B. 3. 207.
Stem bristly, swelled below the joints. Corolla twice the length of the calyx; upper lip nearly straight. *Smith.*
In cultivated ground. — Annual. *July, August.*

4. G. *versicolor* Curtis. Bee-Nettle. E. B. 10. 667.
Stem bristly, swelled below the joints. Corolla thrice the length of the calyx; upper lip tumid; middle lobe of the lower heart-shaped. *Smith.*
G. cannabina *Willd.*
In sandy corn-fields. — Annual. *July, August.*

16. SCUTELLARIA *Linn.*

Calyx short, with both lips entire; a concave scale lying upon the upper lip. *Corolla* longer, curved at the base; the upper lip compressed. vaulted with 2 teeth at the base; the lower broader and emarginate, *Nuts* covered by the closed calyx. *Dec.*

1. S. *galericulata* Linn. Common Skull-cap. E. B. 8. 523.
Leaves lanceolate, crenate, rugged, heart-shaped at the base. Flowers axillary. *Smith.*
About ditches, and other watery situations. — Perennial. *July, August.*

2. S. *minor* Linn. E. B. 8. 524.
Leaves ovate, nearly entire, heart-shaped, and occasionally lobed at the base. Flowers axillary. *Smith.*
On moist heaths. — Perennial. *July, August.*

17. THYMUS.

Calyx striated; the orifice closed with hairs; the limb 2-lipped; the upper lip 3-toothed; the lower bifid, or with 2 bristles. *Corolla* short; the upper lip emarginate; the lower 3-lobed; the middle lobe being broadest and emarginate, or entire. *Nuts* smooth.

* *Calyx campanulate. Middle lobe of the lower lip of the corolla entire.*

1. T. *Serpyllum* Linn. Wild Thyme. E. B. 22. 1514.
Flowers in small heads. Stems recumbent. Leaves flat, ovate, obtuse, entire, fringed at their base. *Smith.*
On heaths and dry mountainous ground. — Perennial. *July, August.*

** *Calyx gibbous at the base. Middle lobe of the lower lip of the corolla nearly entire.*

2. T. *Acinos* Linn. E. B. 6. 411.
Flowers about 6 in a whorl, on simple stalks. Stem branched, ascending. Leaves acute, serrated.
In fields. — Annual. *July, August.*

*** *Calyx cylindrical. Middle lobe of the lower lip of the corolla emarginate.*

3. T. *Calamintha* Scopoli. Calamint. E. B. 24. 1676.
Whorls on forked, many-flowered stalks. Leaves with shallow serratures. Hairs in the mouth of the calyx not prominent. *Smith.*
Melissa Calamintha *Linn.*
By way sides. — Perennial. *July, August.*

4. T. *Nepeta* Smith. E. B. 20. 1414.
Whorls on forked, many-flowered stalks, longer than the adjoining leaf. Leaves serrated. Hairs in the mouth of the calyx prominent. *Smith.*
Melissa Nepeta *Linn.*
On dry banks, and by way sides. — Perennial. *August.*

18. MELITTIS *Linn.*

Calyx large, 3-fid; the upper lip emarginate, larger than the tube of the corolla. *Corolla* twice as long as the calyx; the limb dilated, spreading; the upper lip flat, entire; the lower 3-lobed; the lobes large and unequal. *Nuts* roundish, triangular, shaggy externally.

1. M. *Melissophyllum* Linn. Bastard Balm. E. B. 9. 577.
Calyx with 3 unequal, partly notched, lobes. *Smith.*
In woods and hedges, of the south and west of England. — Perennial. *May, June.*

2. M. *grandiflora* Smith. E. B. 9. 636.
Calyx with 4 nearly equal lobes. *Smith.*
In woods and hedges, in the south-west extremity of England. — Perennial. *May.*

19. PRUNELLA *Linn.*

Calyx 2-labiate, with a naked orifice; the upper lip flat, somewhat truncate, 3-fid; the lower shorter, bifid. *Corolla* with the upper lip concave, entire, or 2-lobed; the lower 3-lobed; the middle lobe being larger than the rest, and emarginate. *Filaments* forked, or 2-toothed at the end; 1 tooth bearing the anther, the other naked. *Nuts* ovate, shining.

1. P. *vulgaris* Linn. Common Self-heal, or Slough-heal. E. B. 14. 961.
All the leaves ovate-oblong, stalked. Teeth of the upper lip of the calyx scarcely discernible. *Smith.*
In meadows and pastures. — Perennial. *July, August.*

20. CLINOPODIUM *Linn.*

Calyx 2-labiate, with a naked orifice; the upper lip 3-fid; the lower

2-parted. *Corolla* with a short tube, gradually widening upwards; the upper lip erect, and emarginate; the lower 3-fid, with a larger emarginate middle lobe. *Dec.*

1. C. *vulgare* Linn. Wild Basil. E. B. 20. 1401.
Whorls bristly. Involucral leaves awl-shaped. Flower-stalks branched. Leaves obscurely serrated. *Smith.*
About hedges, and by road sides. — Perennial. *August.*

21. ORIGANUM *Linn.*

Calyx cylindrical, 5-toothed, when in fruit closed up with hairs. *Corolla* with a compressed tube; the upper lip erect, emarginate; the lower trifid, and nearly equal. *Nuts* roundish.

1. O. *vulgare* Linn. Common Marjoram. E. B. 16. 1143.
Heads of flowers roundish, panicled, crowded, erect. Involucral leaves ovate, smooth. Calyx with 5 acute unequal teeth; throat hairy. *Smith.*
In bushy places, on a limestone or gravelly soil. — Perennial. *July, August.*

Division II. Monochlamydeæ.

ANALYSIS OF THE ORDERS.

Seeds indefinite
  Calyx inferior - - - - - - - - - - 74. Resedaceæ.
  Calyx superior - - - - - - - - - - 77. Aristolochiæ.
Seeds definite
  Calyx superior - - - - - - - - - 66. Santalaceæ.
  Calyx inferior
    Ovarium with more cells than 1
      Fruit drupaceous - - - - - - 76. Empetreæ.
      Fruit dry
        dehiscent - - - - - - - 75. Euphorbiaceæ.
        indehiscent - - - - - - - 79. Ulmaceæ.
    Ovarium with only 1 cell
      Ovule erect, or attached to a funiculus, arising from the base of the cell
        Embryo on the outside of the albumen
          Calyx with an indurated tube - - 72. Scleranthеæ.
          Calyx membranous or succulent
            Embryo inverted - - - 69. Polygoneæ.
            Embryo erect
              Calyx scarious, coloured - 70. Amarantaceæ.
              Calyx green, herbaceous - 71. Chenopodeæ.
        Embryo in the axis of the albumen - 67. Elæagneæ.
      Ovule pendulous
        Cotyledons 2
          Calyx tubular - - - - - - 68. Thymelææ.
          Calyx deeply divided - - - 73. Urticeæ.
        Cotyledons 4 - - - - - - - - 78. Ceratophylleæ.

Order 66. Santalaceæ *R. Br.*

Calyx superior, 4- or 5-cleft, half-coloured, with valvate æstivation.

Stamens 4 or 5, opposite the segments of the calyx, and inserted into their bases.

Ovarium 1-celled, with from 2 to 4 seeds; *ovules* fixed to the top of a central placenta near the summit; *style* 1; *stigma* often lobed.

Fruit 1-seeded, hard and dry, or drupaceous.

Albumen fleshy, of the same form as the seed; *embryo* in the axis, inverted, taper.

Trees or *shrubs*, sometimes *undershrubs* or *herbaceous* plants. *Leaves* alternate, or nearly opposite, undivided, sometimes minute, and resembling stipulæ. *Flowers* in spikes, seldom in umbels, or solitary, small.

1. THESIUM *Linn.*

Calyx 4- or 5-cleft. *Stamens* 4 or 5, opposite the lobes of the calyx. *Fruit* 1-seeded, indehiscent, crowned by the persistent calyx.

1. T. *linophyllum* Linn. Bastard-toadflax. E. B. 4. 247.
Cluster branched. Bracteas 3 together. Leaves linear-lanceolate. Tube of the calyx very short. *Smith.*
T. pratense *Ehr.*
In high, open, chalky pastures. — Perennial. *July.*

Order 67. ELÆAGNEÆ *Juss.*

Flowers diœcious, rarely hermaphrodite.
Male. *Calyx* 4-parted; *stamens* 3, 4, or 8, sessile; *anthers* 2-celled.
Female. *Calyx* inferior, tubular, persistent; the *limb* entire, or 2-4-toothed.
Ovarium superior, simple, 1-celled; *ovulum* solitary, ascending, stalked; *stigma* simple, subulate, glandular.
Fruit crustaceous, enclosed within the calyx become succulent.
Seed erect; *embryo* straight, surrounded by very thin fleshy *albumen*; *radicle* short, inferior; *cotyledons* fleshy.
Trees or *shrubs*, covered with leprous scales. *Leaves* alternate, or opposite, entire, without stipulæ. *Flowers* axillary, often fragrant.

1. HIPPOPHÄE *Linn.*

Flowers diœcious. *Male* catkin-like, tetrandrous. *Female* axillary, solitary. *Calyx* tubular, bifid, and closed at the apex. *Disk* wanting. *Fruit* a nut, contained within a succulent calyx.

1. H. *rhamnoides* Linn. Sea Buck-thorn. E. B. 6. 425.
Leaves linear-lanceolate, scattered. *Smith.*
On sandy cliffs. — Shrub. *May.*

Order 68. THYMELÆÆ *Juss.*

Calyx inferior, tubular, coloured; the limb 4-cleft, seldom 5-cleft, with an imbricated æstivation.
Corolla 0, but sometimes with scales in the orifice.
Stamens definite, inserted in the tube or the orifice often 8, sometimes 4, less frequently 2; when equal in number to the segments of the calyx, or fewer, opposite to them; *anthers* 2-celled, dehiscing lengthwise in the middle.

Ovarium simple, with 1 solitary pendulous ovulum; *style* 1; *stigma* undivided.
Fruit hard, dry, and nut-like, or drupaceous.
Albumen none, or thin and fleshy; *embryo* straight, inverted; *cotyledons* plano-convex; *radicle* short, superior; *plumula* inconspicuous.
Stem shrubby, very seldom herbaceous, with tenacious bark. *Leaves* without stipulæ, alternate, or opposite, entire. *Flowers* capitate or spiked, terminal, or axillary, occasionally solitary. *R. Br.*

1. DAPHNE *Linn.*

Calyx 4-lobed. *Stamens* 8. *Style* short, terminal. *Berry* with 1 cell, and 1 seed. *Dec.*

1. D. *Mezereum* Linn. Common Mezereon. Spurge-olive. E. B. 20. 1381.
Flowers naked on the stem, sessile about 3 together. Leaves lanceolate, deciduous. *Smith.*
In woods, but rare. — Shrub. *March.*

2. D. *Laureola* Linn. Spurge-laurel. E. B. 2. 119.
Clusters axillary, simple, each of about 5 flowers, drooping, shorter than the smooth, obovate-lanceolate, evergreen leaves. Calyx obtuse. *Smith.*
In woods, thickets, and hedges. — Shrub. *March.*

Order 69. Polygoneæ *Juss.*

Calyx divided, inferior, imbricated in æstivation.
Stamens definite, inserted in the bottom of the calyx; *anthers* dehiscing lengthwise.
Ovarium superior, with a single erect ovulum; *styles* or *stigmas* several.
Nut naked, or protected by the calyx.
Seed with farinaceous albumen, rarely with scarcely any; *embryo* inverted, generally on one side; *plumula* inconspicuous.
Herbaceous plants, rarely *shrubs*. *Leaves* alternate, sheathing at the base, or adhering to an intra-foliaceous sheath; the younger revolute. *Flowers* occasionally bisexual, often in racemes. *R. Br.*

1. RUMEX *Linn.*

Calyx 6-parted; the 3 outer segments somewhat cohering at the base; the 3 inner becoming enlarged after flowering. *Stamens* 6. *Styles* 3, reflexed. *Stigmas* 3, cut. *Nut* with 3 sharp angles. *Embryo* on one side. *Radicle* superior.

§ Docks. Juice not acid.

1. R. *maritimus* Linn. E. B. 11. 725.
Lower leaves ovate-lanceolate, acute, flat; upper oblong-lanceolate.

Flowering branches alternate, simple. Whorls dense, many-flowered, all axillary. Inner sepals all acquiring large tubercles, ovate, acuminate, with long setaceous teeth. *Duby*.

In marshes. — Perennial. *July, August.*

2. R. *palustris* Smith. E. B. 27. 1932.

Lower leaves linear-lanceolate, acute, somewhat curled. Flowering branches alternate or twin, simple or divided. Whorls clustered, all axillary. Inner sepals all acquiring tubercles, ovate-lanceolate, nearly acute, each with 3 short teeth. *Duby*.

R. maritimus *Hudson*.

In marshes and ditches. — Perennial. *July, August.*

3. R. *pulcher* Linn. E. B. 22. 1576.

Lower leaves cordate-ovate, bluntish, sometimes fiddle-shaped. Flowering branches alternate. Whorls distant, 1-sided, few-flowered, all axillary. Inner sepals becoming deltoid-ovate, acute or bluntish, serrate, all bearing tubercles, but one in particular.

In pastures. — Perennial. *August.*

4. R. *obtusifolius* Linn. E. B. 28. 1999.

Lower leaves cordate-ovate, obtuse, seldom roundish, slightly curled; the upper ovate-lanceolate, tapering to each end, obtuse, on long stalks. Flowering branches alternate, seldom in pairs, simple. Whorls somewhat clustered; the upper leafless. The inner sepals bearing tubercles, ovate, bluntish, with 2 or 4 short subulate teeth. *Duby*.

In waste ground. — Perennial. *July, August.*

5. R. *glomeratus* Spreng. E. B. 11. 724.

Lower leaves cordate, lanceolate, acute, generally curled at the margin; the upper lanceolate, acuminate. Flowering branches alternate, or rarely double. Whorls distant, many-flowered, axillary. Inner sepals becoming ovate, acute, entire, or with 2 or 3 slight teeth, each bearing a large tubercle. *Duby*.

R. acutus *Linn*.

R. paludosus *With*.

In waste places. — Perennial. *July.*

6. R. *Nemolapathum* Linn.

Lower leaves cordate-lanceolate; the upper lanceolate, undulated, serrated, acute. Flowering branches alternate, simple, straggling. Whorls distant, few-flowered; the upper destitute of leaves. Inner sepals becoming tubercled, ovate, obtuse, and entire. *Duby*.

R. sanguineus β. *Smith*.

By road sides. — Perennial. *July.*

7 R. *sanguineus* Linn. E. B. 22. 1533.

Leaves acute, slightly curled, stalked, stained with crimson; the lower cordate-lanceolate. Flowering branches alternate, simple. Whorls distant, few-flowered; the upper destitute of leaves. Inner sepals becoming ovate-lanceolate, bluntish, and entire, one generally bearing a large tubercle. *Duby*.

By road sides. — Perennial. *July.*

8. R. *crispus* Linn. E. B. 28. 1998.

Leaves wavy, curled, acute; the lower oblong, lanceolate, stalked. Flowering branches alternate, double or triple, simple or divided. Whorls somewhat clustered, many-flowered; the upper destitute of leaves. Inner sepals becoming roundish, cordate, entire, acute, with very large tubercles. *Duby.*

In waste ground. — Perennial. *June, July.*

9. R. *Hydrolapathum* Huds. E. B. 30. 2104.

Leaves lanceolate, acute at each end. Whorls rather crowded, almost destitute of leaves. Inner sepals ovate-oblong, nearly entire, unequally tubercled.

R. Britannica *Hudson.*

R. aquaticus *Smith*; not of Linn.

In ditches. — Perennial. *July, August.*

§ Sorrels. Juice acid.

10. R. *Acetosa* Linn. E. B. 2. 127.

Flowers diœcious. Leaves oblong, arrow-shaped. Permanent sepals tuberculated. *Smith.*

In meadows and pastures. — Perennial. *June.*

11. R. *Acetosella* Linn. E. B. 24. 1674.

Flowers diœcious. Leaves lanceolate, hastate. Permanent sepals without tubercles. *Smith.*

In dry gravelly pastures and fields. — Perennial. *June, July.*

2. OXYRIA *R. Br.*

Flowers hermaphrodite. *Sepals* 4, becoming larger after flowering. *Styles* 2, reflexed. *Nut* 2-edged. *Embryo* in the centre. *Dec.*

1. O. *reniformis* R. Br. Mountain-sorrel. E. B. 13. 910.

Rumex digynus *Linn.*

Rheum digynum *Wahl.*

In mountain bogs, rills, and moist clefts of rocks. — Perennial. *June.*

3. POLYGONUM *Linn.*

Flowers hermaphrodite. *Calyx* monophyllous, divided, persistent, generally petaloid. *Stamens* definite, either equal in number to the segments of the calyx, or twice as many, but generally in part abortive. *Fruit* a 2- or 3-cornered indehiscent monospermous nut. *Meisner.*

* *Styles usually but* 2.

1. P. *amphibium* Linn. E. B. 7. 435.

Styles 2, united half way up. Stamens 5. Spike ovate. *Smith.*

In ponds and ditches. — Perennial. *July, August.*

2. P. *Persicaria* Linn. Spotted Persicaria. E. B. 11. 756.

Styles 2, united half way up. Stamens 6. Clusters dense, ovate-oblong, erect, on smooth stalks. Stipulas fringed. *Smith.*

β. *incanum* Meisner.

Leaves hoary beneath. Flowers pale.

P. incanum *Willd.*

In ditches and watery places. — Annual. *July. August.*

3. P. *lapathifolium* Linn. E. B. 20. 1382.
Styles 2, distinct. Stamens 6. Flower-stalks rough. Stipulas beardless. Seeds concave at each side. *Smith*.
P. pensylvanicum *Curtis*.
P. pallidum *Withering*.
In cultivated ground. — Annual. *July, August.*

4. P. *Hydropiper* Linn. E. B. 14. 989.
Styles 2, united half way up. Stamens 6. Clusters lax, interrupted, drooping. Stem erect. Leaves lanceolate, wavy, without spots. *Smith*.
Common everywhere in ditches. — Annual. *September.*

5. P. *minus* Huds. E. B. 15. 1043.
Styles 2, combined. Stamens 6. Clusters lax, slender, nearly upright. Stem trailing at the base. Leaves linear-lanceolate, flat. *Smith*.
P. intermedium *Ehr.*
On gravelly watery commons. — Annual. *September.*

** *Styles* 3.

6. P. *Bistorta* Linn. Great Bistort, or Snake-weed. E. B. 8. 509.
Stem simple, with a single, spiked cluster of flowers. Leaves ovate, wavy, running down into the footstalks. *Smith*.
In pastures and meadows. — Perennial. *June.*

7. P. *viviparum* Linn. E. B. 10. 669.
Stem simple, with a single, spiked cluster of flowers. Leaves lanceolate, revolute, with copious, prominent marginal veins. *Smith*.
In pastures or moist rocks, in alpine situations. — Perennial. *June, July.*

8. P. *aviculare* Linn. Knot-grass. E. B. 18. 1252.
Flowers axillary. Leaves elliptic-lanceolate, rough-edged. Ribs of the stipulas distant. Stem procumbent, herbaceous. *Smith*.
In waste ground. — Annual. *April—October.*

9. P. *Fagopyrum* Linn. Buck-wheat, or Brank. E. B. 15. 1044.
Leaves heart-arrow-shaped. Stem nearly upright, without prickles. Angles of the fruit even. *Smith*.
In cultivated fields. - Annual. *July, August.*

10. P. *Convolvulus* Linn. Black Bindweed. E. B. 14. 941.
Leaves heart-arrow-shaped. Stem twining, angular. Segments of the calyx bluntly keeled. *Smith*.
In fields and osier grounds. — Annual. *June—September.*

Order 70. Amaranthaceæ *Juss.*

Calyx 3- or 5-leaved, hypogynous, persistent, occasionally with 2 bracteolæ at the base.

Stamens hypogynous, either 5, or some multiple of that number, either distinct or monadelphous, occasionally partly abortive; *anthers* either 2-celled, or 1-celled.

Ovarium single, superior, 1- or few-seeded; the *ovules* hanging from a free central funiculus; *style* 1 or none; *stigma* simple or compound.

Fruit a membranous utricle.

Seeds lentiform, pendulous; *testa* crustaceous; *albumen* central, farinaceous; *embryo* curved round the circumference; *radicle* next the hilum; *plumula* inconspicuous.

Herbs or *shrubs*. *Leaves* simple, opposite or alternate, without stipulæ. *Flowers* scarious, in heads or spikes, usually coloured, occasionally bisexual, generally hermaphrodite. *Pubescence* simple, the hairs divided by internal partitions.

1. AMARANTHUS *Linn.*

Monœcious. *Calyx* 3- or 5-lobed. *Males.* *Stamens* 3 or 5. *Females.* *Styles* 3. *Stigmas* 3. *Fruit* 1-seeded, crowned with the remains of the style, and opening by a transverse incision.

1. A. *Blitum* Linn. E. B. 31. 2212.
Flowers 3-cleft and triandrous, in small lateral tufts. Leaves ovate. Stem diffuse. *Smith.*
Perennial. *May—September.*

Order 71. Chenopodeæ *Vent.*

(Atriplices *Juss.*)

Calyx deeply divided, sometimes tubular at the base, persistent, with an imbricated æstivation.

Stamens inserted into the base of the calyx, opposite its segments, and equal to them in number, or fewer.

Ovarium single, superior, or occasionally adhering to the tube of the calyx, with a single *ovulum* attached to the base of the cavity, and either erect or inverted; *style* in 2 or 4 divisions, rarely simple; *stigmas* undivided.

Fruit membranous, not valvular, sometimes baccate.

Embryo curved round farinaceous albumen; or spiral, or doubled together without albumen; *radicle* next the hilum; *plumula* ncon spicuous.

Herbaceous plants or *undershrubs*. *Leaves* alternate without stipulæ, occasionally opposite. *Flowers* small, occasionally polygamous.

ANALYSIS OF THE GENERA.

Hermaphrodite
  Calyx or the fruit with appendages at the back - - 1. SALSOLA.
  Calyx without appendages
    tubular - - - - - - - - - - 2. SALICORNIA.
    3-4-5-parted
      Fruit loose in the calyx - - - - - - 3. CHENOPODIUM.
      Fruit adhering to the lower half of the calyx - 4. BETA.
Polygamous or monœcious - - - - - - - 5. ATRIPLEX.

---

1. SALSOLA *Linn.* SALTWORT.

Calyx 5-parted, persistent; the segments after flowering producing from their back a scarious appendage. *Stamens* 5. *Stigmas* 2 or 3. *Seed* solitary. *Embryo* in the circumference. *Dec.*

1. S. *Kali* Linn. E. B. 9. 634.
Herbaceous and decumbent. Leaves awl-shaped, spinous-pointed, rough. Calyx with a dilated margin. *Smith.*
On the sea-coast. — Annual. *July.*

2. SALICORNIA *Linn.*

Calyx tubular, ovate, compressed, with 5 inconspicuous teeth. *Stamens* 1 or 2, protruding. *Style* 1, short. *Stigmas* 2, projecting, papillose. *Fruit* a utricle covered by the calyx. *Seed* cochleate.

1. S. *herbacea* Linn. Jointed-glasswort. Marsh Samphire.
E. B. 6. 415.
Stem herbaceous, erect; joints compressed, notched; interstices inversely conical. Spikes tapering upward. *Smith.*
S. annua *Smith.*
On muddy sea-shores. — Annual. *August, September.*

2. S. *procumbens* Smith. E. B. 35. 2475.
Stem herbaceous, procumbent; interstices inversely conical; branches simple. Spikes tapering upward. Stamens 2. *Smith.*
S. herbacea *Ehr.*
In salt marshes. — Annual. *August.*

3. S. *radicans* Smith. E. B. 24. 1691.
Stem woody; procumbent, and taking root at the base. Joints compressed, notched; interstices somewhat cylindrical. Spikes oblong. Stamens 2. *Smith.*
On muddy sea-shores. — Perennial. *September.*

4. S. *fruticosa* Linn. E. B. 35. 2467.
Stem woody, ascending; joints and interstices cylindrical. Spikes nearly sessile, cylindrical, obtuse. *Smith.*
On the sea-shore. — Shrubby. *September.*

3. CHENOPODIUM *Linn.*

Calyx 3-, 4-, or 5-parted, persistent, neither warted nor growing together after flowering. *Stamens* 5, or fewer. *Style* 2-fid. *Stigmas* 2-4. *Fruit* a thin utricle, containing a single polished brittle *seed*.

* *Leaves angular.*

1. Ch. *Bonus Henricus* Linn. E. B. 15. 1033.
Leaves triangular-arrow-shaped, entire. Spikes terminal, compound, leafless. *Smith.*
In waste ground. — Perennial. *May, June.*

2. Ch. *urbicum* Linn. E. B. 10. 717.
Leaves triangular, toothed. Spikes crowded, lobed, very long and straight, approaching the stem, almost leafless. *Smith.*
On dunghills. — Annual. *August, September.*

3. Ch. *rubrum* Linn. E. B. 24. 1721.
Leaves triangular, somewhat rhomboid, deeply toothed and sinuated. Spikes erect, compound, leafy. Fruit very minute. *Smith.*
In waste ground. — Annual. *August.*

4. Ch. *botryodes* Smith E. B. 32. 2247.
Leaves triangular, somewhat toothed; the upper ones bluntish. Spikes erect, compound, rounded, leafy. *Smith.*
In moist sandy places, near the sea. — Annual. *August, September.*

5 Ch. *murale* Linn. E. B. 24. 1722.
Leaves ovate, acute, many-toothed, shining. Spikes aggregate, panicled, cymose, leafless. *Smith.*
In waste ground. — Annual. *August, September.*

6. Ch. *hybridum* Linn. E. B. 27. 1919.
Leaves heart-shaped, pointed, with broad angular teeth. Spikes aggregate, panicled, cymose, divaricated, leafless. *Smith.*
In waste ground. — Annual. *August.*

7. Ch. *album* Linn. Goosefoot. E. B. 24. 1723.
Leaves rhomboid-ovate, jagged, mealy; entire towards the base: upper ones oblong, entire. Fruit quite smooth. *Smith.*
Ch. viride *Linn.*
In waste ground. — Annual. *July, August.*

8. Ch. *ficifolium* Smith. E. B. 24. 1724.
Leaves sinuated, jagged, somewhat hastate; entire towards the base: upper ones oblong, quite entire. Fruit dotted. *Smith.*
Ch. serotinum *Hudson.*
In waste ground. — Annual. *August, September.*

9. Ch. *glaucum* Linn. E. B. 21. 1454.
Leaves all oblong; deeply waved at the margin; glaucous and mealy beneath. Spikes compound, leafless, lobed. Fruit very minutely dotted. *Smith.*
In waste ground. — Annual. *August.*

** *Leaves undivided, entire.*

10. Ch. *olidum* Curtis. E. B. 15. 1034.
Leaves ovate, somewhat rhomboid, entire. Spikes dense, crowded, leafless. *Smith.*
C. Vulvaria *Linn.*
In waste ground. — Annual. *August.*

11. Ch. *polyspermum* Linn. E. B. 21. 1480.
Leaves ovate, obtuse, entire. Stem prostrate. Clusters cymose, divaricated, leafless. *Smith.*
Ch. arrectum *Desm.*
On waste ground. — Annual. *July, August.*

12. Ch. *acutifolium* Smith. E. B. 21. 1481.
Leaves ovate, acute, entire. Stem erect. Clusters spiked, compound, elongated, erect, partly leafy; lower ones somewhat cymose.
C. polyspermum *Curtis.*
In waste ground. — Annual. *July, August.*

13. Ch. *maritimum* Linn. E. B. 9. 633.
Leaves awl-shaped, semicylindrical. Flowers axillary, sessile. *Smith.*
On the sea-shore. — Annual. *July, August.*

14. Ch. *fruticosum* Linn. E. B. 9. 635.
Erect, shrubby. Leaves semicylindrical, bluntish, without spines. *Smith.*
Salsola fruticosa *Linn.*
On the sea-coast. — Shrub. *July, August.*

4. BETA *Linn.*

Calyx 5-parted, half adherent to the ovarium at the base. *Stamens* 5. *Styles* 2. *Fruit* reniform, enveloped in the capsular base of the calyx. *Dec.*

1. B. *maritima* Linn. Sea Beet. E. B. 4. 285.
Stems procumbent. Flowers in pairs. Segments of the calyx entire at the keel. *Smith.*
On the sea-shore. — Perennial. *August.*

5. ATRIPLEX *Linn.*

Polygamous or often monœcious. *Hermaphrodite. Calyx* 5-parted. *Stamens* 5. *Pistillum* usually defective. *Female. Calyx* 2-parted; the segments parallel and close together, uniting after flowering, and forming a cover for the fruit. *Style* bifid. *Fruit* a utricle, with a single brittle seed.

1. A. *portulacoides* Linn. Sea Purslane. E. B. 4. 261.
Stem shrubby, spreading. Leaves opposite, obovate-lanceolate, entire, tapering at the base. Flowers generally completely separated. *Smith.*
On the sea-coast. — Shrub. *July, August.*

2. A. *laciniata* Linn. E. B. 3. 165.
Stem herbaceous, spreading. Leaves trowel-shaped, angular, and toothed; scaly beneath. *Smith.*
On the sea-coast. — Annual. *July.*

3. A. *patula* Linn. E. B. 13. 936.
Stem herbaceous, spreading. Leaves triangular-lanceolate, somewhat halberd-shaped. Calyx of the fruit tuberculated at the sides. *Smith.*
In cultivated, as well as waste ground. — Annual. *June—August.*

4. A. *angustifolia* Smith. E. B. 25. 1774.
Stem herbaceous, spreading. Leaves lanceolate, entire; the lower ones partly 3-lobed. Calyx of the fruit halberd-shaped, slightly warty at the sides. *Smith.*
In waste ground. — Annual. *June—August.*

5. A. *erecta* Huds. E. B. 31. 2223.
Stem herbaceous, erect. Leaves ovate-lanceolate; lower ones sinuated. Calyx of the fruit all over armed with sharp tubercles. *Smith.*
In waste ground. — Annual. *August.*

6. A. *littoralis* Linn. E. B. 10. 708.
Stem herbaceous, erect. Leaves all linear, entire, variously toothed, or sinuated. Calyx of the fruit sinuated; its disk armed with prominent tubercles. *Smith.*
A. serrata *Huds.*
A. marina *Linn.*
In muddy salt marshes. — Annual. *August, September.*

7. A. *pedunculata* Linn. E. B. 4. 232.
Stem herbaceous, zigzag, with spreading branches. Leaves obovate, entire. Seed-bearing flowers stalked, wedge-shaped. *Smith.*
On the sea-coast. — Annual. *August, September.*

Order 72. SCLERANTHEÆ *Link.*

Flowers hermaphrodite.
Calyx 4- or 5-toothed, with an urceolate tube.
Stamens from 1 to 10, inserted into the orifice of the tube.
Ovarium simple, superior, 1-seeded.
Styles 2, or 1, emarginate at the apex.
Fruit a membranous utricle enclosed within the hardened calyx.
Seed pendulous from the apex of a funiculus, which arises from the bottom of the cell; *embryo* cylindrical, curved round farinaceous albumen.
Small *herbs*. *Leaves* opposite, without stipules. *Flowers* axillary, sessile.

1. SCLERANTHUS *Linn.* KNAWEL.

Calyx 5-cleft, persistent, with an urceolate tube. *Stamens* 2, 5, or 10. *Ovarium* 2-seeded. *Fruit* thin, indehiscent, covered by the hardened tube of the calyx.

1. S. *annuus* Linn. E. B. 5. 351.
Calyx of the fruit with spreading, taper, acute segments. Stems spreading. *Smith.*
In sandy fields. — Annual. *July.*

2. S. *perennis* Linn. E. B. 5. 352.
Calyx of the fruit with converging obtuse segments, edged with broad membrane. Stems procumbent. *Smith.*
In sandy fields. — Perennial. *August—October*

Order 73. URTICEÆ *Juss.*

Flowers monœcious or diœcious, scattered or clustered.
Calyx membranous, lobed, persistent.
Stamens definite, distinct, inserted into the base of the calyx; *anthers* curved inwards in æstivation, curving backwards with elasticity when bursting.
Ovarium superior, simple; *ovule* solitary, pendulous; *stigma* sessile, simple.
Fruit a simple indehiscent nut, surrounded either by the membranous or fleshy calyx, sometimes seated upon a dilated fleshy receptacle.
Embryo straight, curved, or spiral, with or without albumen; *radicle* superior.
Trees or *shrubs*. *Leaves* alternate, with stipulæ, hispid or scabrous, either covered with pungent hairs or yielding a milky juice.

1. PARIETARIA *Linn.*

Flowers polygamous, surrounded by an involucrum. *Calyx* 4-parted. *Stamens* 4. *Ovarium* 1. *Style* 1. *Stigma* 1. *Fruit* 1-seeded, covered by the lengthened calyx.

1. P. *officinalis* Linn. Common Wall-pellitory. E. B. 13. 879.
eaves lanceolate-ovate, without lateral ribs at the base. Involucrum 3-flowered, with 7 ovate segments. Stem ascending. *Smith.*
On old walls and rubbish, in sheltered places. — Perennial. *June—Septembc*

2. URTICA *Linn.*

Monœcious, seldom diœcious. *Males* in loose racemes. *Calyx* 4-parted. *Stamens* 4. *Females* in capitate racemes. *Calyx* 2 leaved. *Ovarium* 1. *Stigma* 1. *Fruit* 1-seeded, enclosed in the calyx.

1. U. *pilulifera* Linn. Roman Nettle. E. B. 3. 148.
Leaves opposite, ovate, serrated; with transverse ribs. Fertile flowers in globular heads. *Smith.*

In waste ground near the sea. — Annual. *June, July.*

2. U. *urens* Linn. Small Nettle. E. B. 18. 1236.
Leaves opposite, elliptical, with about 5 longitudinal ribs. Clusters nearly simple. *Smith.*

A troublesome weed. — Annual. *June—October*

3. U. *dioica* Linn. Great Nettle. E. B. 25. 1750.
Leaves opposite, heart-shaped. Clusters much branched, in pairs, mostly diœcious. Roots creeping. *Smith.*

In waste ground. — Perennial. *July, August.*

3. HUMULUS *Linn.*

Diœcious. *Males.* *Calyx* 5-parted. *Stamens* 5. *Females.* Inflorescence a lax membranous cone. *Ovarium* 1. *Styles* 2. *Fruit* 1-seeded. *Embryo* spiral.

1. H. *Lupulus* Linn. Hop. E. B. 6. 427.

In thickets and hedges — Perennial. *July.*

Order 74. RESEDACEÆ. *Lindley.*

Flowers included within a many-parted involucrum, neuter on the outside, hermaphrodite in the centre.

Calyx 1-sided, undivided, glandular.

Barren stamens of the sterile florets linear, petaloid.

Fertile stamens perigynous, definite; *filaments* erect; *anthers* 2-celled, opening longitudinally.

Ovarium sessile, 3-lobed, one-celled, many-seeded with 3 parietal placentæ. *Stigmata* 3, glandular, sessile.

Fruit dry, and membranous, or succulent, opening at the apex.

Seeds several, reniform, attached to 3 parietal placentæ; *embryo* taper, arcuate, without albumen; *radicle* superior.

Herbaceous plants, with alternate *leaves*, the surface of which is minutely papillose.

1. RESEDA *Linn.*

Fruit dry, many-seeded, surrounded by the withered involucrum.

1. R. *Luteola* Linn. Dyer's Rocket. Yellow-weed, or Weld. E. B. 5. 320.
Leaves lanceolate, undivided. Involucrum in four segments.

In waste ground. — Annual. *July.*

2. R. *lutea* Linn. Base Rocket. Wild Mignonette. E. B. 5. 321.
Leaves deeply 3-lobed; lower ones pinnatifid. Involucrum in 6 divisions.

On chalky hills. — Annual, or, in mild winters, perennial. *July, August.*

Order 75. Euphorbiaceæ *Juss.*

Flowers monœcious or diœcious.

Calyx lobed, inferior, with various glandular or scaly internal appendages, sometimes wanting.

Males. *Stamens* definite or indefinite, distinct or monadelphous; *anthers* 2-celled.

Females. *Ovarium* superior, sessile, or stalked, 2- 3- or more-celled; *ovules* solitary, or twin, suspended from the inner angle of their cell; *styles* equal in number to the cells, sometimes distinct, sometimes combined, sometimes none; *stigma* compound, or single with several lobes.

Fruit consisting of 2, 3, or more dehiscent cells, separating with elasticity from their common axis.

Seeds solitary or twin, suspended, with an arillus; *embryo* enclosed in fleshy albumen; *cotyledons* flat; *radicle* superior.

Trees, shrubs, or *herbaceous* plants, often abounding in acrid milk. *Leaves* alternate, simple, rarely compound, with stipulæ. *Flowers* axillary or terminal, usually with bracteæ, sometimes enclosed within an involucrum.

1. EUPHORBIA *Linn.*

Flowers collected in monœcious heads, surrounded by an involucrum, consisting of 1 leaf with five divisions, which have externally 5 glands alternating with them. *Males* naked, monandrous, articulated with their pedicel, surrounding the female, which is in the centre. *Female* naked, solitary. *Ovarium,* stalked. *Stigmas* 3, forked. *Fruit* hanging out of the involucrum, consisting of 3 cells, bursting at the back with elasticity, and each containing 1 suspended seed.

§ *Anisophyllum* Röper.

Leaves with stipulæ. Glands of the involucrum externally supported by membranous processes. Seeds without an arillary caruncula. *Röper.*

1. E. *Peplis* Linn. E. B. 28. 2002.

Branches firm. Leaves oblong, deeply cordate on one side at the base, entire or slightly toothed towards the base, between fleshy and membranous, very smooth. Fruit ovate, 3-cornered, polished and quite smooth. Seeds obovate, somewhat 4-cornered, smooth, whitish. *Röper.*

On the sea-coast. — Annual. *July—September.*

§ *Tithymalus* Tournef.

Leaves without stipulæ. Glands of the involucrum without membranous processes. Seeds without an arillary caruncula. *Röper.*

A. *Glands of the involucrum and cotyledons nearly round.*

a. Seeds netted.

E. *Helioscopia* Linn. E. B. 13. 883.

Annual. Leaves membranous, obovate-cuneate, obtuse, or emarginate, serrated towards the points, smooth, or occasionally with a few hairs. Whorl 5-cleft, rarely 4- or 3-cleft. Ovaria convex at the back, polished, smooth. Seeds obovate, sculptured, brown, not shining. *Röper.*

In fields and waste places. — Annual. *All seasons.*

b. Seeds smooth, or warted with raised points.

3. E. *platyphylla* Linn.

Annual or half-shrubby. Leaves membranous, lanceolate, generally acute, serrulate, smooth, pubescent, or hairy. Whorl 5-cleft, seldom 4- or 3-cleft. Ovaria convex at the back, smoothish or more or less warted, smooth or hairy. Seeds obovate, brown, shining. *Röper.*

β. *stricta* Röper. E. B. 5. 333.

Involucrum generally hairy externally. Whorl generally 5-cleft, seldom 3-cleft. Pericarpium warted, smooth, or hairy. Seeds smooth. *Röper.*

E. stricta *Linn.*

E. Coderiana *Fl. Fr.*

In corn-fields. — Annual. *July, August.*

4. E. *hiberna* Linn. E. B. 19. 1337.

Perennial. Leaves membranous, or between membranous and coriaceous, broadly lanceolate, sessile, obtuse or acute, entire, pubescent or smooth. Whorl 5-cleft, rarely 6-cleft. Ovaria convex at the back, muricated with warts, smooth. Seeds obovate, smooth, somewhat shining, blackish-brown. *Röper.*

In fields, thickets, and woods. — Perennial. *June.*

B. *Glands of the involucrum triangular-lunate, or lunate with 2 horns. Cotyledons linear.*

E. *Esula* Linn. E. B. 20. 1399.

Perennial. Leaves membranous, lanceolate, sessile, bluntish, muricate, entire, or with a few roughish teeth towards the point, nooth. Flowering branches either arranged under the whorl, a sort of multifid false umbel, or occasionally in a 5-cleft whorl. Glands lunate, somewhat 2-horned. Ovaria convex, smooth, and rough, with dots at the back. Seeds obovate, smooth, greyish-brown, not shining. *Röper.*

In shady woods. — Perennial. *July.*

6. E. *Cyparissias* Linn. E. B. 12. 840.

Perennial. Leaves membranous, stiff, linear, sessile, obtuse, or rather pointed, entire, smooth. Flowering branches either arranged under the whorl, in a sort of multifid false umbel, or ocasionally in a 5-cleft whorl. Glands lunate, somewhat 2-horned. Ovaria smooth, convex at the back, roughish with dots. Seeds obovate, smooth, greyish-brown or whitish, not shining. *Röper.*

In groves and thickets. — Perennial. *June, July.*

7. E. *paralias* Linn. E. B. 3. 195.

Perennial. Leaves thickish, leathery, lanceolate, sessile, acute or bluntish, entire, smooth. Whorl 5-cleft or 4- or 3-cleft; the flowering branches occasionally collected beneath the whorl into a sort of multifid false umbel. Glands lunate, somewhat eroded. Ovarium convex at the back, with a longitudinal furrow, covered with extremely minute elevated dots, wrinkled, smooth. Seeds roundish-obovate, smooth or slightly sculptured, cinereous, occasionally with a few brownish spots. *Röper.*

On the sea-coast. — Perennial. *August, September.*

8. E. *segetalis* Linn.

Annual. Leaves membranous, rather stiff, linear-lanceolate, sessile, or tapering into the petiole, acute or rather blunt, mucronate, quite entire, smooth. Whorl 5-cleft. Glands lunate with very long horns. Ovaria convex at the back, and rough with elevated points, smooth. Seeds obovate, whitish, sculptured. *Röper.*

β. *maritima* Röper. E. B. 7. 441.

Bracteæ broad, subcordate, slightly mucronate.

E. Portlandica *Linn.*

On the sea-coast, in the south. — Perennial. *August.*

9. E. *exigua* Linn. E. B. 19. 1336.

Annual. Leaves membranous, rather stiff, linear or linear wedge-shaped, sessile, rather acute, blunt or retuse, slightly mucronate, quite entire, smooth. Whorl trifid, 4-cleft or 5-cleft. Glands with very long horns. Ovaria convex at the back, rough with elevated points, smooth. Seeds obovate-cylindrical, nearly 4-cornered, covered with warted net-work, whitish or brownish ash-colour, not shining. *Röper.*

In corn-fields. — Annual. *July.*

10. E. *Peplus* Linn. E. B. 14. 959.

Leaves membranous, roundish, tapering into the petiole, very blunt, entire, smooth. Whorl trifid, very seldom 5-fid. Glands lunate, with very long horns. Ovaria with a double-winged keel at the back, wrinkled and scabrous, smooth. Seeds obovate-cylindrical, bluntly 6-cornered; 4 of the sides dotted in rows, 2 with a longitudinal furrow; greyish white, not shining. *Röper.*

E. peploides *Gouan.*

A common weed. — Annual. *July, August.*

11. E. *Lathyris* Linn. E. B. 32. 2255.

Biennial. Leaves somewhat coriaceous, linear, sessile, rather acute, or obtuse, mucronate, entire, smooth. Whorl 4-cleft, rarely bifid, still more rarely 5-cleft. Glands lunate, 2-horned; the horns dilated and obtuse. Ovaria convex at the back, with a deep longitudinal furrow, even, smooth. Seeds obovate, truncate at the base, rough, brown, not shining. *Röper.*

In dry thickets. — Biennial. *June, July.*

12. E. *amygdaloides* Linn. E. B. 4. 256.
Perennial and half shrubby. Leaves coriaceous, or between coriaceous and membranous, broadly lanceolate, tapering into the petiole, acute or obtuse, slightly mucronate, entire, pubescent: of the flowering branches connate. Whorl 5-cleft, or the flowering branches collected under the whorl in a 6- or 8-cleft spurious umbel. Glands lunate, 2-horned. Ovaria convex at the back, with extremely minute elevated points, smooth. Seeds roundish-ovate, smooth, brown, not shining. *Röper*
In woods and ditches. — Perennial. *March, April.*

13. E. *Characias* Linn. E. B. 7. 442.
Perennial or half-shrubby. Leaves linear-lanceolate, tapering into the petiole, acute or rather blunt, slightly mucronate, entire, pubescent or smooth: of the flowering branches connate. The flowering branches collected under the whorl in a multifid spurious umbel. Glands lunate, sometimes 2-horned. Ovaria convex at the back, with extremely minute elevated points; when young, woolly, afterwards hairy or nearly smooth. Seeds oblong-ovate, smooth, opaque, blackish. *Röper.*
In bushy mountainous places. — Shrub. *March, April.*

2. MERCURIALIS *Linn.*

Diœcious, or occasionally monœcious. *Calyx* 3-parted. *Males. Stamens* 9-12. *Females. Ovarium* double, with two opposite furrows, and two sterile filaments proceeding from either furrow. *Styles* 2, forked. *Fruit* dry, consisting of 2 cells bursting with elasticity, and containing each 1 seed.

1. M. *perennis* Linn. Perennial Mercury. E. B. 26. 1872.
Stem perfectly simple. Leaves rough. Root creeping. *Smith.*
On banks, and in bushy places. — Perennial. *April, May.*

2. M. *annua* Linn. Annual Mercury. E. B. 8. 559.
Stem cross-branching. Leaves smooth. Root fibrous. Barren flowers in numerous, spiked, alternate tufts. *Smith.*
In waste or cultivated ground. — Annual. *July—September.*

3. BUXUS *Linn.*

Monœcious. Calyx 3- or 4-parted. *Male. Scale* 2-lobed. *Stamens* 4, inserted about the rudiment of an ovarium. *Female. Scales* 3, very small. *Styles* 3. *Stigmas* 3, obtuse. *Fruit* with 3 horns, 3 cells, and 6 seeds. *Dec.*

1. B. *sempervirens* Linn. Common Box-tree. E. B. 19. 1341.
Leaves ovate, convex. Footstalks slightly downy at the edges. Anthers ovate-arrow-shaped. *Smith.*
On dry chalky hills. — Tree, or shrub. *April.*

Order 76. Empetreæ *Nuttall.*

Flowers bisexual.

Sepals 3 or 2, hypogynous, surrounded at the base by imbricated scales.

Stamens equal in number to the sepals, and alternate with them; *anthers* roundish, 2-celled, the cells distinct, bursting longitudinally.

Ovarium superior, seated in a fleshy disk, 3-, 6-, or 9-celled; *ovules* solitary, ascending; *style* 1; *stigma* radiating, the number of its rays corresponding with the cells of the ovarium.

Fruit fleshy, seated in the persistent calyx, 3-, 6-, or 9-celled: the coating of the cells bony.

Seeds solitary, ascending; *embryo* taper, in the axis of fleshy watery albumen; *radicle* inferior.

Small shrubs with heath-like evergreen *leaves* without stipulæ; and minute *flowers* in their axillæ.

1. EMPETRUM *Linn.*

Sepals 3. *Stamens* 3. *Stigma* 6- or 9-cleft. *Fruit* succulent, spherical, with from 6 to 9 seeds.

1. E. *nigrum* Linn. Crow-berry, or Crake-berry. E. B. 8. *526.* Stem and branches procumbent. Leaves slightly elliptical. *Smith.*
On mountainous heaths. — Shrub. *May.*

Order 77. Aristolochiæ *Juss.*

Flowers hermaphrodite.

Calyx superior, tubular, with 3 segments, which are valvate in æstivation, sometimes regular, sometimes very unequal.

Stamens 10 or 12, epigynous, distinct, or cohering to the style and stigmas.

Ovarium inferior, 3- or 6-celled; *ovules* numerous, horizontally attached to the axis; *style* simple; *stigmas* radiating, as numerous as the cells of the ovarium.

Fruit dry or succulent, 3- or 6-celled, many-seeded.

Seeds with a very minute embryo placed in the base of fleshy albumen.

Herbaceous plants or *shrubs*, the latter often climbing. *Leaves* alternate, simple, stalked. *Flowers* axillary, solitary, brown or some dull colour.

1. ASARUM *Linn.* Asarabacca.

Calyx campanulate, 3-lobed. *Stamens* placed upon the ovarium. *Anthers* adnate to the middle of the filaments. *Style* short. *Stigma* stellate, 6-lobed. *Fruit* capsular, 6-celled.

1. A. *europæum* Linn. E. B. 16. 1083.
Leaves two on each stem, kidney-shaped, obtuse. *Smith.*
In mountainous woods. — Perennial. *May.*

2. ARISTOLOCHIA *Linn.*

Calyx tubular, ventricose at the base, dilated at the apex, and lengthened into a strap-like lobe. *Anthers* 6, subsessile, inserted round the base of the style. *Stigma* 6-cleft. *Capsule* with 6 corners and 6 cells.

1. A. *Clematitis* Linn. Birthwort. E. B. 6. 398.
Leaves heart-shaped. Stem erect. Flowers aggregate, upright. Calyx unilateral. *Smith.*
In woods and thickets. — Perennial. *July, August.*

Order 78. Ceratophylleæ *Dec.*

Flowers monœcious.
Calyx inferior, many-parted.
Male. *Stamens* from 12 to 20; *filaments* wanting; *anthers* 2-celled.
Female. *Ovarium* superior, 1-celled; *ovule* solitary, pendulous; *stigma* filiform, oblique, sessile.
Nut 1-celled, 1-seeded, indehiscent, terminated by the hardened stigma.
Seed pendulous, solitary; *albumen* 0; *embryo* with 4 cotyledons, alternately smaller; *plumula* many-leaved; *radicle* superior (*Dec.*)
Floating *herbs*, with multifid, cellular *leaves*.

1. CERATOPHYLLUM *Linn.*

Character that of the order, there being no other genus.

1. C. *demersum* Linn. Hornwort. E. B. 14. 947.
Fruit armed with 3 spines. Segments of the calyx notched at the extremity. *Smith.*
In ditches and fish-ponds. — Perennial. *August, September.*

2. C. *submersum* Linn.
Fruit destitute of spines. Segments of the calyx acute, entire. *Smith.*
ditches. — Perennial. *September.*

Order 79. Ulmaceæ *Mirb.*

Flowers hermaphrodite or polygamous.
Calyx divided, campanulate, inferior.
Stamens definite, inserted into the base of the calyx; erect in æstivation.

Ovarium superior, 2-celled; *ovules* solitary, pendulous; *stigmas* 2, distinct.
Fruit 1 or 2-celled, indehiscent, membranous or drupaceous.
Seed solitary, pendulous; *albumen* none, or in very small quantity; *embryo* with foliaceous cotyledons; *radicle* superior.
Trees or *shrubs* with scabrous, alternate, simple, deciduous leaves, and stipulæ.

1. ULMUS *Linn.*

Calyx campanulate, 4- or 5-toothed, persistent. *Stamens* from 3 to 6. *Ovarium* compressed. *Stigmas* 2, sessile. *Pericarpium* membranous, winged, compressed, 1-seeded.

1. U. *campestris* Linn. Narrow-leaved English Elm. E. B. 27. 1886.
Leaves rhomboid-ovate, acuminate, wedge-shaped, and oblique at the base, always scabrous above, doubly and irregularly serrated, downy beneath, serratures incurved. Branches wiry, slightly corky; when young, bright brown, pubescent. Fruit oblong, deeply cloven, naked.

In hedges in Norfolk, and elsewhere. — Tree. *March* or *April.*

N. B. Of this, the *Hertfordshire elm* of the nurseries is probably a variety.

2. U. *suberosa* Ehr. E. B. 31. 2161.
Leaves nearly orbicular, acute, obliquely cordate at the base, sharply regularly, and doubly serrated; always scabrous above, pubescent below, chiefly hairy in the axillæ. Branches spreading, bright brown, winged with corky excrescences; when young, very hairy. Fruit nearly round, deeply cloven, naked.

In hedges. — Tree. *March.*

3. U. *major* Smith. E. B. 36. 2542.
Leaves ovate-acuminate, very oblique at the base, sharply, doubly, and regularly serrated; always scabrous above, pubescent below, with dense tufts of white hairs in the axillæ. Branches spreading, bright brown, winged with corky excrescences; when young nearly smooth. Fruit obovate, slightly cloven, naked.
U. hollandica *Miller.*

In hedges. — Tree. *March.*

4. U. *carpinifolia.*
Leaves ovate-acuminate, coriaceous, strongly veined, simply crenate, serrate, slightly oblique and cordate at the base, shining, but rather scabrous above, smooth beneath. Branches bright brown, nearly smooth. Fruit . . . . .

Four miles from Stratford on Avon, on the road to Alcester. — Tree.

5. U. *glabra* Miller. E. B. 32. 2248.
Leaves ovate-lanceolate, acuminate, doubly and evenly crenate-serrate, cuneate and oblique at the base, becoming quite smooth above, smooth or glandular beneath, with a few hairs in the axillæ. Branches bright brown, smooth, wiry, weeping. Fruit obovate, naked, deeply cloven.

β. *glandulosa.*

Leaves very glandular beneath.

γ. *latifolia.*

Leaves oblong, acute, very broad.

In woods and hedges; β. near Ludlow; γ. at West Hatch, in Essex. *Mr. Forster.* — Tree. *March.*

N. B. — To this species the Downton elm and Scampston elm of the nurseries probably belong. They, the latter especially, require further examination.

6. U. *stricta.* Cornish Elm.

Leaves obovate, cuspidate, cuneate at the base, evenly and nearly doubly crenate-serrate, strongly veined, coriaceous, very smooth and shining above, smooth beneath, with hairy axillæ. Branches bright brown, smooth, rigid, erect, very compact. Fruit . . . . .

β. *parvifolia.*

Leaves much smaller, less oblique at the base, finely and regularly crenate, acuminate rather than cuspidate.

In Cornwall and North Devon; β. the less common. — Tree.

7. U. *montana* Bauh. Witch Elm. E. B. 27. 1887.

Leaves obovate, cuspidate, doubly and coarsely serrated, cuneate and nearly equal at the base, always exceedingly scabrous above, evenly downy beneath. Branches not corky, cinereous, smooth. Fruit rhomboid-oblong, scarcely cloven, naked.

U. campestris *Willd.*

U. effusa *Sibth.*, not of others.

U. nuda *Ehr.*

U. glabra *Hudson*, according to Smith.

In woods and hedges. — Tree. *March, April.*

N. B. — Of this, the Giant elm and the Chichester elm of the nurseries are varieties. It is often confounded by foreign botanists with U. pedunculata, a totally different species, not found in England. It is very nearly related to he U. rubra of North America, from which it can be scarcely distinguished by the leaves

Division III. Achlamydeæ.

Note. The genus Euphorbia, among Monochlamydeæ, being destitute of calyx and corolla, may, by the student, be referred to some order of this division; but in that genus the absence of floral envelopes is to be ascribed to the excessive development of the involucrum; the other genera of the same order are furnished with calyx. There is a tendency to produce a calyx in Cupuliferæ.

ANALYSIS OF THE ORDERS.

Flowers amentaceous
  Fruit succulent - - - - - - - - 83. Myriceæ.
  Fruit dry
    Stigmas developed. Leaves ribbed
      Fruit naked .. - - - - - - 80. Amentaceæ.
      Fruit in an involucrum - - - - - 81. Cupuliferæ.
    Stigmas wanting. Leaves ribless, resinous - 82. Coniferæ.
Flowers axillary, solitary - - - - - - 84. Callitrichineæ.

Order 80. Amentaceæ *Juss.*

Flowers bisexual; either monœcious or diœcious, amentaceous.

Male. *Stamens* distinct, scarcely ever monadelphous. *Anthers* 2-celled.

Female. *Ovarium* superior, 1- or 2-celled; *ovules* solitary or indefinite, pendulous; *style* single or none; *stigmas* divided.

Fruit membranous and indehiscent, or coriaceous and dehiscent, 1-celled, 1- or many-seeded.

Seeds pendulous, naked or comose; *albumen* none; *embryo* straight or curved; *radicle* superior.

Trees or *shrubs*, with alternate, simple, stipulate, deciduous *leaves*, the veins of which are either reticulated or straight.

§ 1. *Betulineæ* Richard.

Fruit indehiscent, membranous, 2-celled, with solitary ovules. Seeds pendulous, naked. — Trees or shrubs, with leaves having their venæ primariæ running straight from the midrib to the margin.

1. BETULA *Linn.*

Monœcious. *Catkins* cylindrical. *Males.* *Scales* ternate, that in the middle bearing the stamens. *Female.* *Scales* 3-lobed, membranous, deciduous. *Styles* 2. *Ovarium* compressed, 2-celled; 1 cell abortive. *Fruit* membranous, winged, 1-celled.

1. B. *alba* Linn. Common Birch. E. B. 31. 2198.
Leaves ovate, acute, somewhat deltoid, unequally serrated, nearly smooth. Branches erect; when young, pubescent.
In woods. — Tree. *April, May.*

2. B. *pendula* Roth. Weeping Birch.
Leaves ovate, deltoid, acute, unequally serrated, smooth. Branches weeping; when young, smooth and warted.
B. margaritacea *of some.*
B. verrucosa *Ehr.*
In woods. — Tree. *April, May.*

3. B. *nana* Linn. E. B. 33. 2326.
Leaves orbicular, crenate, reticulated with veins beneath. *Smith.*
In spongy bogs, Scotland. — Shrub. *May.*

2. ALNUS *Tourn.*

Monœcious. *Male. Catkins* cylindrical. *Scales* stalked, cordate, with 3 smaller scales beneath them, which are staminiferous at the base. *Female. Catkins* roundish-ovate. *Scales* 2-flowered, coriaceous, persistent. *Ovarium* compressed. *Stigmas* 2. *Fruit* compressed, ovate, 2-celled, 2-seeded.

1. A. *glutinosa* Gærtn. Common Alder. E. B. 21. 1508.
Leaves roundish-wedge-shaped, wavy, serrated, glutinous, rather abrupt; downy at the branching of the veins beneath. *Smith.*
Betula Alnus *Linn.*
In watery meadows. — Tree. *March.*

§ 2. *Salicineæ* Richard.

Fruit 2-valved, 1-celled, many-seeded. Seeds pendulous, comose. — Trees or shrubs, with leaves having their venæ primariæ ramifying within the margin, and forming venæ arcuatæ.

3. SALIX *Linn.*

Flowers diœcious, very seldom monœcious. *Catkins* consisting of imbricated scales. *Males.* Stamens from 2 to 5; sometimes apparently single, in consequence of the cohesion of 2. *Female. Fruit* a 1-celled follicle, with a gland at its base. *Seeds* comose. *Radicle* inferior.

* *Adult leaves serrated, smooth, or nearly so.*

1. S. *triandra* Linn. E. B. 20. 1435.
Leaves linear-oblong, serrated, smooth; rather unequally sloping at the base. Stamens 3. Ovary stalked, ovate, compressed, smooth. Stigmas nearly sessile. *Smith.*
In wet woods and hedges. — Tree. *May and August.*

2. S. *Hoffmanniana* Smith.
Leaves ovate-oblong, serrated, smooth; slightly rounded at the

base. Stamens 3. Ovary stalked, ovate, compressed, smooth. Stigmas nearly sessile. *Smith.*

S. triandra *Hoffm.*

On the banks of rivulets. — Shrub. *May.*

3. S. *lanceolata* Smith. E. B. 20. 1436.

Leaves lanceolate, serrated, smooth; tapering towards each end. Footstalks decurrent. Ovary stalked, ovate, smooth. Style as long as the stigmas. *Smith.*

In low meadows. — Tree. *April, May.*

4. S. *amygdalina* Linn. E. B. 27. 1636.

Leaves ovate, serrated, smooth; rounded and unequal at the base. Stamens 3. Ovary ovate, compressed, smooth; its stalk almost as long as the scale. Stigmas nearly sessile. Young branches furrowed. *Smith.*

On the banks of ditches. — Shrub, or small tree. *April, May;* and again in *August.*

5. S. *pentandra* Linn. Sweet Willow. E. B. 26. 1805.

Leaves ovate, pointed, crenate, glandular, smooth. Footstalks glandular at the summit. Stamens 5 or more, hairy at the base. Ovary ovate, tapering, smooth, nearly sessile. *Smith.*

About rivers, chiefly in the north of England and south of Scotland. — Tree. *June, July.*

6. S. *nigricans* Smith. E. B. 17. 1213.

Leaves elliptic-lanceolate, acute, crenate, smooth, with a downy rib, above; glaucous beneath. Stamens 2, thrice the length of the hairy scales. Ovary lanceolate, downy, on a short downy stalk. *Smith.*

In fens, osier-grounds, woods, and thickets. — Shrub. *April.*

7. S. *phylicifolia* Linn. E. B. 28. 1958.

Leaves elliptic-lanceolate, with wavy serratures, very smooth; glaucous beneath. Stipulas glandular on the inside. Ovary lanceolate, stalked, silky. Style twice the length of the stigmas. Branches trailing. *Smith.*

S. radicans *Smith.*

At Finlarig, Breadalbane, in the Highlands of Scotland. — Shrub. *May.*

8. S. *Borreriana* Smith.

Leaves lanceolate, with shallow even serratures, very smooth; glaucous beneath. Stipulas obsolete. Branches upright. Scales of the catkins acute, shaggy. *Smith.*

In Breadalbane and Glen Nevis. — Shrub. *May.*

9. S. *nitens* Anderson.

Leaves elliptical, acute, unequally serrated; very smooth and glaucous beneath; minutely downy, with a downy mid-rib, above. Stipulas obsolete. Branches spreading. Catkins nearly sessile, with acute, shaggy scales. *Smith.*

In the mountainous parts of Scotland. — Shrub. *April.*

10. S. *Davalliana* Smith.

Leaves obovate-lanceolate, finely serrated, or minutely toothed, tapering at each end, smooth; rather glaucous beneath. Footstalks, mid-rib, and young branches, somewhat downy. Catkins with small rounded scales. Fruit lanceolate, smooth. *Smith.*

S. phylicifolia *Willd.*

In the Highlands of Scotland. — Shrub. *May.*

11. S. *Wulfeniana* Willd.

Leaves obovate or elliptical, somewhat pointed, finely serrated, smooth; glaucous beneath. Catkins dense, with hairy scales, longer than the stalks of the awl-shaped germens. Style longer than the stigmas. *Smith.*

S. phylicifolia *Host.*

In Breadalbane; also by the river-side, near the bridge at Kirby Lonsdale. — Shrub. *April, May.*

12. S. *tetrapla* Walker.

Leaves elliptic-oblong, pointed, unequally serrated, nearly smooth; glaucous, with prominent veins beneath. Stipulas half-heart-shaped. Scales mostly shorter than the hairy stalks of the ovate-oblong smooth ovaries. Style as long as the stigmas. *Smith.*

In Breadalbane. — Shrub. *May.*

13. S. *bicolor* Ehr. E. B. 26. 1806.

Leaves elliptic-oblong, acute, waved and slightly serrated, nearly smooth; glaucous beneath. Footstalks dilated at the base. Stipulas pointed, serrated. Scales obtuse, hairy, half as long as the densely downy, ovate, long-stalked ovary. *Smith.*

S. laurina *Smith.*

In woods and thickets. — Shrub, or small tree. *April, May.*

14 S. *tenuifolia* Smith. E. B. 31. 2186.

Leaves elliptical, acute, serrated, smoothish; glaucous beneath. Stipulas small or none. Scales hairy. Fruit ovate, smooth, on a short smooth stalk. *Smith.*

About the rocky banks of rivers. — Shrub. *May, June.*

15. S. *malifolia* Smith. E. B. 23. 1617.

Leaves elliptic-oblong, toothed, waved, thin and crackling, very smooth. Stipulas heart-shaped, about equal to the footstalks. Scales obovate, bearded. Ovary lanceolate, smooth, on a short smooth stalk. *Smith.*

In thickets. — Shrub. *April.*

16. S. *petiolaris* Smith. E. B. 16. 1147.

Leaves lanceolate, serrated, smooth; glaucous beneath, somewhat unequal at the base. Stipulas lunate, toothed. Catkins lax. Scales hairy, shorter than the stalks of the ovate silky germens. Stigmas divided, sessile. *Smith.*

In osier grounds and swamps. — Shrub, or small tree. *April.*

17. S. *vitellina* Linn. Golden Osier. E. B. 20. 1389.

Leaves lanceolate, acute, with cartilaginous serratures; smooth above, glaucous and somewhat silky beneath. Stipulas minute, lanceolate, deciduous, smooth. Ovary sessile, ovate-lanceolate, smooth. Scales linear-lanceolate, acute, fringed at the base, longer than the pistil. *Smith.*

In osier grounds and swamps. — Tree. *May.*

18. S. *decipiens* Hoffm. White Welsh, or Varnished Willow. E. B. 27. 1937.

Leaves lanceolate, pointed, serrated, very smooth; floral ones partly obovate and recurved. Footstalks somewhat glandular. Ovary tapering, stalked, smooth. Style longer than the cloven stigmas. Branches smooth, highly polished. *Smith.*

In low meadows. — Tree. *May.*

19. S. *fragilis* Linn. E. B. 26. 1807.

Leaves ovate-lanceolate, pointed, serrated throughout, very smooth. Footstalks glandular. Ovary ovate, abrupt, nearly sessile, smooth. Scales oblong, about equal to the stamens and pistils. Stigmas cloven, longer than the style. *Smith.*

In low marshy grounds. — Tree. *April, May.*

20. S. *Russelliana* Smith. Bedford Willow. E. B. 26. 1808.

Leaves lanceolate, tapering at each end, serrated throughout, very smooth. Footstalks glandular, or leafy. Ovary tapering, stalked, longer than the scales. Style as long as the stigmas. *Smith.*

In marshy woods. — Tree. *April, May.*

21. S. *purpurea* Linn. E. B. 20. 1388.

Branches trailing, decumbent. Leaves partly opposite, obovate-lanceolate, serrated, very smooth, narrow at the base. Stamen 1. Stigmas very short, ovate, nearly sessile. *Smith.*

S. monandra *Ehr.*

In low meadows. — Shrub. *March.*

22. S. *Helix* Linn. Rose Willow. E. B. 19. 1343.

Branches erect. Leaves partly opposite, oblong-lanceolate, pointed, slightly serrated, very smooth; linear towards the base. Stamen 1. Style nearly as long as the linear divided stigmas. *Smith.*

S. monandra *Hoffm.*

In marshes. — Tree. *March, April.*

23. S. *Lambertiana* Smith. E. B. 19. 1359.

Branches erect. Leaves partly opposite, obovate-lanceolate, pointed, serrated, smooth, rounded at the base. Stipulas none. Stamen 1. Stigmas ovate, obtuse, notched, very short, nearly sessile. *Smith.*

In low meadows. — Tree. *March, April.*

24. S. *Forbiana* Smith. E. B. 19. 1344.

Branches erect. Leaves alternate, with small stipulas, lanceolate-oblong, with shallow serratures, smooth, rounded at the base; glaucous beneath. Stamen 1. Style nearly as long as the linear divided stigmas. *Smith.*

S. fissa *Relh.*

In meadows and osier-holts. — Shrub. *April.*

25. S. *rubra* Huds. E. B. 16. 1145.

Stamens combined below. Leaves linear-lanceolate, elongated, acute, smooth, with shallow serratures, green on both sides. Stigmas ovate, undivided. *Smith.*

S. fissa *Hoffm.*

S. virescens *Vill.*

In low meadows. — Tree. *April, May.*

26. S. *Croweana* Smith. E. B. 16. 1146.

Stamens combined below. Leaves elliptical, slightly serrated, quite smooth; glaucous beneath. *Smith.*

In swampy meadows. — Shrub. *April, May.*

27. S. *prunifolia* Smith. E. B. 19. 1361.

Leaves broadly ovate, serrated, smooth on both sides, even above, glaucous beneath. Stem erect, much branched. Fruit ovate, shaggy, like the scales, with silky hairs.

S. myrsinites *Lightf.*

On the Highland mountains of Scotland. — Shrub. *April, May.*

28. S. *vacciniifolia* Smith. E. B. 33. 2341.

Leaves lanceolate-ovate, serrated, smooth and even above, glaucous and silky beneath. Fruit ovate, silky. Stems decumbent. *Smith.*

On the Highland mountains of Scotland. — Shrub. *April.*

29. S. *venulosa* Smith. E. B. 19. 1362.

Leaves ovate, serrated, naked, reticulated with prominent veins above, rather glaucous beneath. Fruit ovate, silky. Stem erect, much branched. *Smith.*

In the Highlands of Scotland. — Shrub. *April, May.*

30. S. *myrsinites* Linn. E. B. 19. 1360.

Leaves elliptical, serrated, smooth, veiny, polished on both sides. Young branches hairy. Germens stalked, downy. Fruit awl-shaped. *Smith.*

S. retusa *Dicks.*

S. arbutifolia *Willd.*

In the Highlands of Scotland. — Shrub. *May, June.*

31. S. *Dicksoniana* Smith. E. B. 20. 1390.

Leaves elliptical, acute, slightly toothed, smooth, glaucous beneath. Young branches very smooth. Catkins ovate, short, erect. Ovary stalked, ovate, silky. Stigmas nearly sessile. *Smith.*

S. myrtilloides *Fl. Brit.*

In the Highlands of Scotland. — Shrub. *April.*

32. S. *carinata* Smith. E. B. 19 1363.

Leaves ovate, finely toothed, smooth, minutely veined, folded into a keel. Catkins cylindrical, with rounded, hairy scales. Germen sessile, ovate, silky. *Smith.*

In the Highlands of Scotland. — Shrub. *April.*

33. S. *Arbuscula* Linn. E. B 19. 1366.

Leaves lanceolate, acute, obscurely toothed, smoothish, glaucous beneath, silky when young. Branches downy. Catkins ovate, erect. Ovary stalked, ovate-lanceolate, silky. *Smith.*

In the Highlands of Scotland. — Shrub. *April.*

34. S. *livida* Wahl.

Leaves elliptic-oblong, obscurely toothed, smooth; livid beneath. Stipulas none. Ovary nearly cylindrical, downy; its stalk twice as long as the scale. Stigmas nearly sessile. *Smith.*

In the Lowlands of Scotland. — Shrub. . . . . .

35. S. *herbacea* Linn. E. B. 27. 1907.

Leaves orbicular, serrated, reticulated with veins, very smooth and shining on both sides. Ovary stalked, ovate-lanceolate, smooth. *Smith.*

On the summits of the loftiest mountains of England, Scotland, and Wales. — Shrub. *June.*

** *Adult leaves entire, nearly smooth.*

36. S. *reticulata* Linn. E. B. 27. 1908.

Leaves orbicular, somewhat elliptical, obtuse, entire, coriaceous, with reticulated veins, nearly smooth, glaucous beneath. Ovary sessile, downy. *Smith.*

On the loftiest mountains of Yorkshire, Wales, and Scotland. — Shrub. *June.*

*** *Leaves all shaggy, woolly, or silky.*

37. S. *glauca* Linn. E. B. 26. 1810.

Leaves nearly entire, elliptic-lanceolate; even and nearly smooth above, woolly and snow-white beneath. Footstalks decurrent. Ovary sessile, ovate, woolly. *Smith.*

S. appendiculata *Fl. Dan.*

In the Highlands of Scotland. — Shrub. *May.*

38. S. *Stuartiana* Smith. E. B. 36. 2586.

Leaves nearly entire, ovate-lanceolate, acute; shaggy above, densely silky, somewhat cottony beneath. Style as long as the almost sessile, woolly germen. Stigmas capillary, deeply divided, the length of the style. *Smith.*

In the Highlands of Scotland. — Shrub. *July, August.*

39. S. *arenaria* Linn. E. B. 26. 1809.

Leaves nearly entire, ovate, acute; reticulated and somewhat downy above, veiny and densely woolly beneath. Style as long as the sessile, woolly germen. Stigmas linear, deeply divided, the length of the style. *Smith.*

S. Lapponum *Lightf.*

S. Helvetica *Vill.*

S. limosa *Wahl.*

On mountains in Scotland. — Shrub. *May, June.*

40. S. *lanata* Linn.

Leaves roundish-ovate, pointed, entire, shaggy on both sides, glaucous beneath. Ovary sessile. oblong, smooth. Style four times as long as the blunt, divided stigmas. *Smith.*

S. chrysanthos *Fl. Dan.*

On rocks in the Highlands of Scotland. — Shrub. . . . . .

41. S. *argentea* Smith. E. B. 19. 1364.

Leaves elliptical, entire, somewhat revolute, with a recurved point;

rather downy above, silky and shining beneath, as well as the branches. Stem upright. Ovary ovate-lanceolate, silky; its silky stalk nearly equal to the linear oblong scale. Style not longer than the stigmas. *Smith.*

S. lanata *Roth.*

S. arenaria *Lightf.*

On the sea-shore, among loose blowing sand-banks. — Shrub. *May.*

42. S. *fœtida* Smith. E. B. 28. 1962.

Leaves elliptical, nearly entire, with a recurved point; glaucous and silky beneath. Stem recumbent. Ovary ovate-lanceolate, on a silky stalk nearly equal to the obovate scale. *Smith.*

S. ascendens *Smith.*

S. parvifolia *Smith.*

On moist, heathy, or sandy ground. — Shrub. *May.*

43. S. *repens* Linn. E. B. 3. 183.

Leaves elliptic-lanceolate, straight, somewhat pointed, nearly entire; almost naked above, glaucous and silky beneath. Stipulas none. Stem depressed, with short upright branches. Ovary stalked, ovate, downy. Fruit smooth. *Smith.*

S. depressa *Hoffm.*

On sandy heaths. — Shrub. *May.*

44. S. *fusca* Linn. E. B. 28. 1960.

Leaves elliptic-oblong, acute, straight, flat, with a few glandular teeth; glaucous and silky beneath. Stipulas none. Stem erect, much branched. Ovary sessile, nearly smooth, tapering into an elongated style. *Smith.*

On moist mountainous heaths, in the north. — Shrub. *May.*

45. S. *prostrata* Smith. E. B. 28. 1959.

Leaves elliptic-oblong, convex, somewhat toothed, with a curved point; glaucous, silky, and veiny beneath. Stipulas minute. Stem prostrate, with elongated straight branches. Ovary stalked, ovate, silky. Style shorter than the stigmas. *Smith.*

S. polymorpha *Ehr.*

On heaths and commons. — Shrub. *March, April.*

46. S. *incubacea* Linn.

Leaves elliptic-lanceolate, pointed, straight, nearly entire; convex and smooth above, with prominent reticulated veins; glaucous and silky beneath. Stem recumbent. Catkins ovate-oblong. Stalks of the silky ovary longer than the scales. Fruit smooth. *Smith.*

S. angustifolia *Wulf.*

In sandy meadows. — Shrub. *May.*

47. S. *Doniana* Smith.

Leaves obovate-lanceolate, partly opposite, acute, straight, slightly serrated; livid, and somewhat silky, beneath. Stem and branches erect. Catkins cylindrical. Ovary stalked, silky, longer than the obovate bearded scales. *Smith.*

Scotland. — Shrub. *May.*

48. S. *rosmarinifolia* Linn. E. B. 19. 1365.
Leaves linear-lanceolate, pointed, straight, entire; silky beneath. Stem erect. Catkins ovate, recurved. Ovary stalked, lanceolate, silky. *Smith.*

In moist sandy ground. — Shrub. *April.*

49. S. *cinerea* Linn. E. B. 27. 1897.
Stem erect. Lower leaves entire, upper serrated, obovate-lanceolate; glaucous, downy, and reticulated with veins, beneath. Stipulas half-heart-shaped, serrated. Ovary silky, its stalk half as long as the lanceolate scales. *Smith.*
S. acuminata *Hoffm.*
S. daphnoides *Vill.*

In moist marshy woods. — Tree. *April.*

50. S. *aurita* Linn. E. B. 21. 1487.
Branches trailing. Leaves somewhat serrated, convex, obovate, obtuse, with a small hooked point; hairy, and reticulated with veins, on both sides. Stipulas roundish, convex, toothed. Ovary silky, stalked. Stigmas nearly sessile. *Smith.*
S. uliginosa *Willd.*
S. ulmifolia *Vill.*

In moist upland woods. — Shrub. *April, May.*

51. S. *aquatica* Smith. Sallow. E. B. 20. 1437.
Stem and branches erect. Leaves slightly serrated, obovate-elliptical, minutely downy, flat; rather glaucous beneath. Stipulas rounded, toothed. Ovary silky, stalked. Stigmas nearly sessile. *Smith.*
S. cinerea *Withering.*
S. aurita *Hoffm.*
S. Timmii *Schkuhr.*

In wet hedge-rows. — Shrub, or small tree. *April.*

52. S. *oleifolia* Smith. E. B. 20. 1402.
Stem erect. Branches straight, spreading. Leaves obovate-lanceolate, flat, rather rigid, minutely toothed, acute; glaucous, reticulated, and finely hairy, beneath. Stipulas small, notched, rounded. Catkins oval, nearly half as broad as long. *Smith.*

In woods and hedges. — Tree. *March.*

53. S. *cotinifolia* Smith. E. B. 20. 1403.
Stem erect. Branches spreading, downy. Leaves broadly elliptical, nearly orbicular, slightly toothed; glaucous and downy, with rectangular veins, beneath. Style as long as the linear notched stigmas. *Smith.*
S. spadicea *Villars.*

In thickets and woods. — Shrub. *April.*

54. S. *hirta* Smith. E. B. 20. 1404.
Stem erect. Branches densely hairy. Leaves elliptic-heart-shaped, pointed, finely crenate; downy on both sides. Stipulas half-heart-shaped, flat, toothed, nearly smooth. *Smith.*

In woods and hedges. — Tree. *April, May.*

55. S. *rupestris* Donn. E. B. 33. 2342.
Stem trailing. Leaves obovate, acute, serrated, flat, even, silky on both sides. Stipulas hairy. Branches minutely downy. Ovary stalked, awl-shaped, silky. Style as long as the blunt undivided stigmas. *Smith.*

On rocks, in the Highlands of Scotland. — Shrub. *May.*

56. S. *Andersoniana* Smith. E. B. 33. 2343.
Stem upright. Leaves elliptical, acute, finely notched, slightly downy; paler beneath. Stipulas half-ovate, nearly smooth. Branches minutely downy. Ovary smooth, its stalk almost equal to the scale. Style cloven, longer than the cloven stigmas. *Smith.*

In woods, and on the banks of rivers. — Shrub. *April, May.*

57. S. *Forsteriana* Smith. E. B. 33. 2344.
Stem erect. Branches minutely downy. Leaves elliptic-obovate, acute, crenate, slightly downy; glaucous beneath. Stipulas vaulted. Ovary stalked, awl-shaped, silky. Style as long as the blunt notched stigmas. *Smith.*

In woods, and on the banks of rivers. — Shrub, or small tree. *May.*

58. S. *sphacelata* Smith. E. B. 33. 2333.
Stem erect. Leaves elliptic-obovate, even, veiny, entire, or slightly serrated, downy on both sides, discoloured at the point. Stipulas half-heart-shaped, toothed, erect. Ovary stalked, ovate-lanceolate, silky. Stigmas notched, longer than the style. *Smith.*
S. lanata *Lightf.*

At Finlarig, near the head of Loch Tay. — Tree. *April, May.*

59. S. *caprea* Linn. E. B. 21. 1488.
Stem erect. Leaves roundish-ovate, pointed, serrated, waved, pale and downy beneath. Stipulas somewhat crescent-shaped. Catkins oval. Ovary stalked, ovate, silky. Stigmas nearly sessile, undivided. Fruit swelling. *Smith.*

In woods and hedges. — Tree. *April.*

60. S. *acuminata* Smith. E. B. 20. 1434.
Stem erect. Leaves lanceolate-oblong, pointed, wavy, finely toothed, glaucous and downy beneath. Stipulas half-ovate, then kidney-shaped. Catkins cylindrical. Ovary stalked, ovate, hairy. Style as long as the undivided stigmas. *Smith.*

In woods and hedges. — Tree. *April.*

61. S. *viminalis* Linn. Common Osier. E. B. 27. 1898.
Leaves linear, inclining to lanceolate, elongated, taper-pointed, entire, wavy; snow-white and silky beneath. Branches straight and slender. Ovary sessile. Style as long as the linear undivided stigmas. *Smith.*

In wet meadows. — Tree. *April, May.*

62. S. *Smithiana* Willd. E. B. 21. 1509.
Leaves lanceolate, pointed, slightly wavy, minutely toothed; soft and scarce visibly downy above, whitish and silky beneath. Stipulas

crescent-shaped, minute. Catkins ovate. Ovary stalked. Style shorter than the linear, deeply divided, stigmas. *Smith.*

S. mollissima *Smith.*

In meadows and osier grounds. — Shrub. *April, May.*

63. S. *stipularis* Smith. E. B. 17. 1214.

Leaves lanceolate, pointed, slightly wavy, obscurely crenate; soft and nearly naked above, white and downy beneath. Stipulas half-heart-shaped, stalked, very large. Gland cylindrical. Ovary ovate, nearly sessile, as well as the linear, undivided stigmas. *Smith.*

In osier-holts. — Shrub. *March.*

64. S. *alba* Linn. E. B. 34. 2430.

Leaves elliptic-lanceolate, pointed, serrated, silky on both sides; the lowest serratures glandular. Stamens hairy. Ovary smooth, almost sessile. Stigmas deeply cloven. Scales rounded. *Smith.*

S. cærulea *Smith.*

In moist woods. — Tree. *May;* and often again in *July.*

4. POPULUS *Linn.*

Diœcious. *Catkins* cylindrical with lacerated scales. *Male. Stamens* from 8 to 30, arising out of a little oblique cup. *Female. Fruit* a follicle, almost 2-celled by the rolling inwards of the margins of its two valves. *Seeds* comose. *Radicle* superior.

1. P. *alba* Linn. Abele-tree. E. B. 23. 1618.

Leaves lobed and toothed; somewhat heart-shaped at the base, snow-white and densely downy beneath. Fertile catkins ovate. Stigmas 4. *Smith.*

P. nivea *Willd.*

In woods. — Tree. *March.*

2. P. *canescens* Linn. White Poplar. E. B. 23. 1619.

Leaves roundish, deeply waved, toothed; hoary and downy beneath. Fertile catkins cylindrical. Stigmas 8. *Smith.*

In wet meadows, or on dry heaths. — Tree. *March.*

3. P. *tremula* Linn. Aspen. E. B. 27. 1909.

Leaves nearly orbicular, toothed, smooth on both sides. Footstalks compressed. Young branches hairy. Stigmas 4, erect, auricled at the base. *Smith.*

In woods. — Tree. *March, April.*

4. P. *nigra* Linn. Black Poplar. E. B. 27. 1910.

Leaves deltoid, pointed, serrated; smooth on both sides. Catkins all lax and cylindrical. Stigmas 4, simple, spreading. *Smith.*

β. *viridis.*

Branches green. Leaves broader.

In watery places; β. in Norfolk. — Tree. *March.*

Order 81. Cupuliferæ *Richard.*

Quercineæ *Juss.* Corylaceæ *Mirb.*

Flowers bisexual; males amentaceous; females aggregate or amentaceous.

Male. *Stamens* 5 to 20, inserted into the base of the scales, generally distinct.

Female. *Ovaries* crowned by the rudiments of a superior calyx, seated within a coriaceous involucrum (*cupule*) of various figure, with several cells and several ovules, the greater part of which are abortive; *ovules* twin or solitary, pendulous; *stigmata* several, subsessile, distinct.

Fruit a bony or coriaceous 1-celled nut, more or less enclosed in the involucrum.

Seeds solitary, 2 or 3, pendulous; *embryo* large, with plano-convex fleshy cotyledons, and a minute superior radicle.

Trees or shrubs. *Leaves* with stipulæ, alternate, simple, with veins proceeding straight from the midrib to the margin.

1. FAGUS *Linn.*

Monœcious. *Males.* *Catkins* pendulous, globose, dense. *Calyx* 6-lobed. *Stamens* 8. *Females* 2, inclosed in a spiny 4-lobed involucrum. *Stigmas* 3. *Ovarium* 3-cornered, 3-celled. *Nut* by abortion 1-celled, 1- or 2-seeded.

1. F. *sylvatica* Linn. Common Beech. E. B. 26. 1846.
Leaves ovate, obsoletely serrated. Prickles of the outer calyx simple. Stigmas 3. *Smith.*
In woods. — Tree. *April, May.*

2. CASTANEA *Gærtn.*

Polygamous. *Male.* *Catkins* very long, with irregular clusters of flowers. *Stamens* from 5 to 20. *Hermaphrodite.* *Involucrum* generally 3-flowered, 4-lobed, spiny. *Stamens* 12, abortive. *Ovarium* 6-celled, with 2 ovules in each cell. *Styles* 6. *Nut* 1-celled, with from 1 to 3 seeds.

1. C. *vesca* Gærtn. Sweet Chesnut. E. B. 13. 886.
Leaves oblong-lanceolate, acuminate, with mucronate serratures, smooth on each side.
Fagus Castanea *Linn.*
Castanea vulgaris *Dec.*
In woods. — Tree. *May.*

3. QUERCUS Linn.

Monœcious. *Male.* *Catkin* lax and pendulous. *Stamens* rom 5 to 10. *Female.* *Involucrum* cup-shaped, covered with scales.

Ovarium with 3 cells, 2 of which are abortive. *Stigmas* 3. *Acorn* 1-celled, 1-seeded, seated in the cup-shaped involucrum.

1. Q. *Robur* Linn. Common British Oak. E. B. 19. 1342.
Leaves deciduous, oblong, wider towards the extremity; their sinuses rather acute, lobes obtuse. Fruit-stalks elongated. *Smith.*
Q. pedunculata *Willd.*
Q. femina *Withering.*
In woods and hedges, everywhere. — Tree. *April.*

2. Q. *sessiliflora* Salisb. E. B. 26. 1845.
Leaves on elongated stalks, deciduous, oblong, with opposite, acute sinuses. Fruit sessile. *Smith.*
Q. sessilis *Ehr.*
Q. Robur *Willd.*
In woods, less common than the foregoing. — Tree. *April.*

4. CORYLUS *Linn.*

Monœcious. Male. Catkins cylindrical, with 3-lobed scales, the middle lobe of which covers the 2 lateral ones. *Stamens* 8. *Anthers* 1-celled. *Female. Flowers* numerous, enclosed in a scaly bud. *Stigmas* 2. *Nut* enclosed in a lacerated involucrum.

1. C. *Avellana* Linn. Common Hazel-nut. E. B. 11. 723.
Stipulas ovate, obtuse. Leaves roundish, heart-shaped, pointed. Young branches hairy. Calyx shorter than the nut. *Smith.*
In hedges and coppices, everywhere. — Tree. *March, April.*

5. CARPINUS *Linn.*

Monœcious. Male. Catkins long, cylindrical. *Scales* ciliated at the base. *Stamens* from 8 to 14, somewhat bearded at the apex. *Females. Cones* lax, membranous. *Involucrum* scale-shaped, 3-lobed, 2-flowered. *Ovarium* with 2 cells, of which 1 is abortive. *Stigmas* 2. *Nut* long.

1. C. *Betulus* Linn. Hornbeam. E. B. 29. 2032.
Bracteas of the fruit flat, oblong, serrated, with two lateral lobes. *Smith.*
In woods and hedges. — Tree. *May.*

Order 82. Coniferæ *Juss.*

Flowers monœcious or diœcious, amentaceous.
Males in deciduous, scarious catkins; *stamens* distinct, adhering to the scales, with 1- or many-celled *anthers.*
Females either in cones, or solitary, and surrounded by imbricated scales.
Ovules erect, naked, sometimes seated in an envelope, which in Taxus becomes succulent.

Nuts either solitary and naked, or enclosed within the hardened scales of a woody cone. *Embryo* straight in the axis of a fleshy albumen. *Cotyledons* subulate, from 2 to 10 in number; *radicle* superior.

Trees abounding in resin; their *wood* consisting of longitudinal cellular tissue, intermixed with woody fibre; the coats of the cellules covered with transparent spherules marked with a coloured central point.

Leaves generallv acerose and persistent, sometimes broad, very rarely deciduous.

1. PINUS *Linn.*

Monœcious. *Males. Catkins* with the scales each bearing 2 1-celled anthers at the ends. *Females. Catkins* with acuminate scales. *Ovaries* 2. *Cones* with oblong, clavate, woody scales, with an angular termination. — Leaves *two or more from the same sheath.*

1. P. *sylvestris* Linn. Scotch Fir. E. B. 35. 2460.
Leaves rigid, in pairs. Young cones stalked, recurved. Crest of the anthers very small. *Smith.*

In the Highlands of Scotland. — Tree. *May.*

Obs. There are several strongly-marked varieties, or perhaps species, confounded under this name. They were distinguished by the late Mr. George Don, and deserve re-examination by some Scottish botanist, who has opportunities of observing them on their native hills.

2. JUNIPERUS *Linn.*

Diœcious or monœcious. *Males. Catkins* ovate, with 4-8 1-celled anthers. *Females. Cone* round, consisting of 3 fleshy scales growing together and enclosing 3 bony nuts.

1. J. *communis* Linn. Common Juniper. E. B. 16. 1100.
Leaves 3 in each whorl, tipped with a spine, spreading, longer than the ripe fruit. Stem erect. *Smith.*

On hills and heathy downs, especially where the soil is chalky. — Shrub. *May.*

2. J. *nana* Willd.
Leaves 3 in each whorl, tipped with a spine, somewhat imbricated, curved, the length of the oval ripe fruit. Stem recumbent. *Smith.*

On mountains. — Shrub. *May.*

3. TAXUS *Linn.*

Flowers diœcious or monœcious, surrounded by scales. *Males. Stamens* 8 or 10, monadelphous. *Females. Nut* enclosed in a succulent cup.

1. T. *baccata* Linn. Common Yew. E. B. 11. 746.
Leaves linear, distichous. Fruit roundish.

In mountainous woods, and on the ledges of limestone cliffs. — Tree. *March, April.*

2. T. *fastigiata.* Irish Yew.
Leaves linear, scattered, crowded. Fruit oblong.

In Ireland, about the Giant's Causeway. — Tree. *March, April.*

Order 83. Myriceæ *Rich.*

Flowers bisexual, amentaceous.
Males. *Stamens* 1 or several, each with an hypogynous scale. *Anthers* 2-celled, opening lengthwise.
Females. *Ovarium* 1-celled, surrounded by several hypogynous scales; *ovulum* solitary, erect, with a foramen in its apex; *stigmas* 2, subulate.
Fruit drupaceous, covered with waxy secretions; formed of the hypogynous scales of the ovarium become fleshy and adherent.
Seed solitary, erect; *embryo* without albumen; *cotyledons* 2, planoconvex; *radicle* short, superior.
Shrubs with resinous glands and dots. *Leaves* alternate, simple.

1. MYRICA *Linn.*

Diœcious. *Catkins* ovate, with lunate scales. *Males.* *Stamens* 4-6. *Anthers* 4-valved. *Females.* *Ovarium* 1. *Stigmas* 2. *Drupe* 1-celled, 1-seeded.

1. M. *Gale* Linn. Sweet Gale, or Dutch Myrtle. E. B. 8. 562.
Leaves lanceolate, serrated, tapering and entire at the base. Scales of the catkins pointed. *Smith.*
In bogs. — Shrub. *May.*

Order 84. Callitrichineæ *Link.*

Flowers usually bisexual, monœcious, naked, with 2 fistular coloured bracteæ.
Stamen single; *filament* filiform, furrowed along the middle; *anther* reniform, 1-celled, 2-valved; the valves opening fore and aft.
Ovarium solitary, 4-cornered, 4-celled; *ovules* solitary, peltate; *styles* 2, right and left, subulate; *stigmas* simple points.
Fruit 4-celled, 4-seeded, indehiscent.
Seeds peltate; *embryo* inverted in the axis of fleshy *albumen*; *radicle* very long, curved, superior; *cotyledons* very short.
Small aquatic *herbaceous* plants, with opposite, simple, entire *leaves*. *Flowers* axillary, solitary, very minute.

N. B.—The affinity of this order to other Dicotyledones appears to be of precisely the same nature as that borne by Lemna to Monocotyledones. They each exhibit the lowest degree of organization known in their respective classes. I can by no means agree in the definition of the genus proposed by M. Decandolle; but, at the same time, I fully assent to Mr. Brown's opinion of its affinity with Halorageæ, although I do not place it in the same order.

---

1. CALLITRICHE *Linn.* WATER-STARWORT

The character that of the order, there being no other genus.

1. C. *verna* Linn. E. B. 11. 722.
Leaves triple-ribbed; the uppermost crowded, obovate. Margin of the fruit obtuse. *Smith.*

In ditches and slow streams, everywhere. — Annual. *April, May.*

2. C. *autumnalis* Linn. E. B. 11. 722
Leaves linear, abrupt, single-ribbed, uniform. Margin of the fruit membranous. *Smith.*

In clear pools and lakes. — Annual. *June—October.*

Subclass II. Monocotyledones.

Trunk cylindrical, formed of bundles of woody fibre, intermixed with cellular tissue; with no distinction of wood, bark, or pith, and destitute of medullary rays; increasing by the addition of new matter to the centre. Leaves generally sheathing at the base, and not articulated with the stem, always alternate, with parallel simple veins connected by others which traverse the space between them. Flowers usually with a ternary division of the floral envelopes; the calyx and corolla either distinct, or confounded together; in this case, the two taken together are called the perianthium. Embryo with only one cotyledon, or, if with two, then the accessory one is imperfect and alternate with the other; radicle enclosed within the substance of the embryo, through which it bursts when germinating.

DIVISIONS.

1. Petaloideæ. Flowers having a regular perianthium; or if destitute of one, naked.
2. Glumaceæ. Flowers destitute of a perianthium, and composed of imbricated alternate bracteæ.

Division I. Petaloideæ.

ANALYSIS OF THE ORDERS.

Ovarium superior
- Flowers solitary, or in spikes. (The inflorescence fully developed). Embryo undivided
  - Anthers turned inwards
    - Placentæ in the centre
      - Embryo on the outside the albumen 100. Restiaceæ.
      - Embryo in the inside the albumen
        - Perianthium dry, glumaceous - 101. Junceæ.
        - Perianthium coloured and petaloid
          - Testa black, brittle - - - 97. Asphodeleæ.
          - Testa membranous - - 98. Smilaceæ.
          - Testa spongy and dilated - - 96. Liliaceæ.
    - Placentæ parietal - - - - - - 99. Butomeæ.
  - Anthers turned outwards
    - Ovarium many-seeded - - - - 94. Melanthaceæ.
    - Ovarium 1- or 2-seeded
      - Sepals and calyx distinct - - - 90. Alismaceæ.
      - Sepals and calyx confounded - - 89. Juncagineæ.
- Flowers in a spadix, or solitary. (The inflorescence contracted). Embryo with a lateral slit for the emission of the plumula
  - Perianthium wanting
    - Flowers in a spadix. Stems leafy - - 85. Aroideæ.
    - Flowers solitary. Stems leafless, floating 88. Pistiaceæ.

Periantbium present
Embryo with albumen. Leaves rigid - - 86. TYPHACEÆ.
Embryo without albumen. Leaves membranous - - - - - - 87. FLUVIALES.
Ovarium inferior
Anthers turned outwards - - - - - - 92. IRIDEÆ.
Anthers turned inwards
Stamens united in a column - - - - - 93. ORCHIDEÆ.
Stamens distinct
Calyx and petals distinct. Seed without albumen - - - - - - 91. HYDROCHARIDEÆ.
Calyx and petals confounded. Seed with albumen - - - - - - 95. AMARYLLIDEÆ.

Order 85. AROIDEÆ *Juss.*

Flowers monœcious, arranged upon a spadix, occasionally surrounded by a few scales.

Perianthium wanting.

Males. *Stamens* definite or indefinite, hypogynous; *anthers* very short, 1- or 2-celled, ovate, turned outwards.

Females. *Ovarium* superior, 1-celled, very seldom 3-celled, and many-seeded; *ovules* erect, or pendulous, or parietal; *stigmas* sessile, as many as the cells.

Fruit succulent or dry, not opening.

Seeds solitary or numerous; *embryo* in the axis of fleshy or mealy albumen, straight, taper, with a cleft in one side, in which the plumula lies; *radicle* obtuse, usually next the hilum; occasionally at the opposite extremity.

Herbaceous plants or *shrubs*, stemless, or arborescent, or climbing by means of aerial roots. *Leaves* either with parallel or branching veins; often cordate. *Spadix* generally enclosed in a *spathe*.

1. ARUM *Linn.*

Spadix naked at the apex, enclosed in a *spathe*. *Flowers* naked, the males crowded about the middle of the spadix; the females seated at the base. *Berry* 1-celled, many-seeded.

1. A. *maculatum* Linn. Cuckow-pint, or Wake Robin.
E. B. 19. 1298.

Stem none. Leaves halberd-shaped, entire. Common stalk of the flowers club-shaped, obtuse. *Smith.*

In groves and hedge banks. — Perennial. *May.*

2. ACORUS *Linn.*

Spathe wanting. *Perianthium* persistent, globose, 6-parted. *Stamens* 6, opposite the divisions of the perianthium. *Ovary* globose, 3-celled, many-seeded. *Stigma* sessile. *Fruit* dry.

1. A. *Calamus* Linn. Sweet Flag. E. B. 5. 356.

Leafy summit of the flower-stalk rising high above the spadix. *Smith.*

In watery places. — Perennial. *June.*

Order 86. Typhaceæ *Juss.*

Flowers monœcious, arranged upon a naked spadix.
Sepals 3, or more.
Petals wanting.
Males. *Stamens* 3 or 6; *anthers* wedge-shaped, attached by their base to long filaments.
Females. *Ovary* single, superior, 1-celled; *ovulum* solitary, pendulous; *style* short; *stigmas* 1 or 2, simple, linear.
Fruit dry, not opening, 1-celled, 1-seeded.
Embryo in the centre of *albumen*, straight, taper, with a cleft in one side, in which the plumula lies; *radicle* next the hilum.
Herbaceous plants, growing in marshes or ditches. *Leaves* rigid, ensiform, with parallel veins. *Spadix* without a spathe.

1. TYPHA *Linn.*

Spikes cylindrical. *Males.* *Sepals* 3, imperfect. *Stamens* 3, united at the base into one. *Females.* *Sepals* several, filiform, surrounding the stalk of the fruit.

1. T. *latifolia* Linn. Bulrush, Cat's-tail, or Reed-mace. E. B. 21. 1455.
Leaves somewhat convex beneath. Catkin continuous. Receptacle hairy. *Smith.*
Typha major *Curtis.*
In ponds, ditches, and slow streams. — Perennial. *July.*

2. T. *angustifolia* Linn. E. B. 21. 1456.
Leaves slightly semicylindrical; channelled above. Barren catkin separated from the fertile one. Receptacles scaly. *Smith.*
Typha minor *Curtis.*
In pools and ditches. — Perennial. *June, July.*

3. T. *minor* Smith. E. B. 21. 1457.
Leaves linear, convex beneath. Catkins a little distant; barren one leafy; fertile short and turgid; often interrupted. Anthers nearly solitary. Receptacle naked. *Smith.*
In marshes, but rare. — Perennial. *July.*

2. SPARGANIUM *Linn.* Bur-reed.

Spikes round. *Sepals* 3. *Stamens* 6; *anthers* wedge-shaped. *Fruit* sessile, turbinate, without bristles at the base.

1. S. *ramosum* Hudson. E. B. 11. 744.
Leaves triangular at the base, with concave sides. Common flower-stalk branched. Stigmas linear. *Smith.*
S. erectum *Linn.*
In ditches, and the margins of ponds and rivers. — Perennial. *July, August.*

2. S. *simplex* Hudson. E. B. 11. 745.
Leaves triangular at the base, with flat sides. Common flower-stalk simple. Stigma linear. *Smith.*

S. superaxillare *Ehr.*

In pools and ditches. — Perennial. *July, August.*

3. S. *natans* Linn. E. B. 4. 273.

Leaves floating, flat; concave at the base. Common flower-stalk simple. Stigma ovate, very short. Head of barren flowers mostly solitary. *Smith.*

In muddy fens, or slow rivers. — Perennial. *July.*

Order 87. Fluviales *Ventenat.*

Potameæ *Juss.*

Flowers hermaphrodite or bisexual.

Perianthium of 2 or 4 pieces, often deciduous, rarely wanting.

Stamens definite, hypogynous.

Ovary 1 or more, superior; *stigma* simple; *ovule* solitary, pendulous.

Fruit dry, not opening, 1-celled, 1-seeded.

Seed pendulous; *albumen* none; *embryo* having a direction contrary to that of the seed, with a lateral cleft for the emission of the plumula.

Water-plants. Leaves very vascular, with parallel veins. *Flowers* inconspicuous, usually arranged in terminal spikes.

1. POTAMOGETON *Linn.*

Sepals 2. *Petals* 2. *Stamens* 4, opposite the sepals and petals; anthers nearly sessile. *Ovaries* 4, alternate with the stamens; *ovules* solitary, suspended. *Nuts* 4, compressed. *Seed* suspended, arcuate, more or less spiral. *Albumen* none. *Plumula* dorsal. — Floating *plants*, with pellucid leaves.

Tribe i. *Leaves opposite.*

1. P. *densus* Linn. E. B. 6. 397.

Leaves opposite. *Schlecht.*

P. oppositifolius *Dec.*

P. setaceus *Lam.*

P. serratus *Linn.*

P. pauciflorus *Lam.*

In ditches. — Perennial. *June*

Tribe ii. *Leaves sheathing; their stipules adnate to the petiole.*

2. P. *pectinatus** Linn. E. B. 5. 323.

Leaves sheathing, 1-ribbed. Nuts very large, keeled at the back. *Schlecht.*

---

* The following species is not yet recorded as British, but is probably to be found in these islands:

P. *filiformis* Persoon.

Leaves sheathing, 1-ribbed. Nut small, not keeled at the back. *Schlecht.*

P. marinum *Linn.*
P. interruptus *Kitaib.*
P. Vaillantii *R. & S.*
P. tenuifolium *Humb. & Bonpl.*
In rivers and ponds. — Perennial. *July.*

Tribe iii. *Leaves all linear.*

3. P. *pusillus* * Linn. E. B. 3. 215. & 6. 418.
Leaves 3- 5-ribbed, with a few obsolete veins. Spikes somewhat interrupted, on long stalks. *Schlecht.*
P. compressus *Linn.*
P. acutifolium *Presl.*
P. gramineum *Merat.*
P. denticulatum *Link.*
In ditches. — Perennial. *July.*

4. P. *obtusifolius* † Mertens & Koch. E. B. 32. 2253.
Leaves 3-ribbed, blunt, with a few obsolete veins. Spikes ovate, on short stalks. *Schlecht.*
P. gramineum *Smith.*
In ponds and ditches. — Perennial. *July.*

5. P. *zosterifolius* Schumacher.
Leaves many-ribbed, acuminate. Spikes cylindrical, on long stalks. *Schlecht.*
P. cuspidatum *Schrad.*
P. complanatus *Willd.*
P. compressum *Teesdale in L. Trans.*
In ponds. — Perennial. *July.*

Tribe iv. *Leaves dilated, all under water.*

6. P. *crispus* Linn. E. B. 15. 1012.
Leaves linear-oblong, 3-ribbed, evidently serrulated. Nuts with long beaks. *Schlecht.*
P. serratus *Lam.*
In ditches. — Perennial. *June, July.*

7. P. *perfoliatus* ‡ Linn. E. B. 3. 168.
Leaves ovate, ribbed, amplexicaul; scabrous at the margin, as long as the peduncle. *Schlecht.*
P. Löselii *R. & S.*
In ponds. — Perennial. *July, August.*

---

* The following species, distinguished from this by Schlechtendahl, is probably British:
P. *trichoides* Schlecht.
Leaves setaceous, 1-ribbed, veinless. Spikes somewhat interrupted, on long stalks. *Schlecht.*

† With this has been confounded the following, which may be British:
P. *acutifolius* Link.
Leaves many-ribbed, acute. Spike ovate, on short stalks. *Schlecht.*

‡ Search should be made for
P. *prælongus* Wulff.
Leaves oblong, ribbed, half-amplexicaul; smooth at the margin; navicular at the apex. *Schlecht.*
Found in very deep waters, with stems more than 8 feet long.

8. P. *lanceolatus Smith.* E. B. 28. 1985.

Leaves lanceolate, membranous, flat, entire; contracted at the base; with chain-like reticulations near the ribs. Spikes ovate, dense, of few flowers. *Smith.*

In Anglesey and Scotland. — Perennial. *July, August.*

Tribe v. *Floating leaves coriaceous, accessory.*

9. P. *Proteus* Schlecht.

Spikes cylindrical, on long stalks. Peduncle thickened. *Schlecht.*

α. lucens Schl. E. B. 6. 376.

Leaves lanceolate or oval, very large, distinctly mucronate, subsessile, all submersed. *Schl.*

P. lucens *Linn.*
P. acuminatum *Schum.*

β. heterophyllus Schl. E. B. 18. 1285.

Leaves small; the lower lanceolate, sessile, acute; the upper often on long stalks, floating, coriaceous. *Schl.*

P. heterophyllum *Schreb.*
P. hybridum *Petagn.*
P. augustanum *Balb.*
P. distachyum *Bellard.*

In pools. — Perennial. *June — September.*

10. P. *rufescens* Schrader. E. B. 18. 1286.

Leaves tapering to the base, obtuse at the end; those which are under water thin, with thin ribs; those above water generally floating and stalked, as long as the peduncles. *Schlecht.*

P. fluitans *Smith.*
P. obscurum *Dec.*
P. annulatum *Bellard.*
P. alpinum *Balb.*

In ponds. — Perennial. *July, August.*

Tribe vi. *Leaves floating, coriaceous, all necessary.*

11. P. *oblongus* Viviani.

Lower leaves floating, or all swimming (without leafless petioles) Nuts small, blunt at the back. *Schlecht.*

P. Plantago *Batard.*
P. polygonifolius *Pour.*
P. parnassifolius *Schrad.*
P. uliginosum *Bonnigh.*
P. affine *Ib.*

In ditches. — Perennial. *July.*

12. P. *natans* Linn. E. B. 26. 1822.

Lower petioles leafless, elongated. Nuts large, keeled at the back. *Schlecht.*

In ditches. — Perennial. *July.*

Also for

P. *coloratus* Hornemann.

Leaves stalked; all membranous and transparent. Spikes on long stalks, cylindrical, many-flowered. Nuts very small. *Schlecht*

Common in the north of France.

2. ZOSTERA *Linn.*

Flowers monœcious or diœcious, arranged in a unilateral manner within the base of a leaf. *Style* bifid. *Nuts* 1-seeded.

1. Z. *marina* Linn. Grass-wrack. E. B. 7. 467.
Leaves entire, obscurely 3-ribbed. Stem slightly compressed. *Smith*.
In creeks and ditches of salt water. — Perennial. *August, September.*

3. RUPPIA *Linn.*

Flowers hermaphrodite, distichous, arranged upon a solitary spadix. *Sepals* 2, deciduous. *Anthers* 4, sessile, reniform, 1-celled. *Ovaries* 4, acquiring pedicels as they approach maturity. *Fruit* dry, ovate, 1-seeded crowned by the persistent stigma.

1. R. *maritima* Linn. E. B. 2. 136.
In salt-water ditches. — Perennial ? *August, September.*

4. ZANNICHELLIA *Linn.*

Flowers solitary, monœcious. *Males.* *Stamen* single, naked, placed at the base of the female flower on the outside. *Female.* *Perianthium* campanulate. *Ovaries* 2-6. *Fruit* dry, 1-seeded, sessile, compressed, gibbous, crenated outwardly.

1. Z. *palustris* Linn. Horned-pondweed. E. B. 26. 1844.
Anther of 4 cells. Stigmas entire. *Smith*
In ponds and ditches. — Annual. *July.*

Order 88. PISTIACEÆ *Richard.* — *Lindley in Hooker's Flora Scot* 2. 191.

Flowers 2, naked, enclosed in a spatha.
Male. *Stamens* definite.
Female. *Ovarium* 1-celled, with 1 or more erect *ovules*; *style* short; *stigma* simple.
Fruit membranous or capsular, not opening, 1 or more seeded.
Seeds with a striated testa, marked with a distinct raphe and chalaza; *embryo* in the axis of fleshy albumen, with a contrary direction to that of the seed, having a lateral cleft for the emission of the *plumule*.
Floating plants, with very vascular, lenticular, or lobed *stems*. *Leaves* none. *Flowers* appearing from the margin of the stems.

1. LEMNA *Linn.* DUCK-WEED.

Spatha membranous, inflated. *Stamens* 1 or 2. *Utricle* 1-celled, — Floating minute plants, with lenticular stems, and no leaves.

1. L. *trisulca* Linn. E. B. 13. 926.
Leaves stalked, elliptic-lanceolate, proliferous.
In clear still waters. — Annual. *June.*

2. L. *minor* Linn. E. B. 16. 1095.
Leaves obovate, flattish above and beneath. Roots solitary.
In ponds and ditches. — Annual. *June, July.*

3. L. *gibba* Linn. E. B. 18. 1233.
Leaves obovate; slightly convex above; hemispherical beneath. Roots solitary.
In ponds and ditches. — Annual. *June, July.*

4. L. *polyrrhiza* Linn. E. B. 35. 2458.
Leaves roundish-obovate; convex beneath. Roots clustered.
In ditches and standing pools. — Annual. *Flower not observed in Britain.*

Order 89. JUNCAGINEÆ *Richard.*

Sepals and *petals* both herbaceous.
Stamens 6.
Ovaries 3 or 6, superior, cohering firmly; *ovules* 1 or 2, approximated at their base, erect.
Fruit dry, indehiscent, 1 or 2-seeded.
Seeds erect; *albumen* wanting; *embryo* having the same direction as the seed, with a lateral cleft for the emission of the plumule.
Herbaceous bog-plants. *Leaves* ensiform, with parallel veins. *Flowers* in naked spikes, inconspicuous.

1. TRIGLOCHIN *Linn.*

Sepals and *petals* deciduous, herbaceous. *Anthers* 6, nearly sessile. *Fruit* consisting of from 3 to 6 cohering pieces, each of which is 1-seeded.

1. T. *palustre* Linn. Arrow-grass. E. B. 6. 366.
Capsule nearly linear, of 3 cells; tapering at the base. Root fibrous. *Smith.*
In wet boggy meadows. — Perennial. *June, July.*

2. T. *maritimum* Linn. E. B. 4. 255.
Capsule ovate, of 6 cells. *Smith.*
In salt marshes. — Perennial. *May—August.*

2. SCHEUCHZERIA *Linn.*

Sepals and *petals* permanent, brown. *Anthers* 6, long, upon capillary filaments. *Fruit* consisting of 3, roundish, spreading pieces, each of which has 2 valves, and contains 1 or 2 seeds.

1. S. *palustris* Linn. E. B. 26. 1801.
In Lakeby Car, near Boroughbridge in Yorkshire. — Perennial. *June.*

Order 90. ALISMACEÆ *Juss.*

Sepals 3, herbaceous.
Petals 3, petaloid.
Stamens definite or indefinite.
Ovaries superior, several, 1-celled; *ovules* solitary, or 2, attached to the suture, at a distance from each other.
Styles and *stigmas* the same number as the ovaries.
Fruit dry, not opening, 1- or 2-seeded.
Seeds without albumen; *embryo* shaped like a horse-shoe, undivided with the same direction as the seed.
Floating plants. *Leaves* with parallel veins.

1. ALISMA *Linn.* WATER-PLANTAIN.

Stamens 6. *Ovaries* from 6 to 25. *Nuts* distinct, generally 1-seeded, deciduous, indehiscent.

1. A. *Plantago* Linn. E. B. 12. 837.
Leaves ovate, acute. Capsules obtusely triangular. *Smith.*
Alisma lanceolata *With.*
In pools and ditches. — Perennial. *July.*

2. A. *Damasonium* Linn. E. B. 23. 1615.
Leaves oblong; heart-shaped at the base. Styles 6. Capsules tapering. *Smith.*
In ditches and pools. — Perennial. *June, July.*

3. A. *natans* Linn. E. B. 11. 775.
Leaves elliptical, obtuse. Flower-stalks simple. Capsules striated. *Smith.*
In the lakes of North Wales and Cumberland. — Perennial. *July, August.*

4. A. *ranunculoides* Linn. E. B. 5. 326.
Leaves linear-lanceolate. Capsules angular, acute, numerous, in a globular head. Stem none. *Smith.*
In swamps and turfy bogs. — Perennial. *August.*

5. A. *repens* Cav.
Leaves lanceolate. Capsules compressed, acute, numerous, in a globular head. Stems prostrate, creeping *Smith.*
On the margins of lakes in North Wales. — Perennial. *September, October.*

2. SAGITTARIA *Linn.*

Monœcious. *Male.* *Stamens* about 24. *Female.* *Ovaries* numerous, seated upon a globose receptacle. *Nuts* compressed, bordered, 1-seeded.

S. *sagittifolia* Linn. Arrow-head. E. B. 2. 84.
Leaves arrow-shaped, acute. *Smith.*
In ditches and rivers. — Perennial. *July, August.*

Order 91. HYDROCHARIDEÆ *Juss.*

Flowers hermaphrodite or diœcious.
Sepals 3, herbaceous.
Petals 3, petaloid.
Stamens definite or indefinite
Ovary single, inferior; *stigmas* several; *ovules* indefinite, parietal.
Fruit dry or succulent, indehiscent, with 1 or more cells.
Seeds without albumen; *embryo* undivided.
Floating plants. *Leaves* with parallel or branched veins

1. STRATIOTES *Linn.*

Spathe compressed, persistent, deeply parted in two, keeled, 1-flowered. *Perianthium* tubular; *sepals* 3, minute, green; *petals* 3, large. *Stamens* about 20, inserted into the apex of the tube, or margin of the ovarium. *Styles* 6, bifid. *Fruit* fleshy, taper-pointed, 6-cornered, 6-celled. *Seeds* somewhat angular, attached to the dissepiments.

1. S. *aloides* Linn. Water Aloe, or Water-soldier. E. B. 6. 379. Leaves sword-shaped, channelled, with a prominent rib, and sharp marginal prickles. *Smith.*

In deep fen ditches and pools. — Perennial. *July.*

2. HYDROCHARIS *Linn.*

Diœcious. *Male. Spathe* 2-parted, 3-flowered. *Perianthium* of 6 pieces. *Stamens* 12, or by abortion 9, placed in a triple order upon the rudiments of an abortive ovary. *Female. Spathe* sessile, 1-flowered. *Perianthium* like that of the male, with 6 filiform abortive stamens. *Stigmas* 6, wedge-shaped, bifid. *Fruit* leathery, 6-celled, many-seeded.

1. H. *Morsus ranæ* Linn. Frog-bit. E. B. 12. 808.

In ditches and streams. — Perennial. *July.*

Order 92. IRIDEÆ *Juss.*

Perianthium superior, petaloid, in six parts, sometimes irregular, deciduous; the 3 petals occasionally abortive.
Stamens 3, inserted into the sepals; *filaments* distinct or connate; *anthers* turned outwards.
Ovarium 3-celled, many-seeded; *style* 1 or 3, united at the base and petaloid; *stigmas* either simple or 3-lobed.
Capsule 3-celled, 3-valved, with a loculicidal dehiscence.
Seeds attached to the axis of the fruit; *albumen* horny or densely fleshy; *embryo* included, undivided.

Herbaceous plants, very seldom undershrubs. *Roots* tuberous or fibrous. *Leaves* equitant, distichous. *Bracteæ* usually spathaceous. *Flowers* brightly coloured.

1. IRIS *Linn.*

Perianthium 6-parted; the *sepals* larger and spreading, the petals smaller and erect. *Stamens* distinct, opposite the sepals. *Styles* 3, very large, petaloid, opposite the sepals, and incumbent upon the stamens.

1. I. *Pseud-acorus* Linn. E. B. 9. 578.
Corolla beardless; inner segments smaller than the stigmas. Leaves sword-shaped. Seeds angular. *Smith.*
In ditches and rivers. — Perennial. *July.*

2. I. *fœtidissima* Linn. Gladwyn. Roast-beef plant. E. B. 9. 596.
Corolla beardless; inner segments spreading. Stem with 1 angle. Leaves sword-shaped. Seeds globose.
In groves, thickets, and under hedges. — Perennial. *May.*

2. TRICHONEMA *Ker.*

Perianthium in 6 deep equal segments; longer than the tube. *Filaments* downy. *Stigmas* very slender, deeply cloven.

1. T. *Bulbocodium* Ker. E. B. 36. 2549.
Leaves linear, channelled, recurved, longer than the flower-stalks. *Smith.*
Ixia Bulbocodium *Linn.*
On grassy hillocks in Guernsey. — Perennial. *March, April.*

3. CROCUS *Linn.*

Perianthium with a slender tube twice as long as the limb. *Limb* 6-parted, equal, inflated, erect. *Stigmas* 3, convolute, many-lobed.

1. C. *sativus* Linn. Saffron Crocus. E. B. 5. 343.
Stigma prominent laterally, in 3 deep, linear, notched segments. *Smith.*
C. officinalis *Hudson.*
C. autumnalis *E. Bot.*
In meadows and pastures. — Perennial. *September.*

2. C. *vernus* Willd. E. B. 5. 344.
Stigma within the flower, in 3 short wedge-shaped jagged lobes. Tube hairy at the mouth. *Smith.*
In meadows about Nottingham. — Perennial. *March.*

3. C. *nudiflorus* Smith. E. B. 7. 491.
Stigma within the flower, in 3 deeply-laciniated tufted segments. Flower unaccompanied by leaves. *Smith.*
In sandy meadows between Nottingham castle and the Trent. — Perennial. *October.*

N. B. — C. reticulatus, admitted as a British plant by Sir James Smith, is not more wild in Sir Henry Bunbury's park, than C. luteus, which is found in the same place.

Order 93. ORCHIDEÆ *Juss.*

Perianthium superior, ringent.
Sepals 3, usually coloured, of which the odd one is uppermost in consequence of a twisting of the ovarium.
Petals 3, usually coloured, of which 2 are uppermost in consequence of a twisting of the ovarium, and 1, called the *lip*, undermost; this latter is frequently lobed, always of a different form from the others, and very often spurred at the base.
Stamens 3, united in a central column, the 2 lateral usually abortive, the central perfect, or the central abortive, and the 2 lateral perfect; *anther* either persistent or deciduous, 2- or 4- or 8-celled; *pollen*, either powdery or cohering in definite or indefinite waxy masses, either adhering to a gland or loose in their cells.
Ovarium 1-celled, with 3 parietal placentæ; *style* forming part of the column of the stamens; *stigma*, a viscid space in front of the column, communicating directly with the ovarium by a distinct open canal. *Impregnation* taking effect by absorption from the pollen masses through their gland into the stigmatic canal.
Capsule inferior, bursting with 3 valves and 3 ribs.
Seeds parietal, very numerous; *testa* loose, reticulated, contracted at each end; *albumen* none; *embryo*, a solid undivided fleshy mass.

ANALYSIS OF THE GENERA.

Pollen simple, or consisting of granules in a slight state of cohesion
Anther parallel with the stigma - - - - i. NEOTTIEÆ.
Anther terminal, like a lid - - - - - - - ii. ARETHUSEÆ.
Pollen cohering in grains or masses, which are indefinite in number, and waxy - - - - - } iii. OPHRYDEÆ.
Pollen cohering in grains or masses, which are definite in number, and waxy - - - - - - } iv. MALAXIDEÆ.
Lateral anthers fertile; intermediate sterile and petaloid v. CYPRIPEDIEÆ.

NEOTTIEÆ
Lip entire
saccate - - - - - - - - - - 1. GOODYERA.
shovel-shaped - - - - - - - - - 2. SPIRANTHES.
Lip lobed
Anther naked - - - - - - - - 3. NEOTTIA.
Anther enclosed in a hood - - - - - 4. LISTERA.

ARETHUSEÆ
Lip spurred - - - - - - - - - - 5. CORALLORHIZA.
Lip not spurred - - - - - - - - - 6. EPIPACTIS.

OPHRYDEÆ
Lip spurred
Glands of the pollen masses enclosed in a pouch
two glands - - - - - - - - - 7. ORCHIS.
one gland - - - - - - - - - 8. ANACAMPTIS.
Glands of the pollen masses naked
Lobes of the anther parallel - - - 9. GYMNADENIA.
Lobes of the anther diverging - - - 10. PLATANTHERA.

Lip without a spur
Glands of the pollen masses enclosed in a pouch
Pouch single - - - - - - - 11. ACERAS.
Pouches 2 - - - - - - - 12. OPHRYS.
Glands of the pollen masses naked - - - 13. HERMINIUM.

MALAXIDEÆ
Lip uppermost. Pollen masses 2 - - - - 14. MALAXIS.
Lip undermost. Pollen masses 4 - - - - 15. LIPARIS.

CYPRIPEDIEÆ - - - - - - - - - - - 16. CYPRIPEDIUM

Tribe i. *Neottieæ* Lindley.

1. GOODYERA *R. Br.*

Sepals spreading, ovate, herbaceous. *Petals* erect; *lip* saccate, entire. *Column* taper, distinct, with 2 teeth at the apex. *Stigma* prominent, roundish.

1. G. *repens* R. Br. E. B. 5. 289.
Leaves ovate. Spike spiral. Point of the lip elongated, deflexed. *Smith.*
Neottia repens *Swartz.*
Satyrium repens *Linn.*
In mossy woods in Scotland. — Perennial. *July.*

2. SPIRANTHES *Rich.*

Sepals coloured, and *petals* converging, parallel with the lip; *lip* shovel-shaped, unguiculate, with two fleshy projections at the base. *Column* taper, club-shaped, distinct, with 2 teeth at the apex. *Stigma* prominent, rostrate.

1. S. *autumnalis* Rich. Ladies' Traces. E. B. 8. 541.
Leaves ovate, stalked. Spike twisted, unilateral. Bracteas downy, tumid. Lip ovate, entire. *Smith.*
Neottia spiralis *Swartz.*
Ophrys spiralis *Linn.*
In open pastures, on a chalky or gravelly soil. — Perennial. *August, September.*

2. S. *gemmipara.*
Leaves lanceolate, as tall as the stalk. Spike 3-ranked, twisted. Bracteas smooth. *Smith.*
Neottia gemmipara *Smith.*
Near Castletown, opposite to Bearhaven on the northern side of Bantry Bay, county of Cork, in small quantities. — Perennial. *July.*

3. NEOTTIA *Linn. act. ups.* 1740.

Sepals and *petals* brown, converging; *lip* dependent, 2-lobed, concave at the base. *Column* taper, erect, 4 times as long as the stigma. *Stigma* distinctly 2-lipped; the upper lip narrower than the lower. *Anther* naked. — A leafless brown parasite, with succulent clustered roots.

1. N. *Nidus avis.* E. B. 1. 48.
Leaves none. Stem clothed with sheathing scales. Lip with 2 spreading lobes.
Epipactis nidus avis *Swartz.*
Ophrys nidus avis *Linn.*

In shady woods, especially beech, on a chalky or loamy soil. — Perennial. *May, June.*

4. LISTERA *R. B*

Sepals and *petals* herbaceous, spreading; *lip* dependent, 2-lobed. *Column* taper, erect, much shorter than the stigma. *Stigma* plane, ovate. *Anther* covered by a hood proceeding from the back of the column. — Leafy herbaceous plants, with fascicled fibrous roots.

1. L. *ovata* R. Br. Twayblade. E. B. 22. 1548.
Leaves elliptical, opposite. Lip with 2 linear-oblong, nearly parallel, lobes.
Ophrys ovata *Linn.*

In groves and thickets. — Perennial. *June.*

2. L. *cordata* R. Br. E. B. 5. 358.
Leaves heart-shaped, opposite. Lip with 4 lobes.
Ophrys cordata *Linn.*

On turfy mountainous moors in the north. — Perennial. *July.*

Tribe ii. *Arethuseæ* Lindley.

5. CORALLORHIZA *Haller.*

Sepals and *petals* more or less coloured, spreading; the lower sepals cohering at the base; *lip* more or less lobed, producing from the base a spur, which is more or less adherent to the ovary. *Column* plano-convex, entire. *Pollen masses* spherical. *Anther* round, 2-celled, with 2 anterior valves.

1. C. *innata* R. Br. Coral-root. E. B. 22. 1547.
Spur short, not distinct from the slightly 3-lobed lip. Root copiously branched. *Smith.*
Cymbidium corallorhizum *Swartz.*
Ophrys corallorhiza *Linn.*

In marshy umbrageous woods in Scotland. — Perennial. *May, June.*

6. EPIPACTIS *Swartz.* HELLEBORINE.

Sepals and *petals* spreading, or converging, more or less coloured; *lip* inflated at the base, either entire or with 3 lobes, of which the middle one is articulated with the others. *Column* plano-convex, with 2 teeth at the apex. *Pollen masses* acuminate. *Anther* 2-celled.

1. E. *latifolia* Swartz. E. B. 4. 269.
Leaves ovate, clasping the stem. Lower bracteas longer than the

drooping flowers. Lip shorter than the sepals, entire, with a minute point. Ovary downy.
Serapias latifolia *Linn.*

In shady mountainous woods and thickets. — Perennial. *July, August.*

2. E. *purpurata* Smith.
Leaves ovate-lanceolate. Bracteas linear, all twice as long as the flowers. Lip shorter than the sepals, entire. Ovary downy. *Smith.*

Parasitical on the stump of a maple or hazel, in a wood near the Noris farm, at Leigh, Worcestershire. — Perennial. *June.*

3. E. *palustris* Swartz. E. B. 4. 270.
Leaves lanceolate, clasping the stem. Flowers drooping. Lip rounded, obtuse, crenate, as long as the petals, with a notched protuberance on the disk. *Smith.*
Serapias palustris *Scop.*

In watery places or swampy meadows. — Perennial. *July, August.*

4. E. *grandiflora* Smith. E. B. 4. 271.
Leaves elliptic-lanceolate. Bracteas longer than the smooth ovary. Flowers sessile, erect. Lip abrupt, shorter than the sepals, with elevated lines on the disk.
Epipactis pallens *Swartz.*
Serapias grandiflora *Linn.*
S. longifolia *Hudson.*
S. lancifolia *Murray.*
In woods and thickets. — Perennial. *June.*

5. E. *ensifolia* Swartz. E. B. 7. 494.
Leaves lanceolate, pointed. Bracteas minute, much shorter than the smooth ovary. Flowers sessile, erect. Lip abrupt, half as long as the sepal, with elevated lines on the disk.
E. xiphophylla *Swartz.*
Serapias ensifolia *Murray.*
S. grandiflora *Fl. Dan.*

In mountainous woods, but rarely. — Perennial. *May, June.*

6. E. *rubra* Swartz. E. B. 7. 437.
Leaves lanceolate. Bracteas longer than the downy ovary. Flowers sessile, erect. Lip tapering to a point, with elevated undulating lines on the disk. *Smith.*
Serapias rubra *Linn.*

In stony mountainous woods. — Perennial *June, July.*

Tribe iii. *Ophrydeæ* Lindley.

7. ORCHIS *Linn.*

Sepals and *petals* ringent, coloured; *lip* lobed, spurred at the base. *Pollen masses* with 2 glands, enclosed in a common pouch.

* *Knobs of the root roundish, undivided.*

1. O. *Morio* Linn. E. B. 29. 2059.
Knobs of the root oval. Lip 4-cleft, somewhat crenate; spur obtuse, ascending. Sepals many-ribbed, converging.
In meadows and pastures. — Perennial. *May, June.*

2. O. *mascula* Linn. E. B. 9. 631.
Knobs of the root oval. Lip 4-cleft, crenate; spur obtuse. Sepals 3-ribbed; two lateral ones reflexed upwards.
In pastures. — Perennial. *April, May.*

3. O. *ustulata* Linn. E. B. 1. 18.
Knobs of the root oval. Lip 4-lobed, rough with small points. Spur obtuse, not half the length of the ovary. Sepals converging. Leaves lanceolate.
On dry, open, chalky downs. — Perennial. *June.*

4. O. *fusca* Jacq. E. B. 1. 16.
Knobs of the root oval. Lip 5-lobed, dilated, rough. Spur obtuse, not half the length of the ovary. Sepals converging, blunt-pointed. Leaves elliptic-oblong.
O. purpurea *Hudson.*
O. militaris *E. Bot.*
O. moravica *Jacq.*
On chalky bushy hills, chiefly in Kent. — Perennial. *May.*

5. O. *militaris* Linn. Military Orchis.
Knobs of the root oval. Lip 5-lobed, downy; 2 middle lobes dilated, rounded. Spur obtuse, not half the length of the ovary. Sepals converging, taper-pointed.
On chalky hills. — Perennial. *May.*

6. O. *tephrosanthos* Villars. Monkey Orchis. E. B. 27. 1873.
Knobs of the root oval. Lip downy, in 5 lobes; 4 of them equal, linear, entire. Spur obtuse, not half the length of the ovary. Sepals converging, taper-pointed.
On chalky hills. — Perennial. *May.*

7. O. *hircina* Scopoli. Lizard Orchis. E. B. 1. 24.
Knobs of the root globose. Lip downy, in 3 linear segments; the middle one very long, twisted, notched at the end. Sepals converging.
Satyrium hircinum *Linn.*
In pastures and bushy places. — Perennial. *July.*

** *Knobs of the root palmate.*

8. O. *latifolia* Linn. E. B. 33. 2308.
Knobs imperfectly palmate. Lip convex, crenate, slightly 3-cleft; spur conical. Bracteas longer than the flowers. Stem hollow. *Smith.*
In marshes and moist meadows. — Perennial. *May, June.*

9. O. *maculata* Linn. E. B. 9. 632.
Knobs palmate, spreading. Lip flat, crenate, 3-lobed; spur cylin-

drical, rather shorter than the ovary. Bracteas shorter than the flowers.

In meadows and woods. — Perennial. *June, July.*

8. ANACAMPTIS *Rich.*

Sepals and *petals* ringent, coloured; *lip* lobed, with 2 projecting plates along its middle, spurred at the base. *Pollen masses* with 1 gland, enclosed in a pouch.

1. A. *pyramidalis* Rich. E. B. 2. 110.
Knobs of the root oval. Lip in 3 equal entire lobes, with 2 protuberances above; spur long and slender. *Smith.*
Orchis pyramidalis *Linn.*
On grassy hills or banks. — Perennia . *July.*

9. GYMNADENIA *R. Br.*

Sepals and *petals* converging, coloured; *lip* coloured, lobed, spurred. *Lobes* of the *anther* parallel. *Pollen masses* with 2 naked glands.

1. G. *conopsea* R. Br. E. B. 1. 10.
Knobs palmate. Lip in 3 entire equal lobes; spur very slender, twice as long as the ovary. Sepals widely spreading.
Orchis conopsea *Linn.*
In rather moist meadows and pastures. — Perennial. *June.*

10. PLATANTHERA *Rich.*

Sepals spreading or converging, coloured or herbaceous. *Petals* of the same figure as the sepals, coloured or herbaceous; *lip* entire or 3-lobed, with a spur at the base. *Column* very much compressed, often lengthened at the base, in front, into 2 projecting processes. *Lobes* of the *anther* diverging, not distinct from the processes of the column. *Pollen masses* with 2 naked glands.

1. P. *bifolia* Linn. Butterfly Orchis. E. B. 1. 22.
Knobs of the root oval, taper-pointed. Lip lanceolate, entire, about half the length of its very long spur. Lateral sepals spreading downwards.
Orchis bifolia *Linn.*
Habenaria bifolia *R. Br.*
In groves and thickets. — Perennial. *June.*

2. P. *albida* E. B. 8. 505.
Knobs tapering, clustered, undivided. Lip in 3 deep acute lobes, the middle one largest; spur one-third the length of the ovary.
Orchis albida *Swartz.*
Satyrium albidum *Linn.*
Habenaria albida *R. Br.*
In grassy mountain pastures. — Perennial. *June.*

3. P. *viridis.* Frog Orchis. E. B. 2. 94.
Knobs tapering, clustered, divided. Lip linear, with 3 teeth; the middle one smallest. Spur very short, slightly cloven. *Smith.*
Orchis viridis *Swartz.*

Satyrium viride *Linn.*
Habenaria viridis *R. Br.*

In moist pastures and meadows, especially on gravelly or stony ground. — Perennial. *June, July.*

11. ACERAS *R. Br.*

Sepals and *petals* helmet-shaped, herbaceous; *lip* coloured, lobed, hanging down, not spurred. *Pollen masses* with 2 glands enclosed in a common pouch.

1. A. *anthropophora* R. Br. Green Man-orch's. E. B. 1. 29.
Lip longer than the ovary.
Ophrys anthropophora *Linn.*

In chalk pits and on banks, on a chalky soil. — Perennial. *June.*

12. OPHRYS *Linn.*

Sepals spreading, coloured or herbaceous. *Petals* much smaller than the sepals, generally coloured; *lip* convex, not spurred, more or less lobed, usually hairy, and figured. *Pollen masses* with 2 glands, each enclosed in a separate pouch.

1. O. *muscifera* Huds. Fly Orchis. E. B. 1. 64.
Lip twice as long as the calyx, flat, with 4 expanded lobes, somewhat downy; the disk polished. Petals linear, smooth.
Ophrys myodes *Swz.*

In chalky pastures, or in meadows. — Perennial. *June.*

2. O. *apifera* Huds. Bee Orchis. E. B. 6. 383.
Lip the length of the calyx, tumid, with 5 reflexed marginal lobes; the terminal one awl-shaped; the rest hairy above. Sepals coloured. Petals ciliated.

In meadows and pastures. — Perennial. *July.*

3. O. *aranifera* Huds. Spider Orchis. E. B. 1. 65.
Lip the length of the calyx, tumid, hairy, rounded, emarginate, with 4 shallow, reflexed, marginal lobes. Sepals herbaceous. Petals linear, smooth.
Ophrys fucifera *Curtis.*

In dry chalky or gravelly pastures. — Perennial. *April.*

4. O. *fucifera* Smith. Drone Orchis.
Lip longer than the calyx, obovate, hairy, undivided, with a spreading wavy margin. Column bluntly pointed, incurved. Petals roughish; ovate at the base. *Smith.*

On chalky hillocks and banks, in Kent. — Perennial. *May, June.*

5. O. *arachnites* Willd. Late Spider Orchis.
Lip longer than the calyx, dilated, somewhat tumid, with 5 shallow, inflexed, marginal lobes; the terminal one flattened. Sepals coloured. Petals deltoid, downy.
Orchis arachnites *Scop.*

In chalky pastures. — Perennial. *July.*

13. HERMINIUM *R. Br.*

Sepals and *petals* herbaceous, spreading; *lip* short, lobed, not spurred. *Lobes* of the *anther* parallel. *Pollen masses* with 2 naked glands.

1. H. *monorchis* R. Br. Green Musk-orchis. E. B. 1. 71.
Radical leaves 2, lanceolate. *Br.*
Ophrys monorchis *Linn.*
On chalky banks and hillocks. — Perennial. *June, July.*

Tribe iv. *Malaxideæ* Lindley.

14. MALAXIS *Linn.*

Sepals herbaceous, ovate, spreading. *Petals* herbaceous, reflexed; *lip* uppermost, much smaller than the sepals, and similar in size and figure to the petals. *Column* very short. *Pollen masses* 2.

1. M. *paludosa* Linn. Bog-orchis. E. B. 1. 72.
Leaves about 4, spatulate; rough at the tip. Stalk with 5 angles. Lip entire, concave, erect, acute, half the length of the calyx.
Ophrys paludosa *Linn.*
In spongy turfy bogs. — Perennial. *July.*

15. LIPARIS *Richard.*

Sepals more or less herbaceous, spreading. *Petals* linear, spreading; *lip* undermost, dilated, much larger than the sepals. *Column* nearly as long as the sepals. *Pollen masses* 4.

1. L. *Lœselii* Rich. E. B. 1. 47.
Leaves 2, elliptic-lanceolate. Stalk triangular. Lip channelled undivided, recurved, longer than the calyx.
Malaxis Löselii *Swartz.*
Cymbidium Löselii *Swartz.*
Ophrys Löselii *Linn.*
O. liliifolia *Hudson.*
O. paludosa *Fl. Dan.*
On sandy bogs, among rushes. — Perennial. *July.*

Tribe v. *Cypripedieæ* Lindley.

16. CYPRIPEDIUM *Linn.*

Lip inflated, sometimes saccate. *Column* terminated at the back by a petaloid lobe representing a barren stamen, and dividing the anthers. The 2 anterior *sepals* often united. *R. Br.*

1. C. *Calceolus* Linn. Ladies' Slipper. E. B. 1. 1.
Stem leafy. Appendage to the column elliptical, obtuse, channelled. Lip somewhat compressed, shorter than the petals. *Smith.*
In mountainous woods and thickets in the north of England. — Perennial. *June.*

Order 94. MELANTHACEÆ *R. Brown.*

Colchicaceæ *Decandolle.*

Perianthium inferior, petaloid, in 6 pieces, or, in consequence of the cohesion of their claws, tubular; the pieces generally involute in æstivation.

Stamens 6; *anthers* mostly turned outwards.

Ovarium 3-celled, many-seeded; *style* trifid or 3-parted; *stigmas* undivided.

Capsule generally divisible into 3 pieces; sometimes with a loculicidal dehiscence.

Seeds with a membranous testa; *albumen* dense, fleshy. *R. Br.*

Roots bulbous or fibrous. *Leaves* sheathing at the base, with parallel veins. *Flowers* either arising from under the surface of the ground, or arranged upon tall leafy stems in large panicles; or disposed in spikes or racemes upon a naked scape.

1. COLCHICUM *Linn.*

Perianthium tubular, long, with a campanulate 6-parted limb. *Stamens* inserted in the orifice of the tube. *Anthers* oblong, versatile. *Ovary* 1. *Styles* 3, very long. *Follicles* 3, inflated, erect, united at the base, many-seeded.

1. C. *autumnale* Linn. Meadow-saffron. E. B. 2. 133.
Leaves flat, lanceolate, erect. Segments of the corolla oblong. *Smith.*

In moist rich meadows. — Perennial. *September.*

2. TOFIELDIA *Huds.*

Perianthium 6-parted, with 3 minute bracteæ at its base. *Stamens* smooth. *Follicles* 3 or 6, united at the base, many-seeded.

1. T. *palustris* Huds. Scottish Asphodel. E. B. 8. 536.
Flowers in an ovate head. Stem smooth, thread-shaped, leafless. Petals obovate, obtuse. Ovaries roundish.
T. borealis *Wahl.*
Helonias borealis *Willd.*
Anthericum calyculatum *Linn.*
Narthecium calyculatum *Lam.*

In the black boggy margins of pools and rills. — Perennial. *August.*

Order 95. AMARYLLIDEÆ *R. Br.*

Perianthium superior, in 6 parts, regular, with an imbricated æstivation; the 3 sepals overlapping the petals.

Stamens 6, inserted upon the segments of the perianthium; *filaments* sometimes connate at the base; *anthers* turned inwards.

Ovary 3-celled, either many-seeded or few-seeded; in the latter case the ovules ascending. *Style* 1; *stigma* with 3 lobes.

Fruit 3-celled, either capsular, with 3 loculicidal valves and many seeds; or succulent, with from 1 to 3 seeds.

Seeds neither black nor crustaceous; the *testa* often extremely fleshy; *albumen* fleshy; *embryo* straight, entire, pointing to the hilum.

Roots either fibrous or bulbous. *Flowers* usually with bright colours. *L aves* with parallel minute veins.

1. GALANTHUS *Linn.*

Perianthium in 6 pieces; the *petals* twice as short as the sepals, and emarginate. *Stigma* simple.

1. G. *nivalis* Linn. Snowdrop. E. B. 1. 19.
Leaves not plaited.
In meadows, orchards, woods, hedges, and on the banks of rivers. — Perennial. *February.*

2. LEUCOJUM *Linn.*

Perianthium with a short tube, and a campanulate, equal limb formed of 6 pieces, which are thickish at the apex. *Stigma* simple.

1. L. *æstivum* Linn. Snowflake. E. B. 9. 621.
Flowers several. Style club-shaped. *Smith.*
In moist meadows, and marshes near rivers. — Perennial. *May.*

3. NARCISSUS *Linn.*

Perianthium funnel-shaped, with a spreading, 6-parted limb, surrounded at the orifice of the tube by a cup. *Stamens* 6, inserted in the tube, and concealed within the cup.

1. N. *poeticus* Linn. E. B. 4. 275.
Flowers mostly solitary. Crown very short, depressed; membranous and crenate at the margin. Leaves bluntlv keeled· their edges reflexed.
In heathy, elevated, open fields. — Perennial. *May.*

2. N. *biflorus* Curt. Primrose peerless. E. B. 4. 276.
Flowers in pairs. Crown very short, depressed; membranous and crenate at the margin. Leaves acutely keeled; their edges inflexed.
N. poeticus *Hudson*
In sandy fields. — Perennial. *April, May.*

3. N. *Pseudo-narcissus* Linn. Daffodil. E. B. 1. 17.
Flowers solitary Crown bell-shaped, erect, crisped, with 6 marginal segments; its length equal to that of the ovate petals.
In rather moist woods and thickets. — Perennial. *March.*

Order 96. Liliaceæ *Juss.*

Tulipaceæ *Dec.* Hemerocallideæ *R. Br.*

Perianthium inferior, in 6 pieces, coloured, regular; occasionally with a tube.

Stamens 6, inserted into the pieces of the perianthium.

Ovary superior, 3-celled, many-seeded; *stigma* simple, or 3-lobed.

Fruit dry, capsular, 3-celled, many-seeded, with a loculicidal dehiscence.

Seeds flat, packed one upon another in 1 or 2 rows, with a spongy, dilated, often winged integument; *embryo* with the same direction as the seed, in the axis of fleshy *albumen*.

Roots scaly bulbs. *Leaves* with parallel veins, either lanceolate or cordate. *Flowers* large, usually with bright colours, often solitary.

1. FRITILLARIA *Linn.*

Perianthium campanulate, of 6 pieces, with an oval honey-pore at their base. *Stigmas* 3. *Seeds* flat.

1. F. *Meleagris* Linn. Common Fritillary. Chequered Daffodil. Snake's-head. E. B. 9. 622.

All the leaves alternate, linear-lanceolate, pointed. Stem single-flowered. Honey-pore linear. Points of the perianthium inflexed. *Smith.*

In moist meadows and pastures. — Perennial. *April.*

2. TULIPA *Linn.*

Perianthium campanulate, of 6 pieces, without honey-pores at the base. *Stigmas* 3, thick, sessile. *Capsule* oblong, 3-cornered. *Seeds* flat.

1. T. *sylvestris* Linn. Wild Tulip. E. B. 1. 63.

Flower solitary, a little drooping. Leaves lanceolate. Stigma triangular, abrupt. Stamens hairy at the base. *Smith.*

In chalk pits. — Perennial. *April.*

Order 97. Asphodeleæ *R. Br.*

Calyx and *corolla* forming a 6-parted or 6-cleft, petaloid, regular *perianthium*.

Stamens 6, inserted upon the perianthium, or hypogynous; the 3 opposite the sepals sometimes either unlike the rest or wanting.

Ovarium superior, 3-celled, with 2- or many-seeded cells; *ovules* when 2 ascending. *Style* 1. *Stigma* entire or with 3 short lobes.

Fruit mostly a 3-celled, 3-valved *capsule*, with a loculicidal dehiscence; occasionally succulent, and sometimes 3-parted.

Seeds with a testa, which is black, brittle, and crustaceous; *albumen* fleshy; *embryo* included.

Herbaceous plants, or occasionally *trees*, with bulbous or fascicled roots. *Leaves* with parallel veins. *Peduncles* articulated in the middle. *Flowers* coloured.

ANALYSIS OF THE GENERA.

Fruit succulent - - - - - - - - - 1. ASPARAGUS.
Fruit capsular
  Perianthium spreading open
    Inflorescence an umbel
      Stigma simple - - - - - - - - 2. ALLIUM.
      Stigma gaping - - - - - - - - 3. GAGEA.
    Inflorescence a spike or raceme
      Perianthium withering - - - - - 4. ORNITHOGALUM.
      Perianthium deciduous
        Filaments bearded - - - - - 5. ANTHERICUM.
        Filaments smooth - - - - - 6. SCILLA.
  Perianthium tubular
    6-toothed - - - - - - - - - - 7. MUSCARI.
    6-parted - - - - - - - - - - 8. HYACINTHUS.

---

1. ASPARAGUS *Linn.*

Perianthium 6-parted, spreading, equal, deciduous. *Stamens* 6, inserted in the base of the sepals and petals. *Filaments* subulate, smooth. *Anthers* peltate, erect. *Ovarium* with 2-seeded cells. *Style* short, with 3 furrows. *Stigma* 3-lobed. *Berry* round, with from 1 to 3 cells, and few seeds. *Embryo* out of the centre. *R. Br.*

1. A. *officinalis* Linn. Asparagus. E. B. *5*. 339.
Stem herbaceous, round, erect, without prickles. Leaves bristle-shaped, flexible. Stipulas mostly solitary. *Smith.*

On the sea-coast. — Perennial. *June, July.*

2. ALLIUM *Linn.*

Perianthium 6-parted, spreading. *Stigma* simple. *Capsule* 3-angular, the cells deeply parted in two, separating from a permanent filiform axis. — Flowers *in terminal umbels, with two herbaceous* bracteæ.

* *Stem leafy. Leaves flat.*

1. A. *Ampeloprasum* Linn. E. B. 24. 1657.
Umbel globose, without bulbs. Stem leafy below. Leaves flat. Three alternate stamens deeply 3-cleft. Keel of the petals rough. *Smith.*

In open hilly places. — Perennial. *August.*

2. A. *arenarium* Linn. E. B. 19. 1358.
Umbel globose, bearing bulbs. Stem leafy below. Leaves flat, with cylindrical sheaths. Bracteas obtuse. Three alternate stamens dilated, 3-cleft. Keel of the petals roughish. *Smith.*

In mountainous woods and fields. — Perennial. *July.*

3. A. *carinatum* Linn. E. B. 24. 1658.
Umbel lax, bearing bulbs. Leaves flat. Bracteas tapering, longer than the umbel. Stamens simple, awl-shaped. *Smith.*
On mountains and rocks in the north. — Perennial. *July.*

** *Stem leafy. Leaves somewhat cylindrical.*

4. A. *oleraceum* Linn. E. B. 7. 488.
Umbel lax, bearing bulbs. Leaves semicylindrical, tubular, rough; channelled above; ribbed beneath. Bracteas pointed, longer than the umbel. Stamens simple, awl-shaped. *Smith.*
In corn-fields and their borders. — Perennial. *July.*

5. A. *vineale* Linn. E. B. 28. 1974.
Umbel spherical, bearing bulbs. Leaves cylindrical, smooth. Three alternate stamens deeply 3-cleft. *Smith.*
In pastures and waste ground. — Perennial. *July.*

*** *Stalk radical, naked.*

6. A. *ursinum* Linn. Ramsons. E. B. 2. 122.
Stalk naked, semicylindrical. Leaves elliptic-lanceolate, stalked. Umbel level-topped. Stamens simple. *Smith.*
In moist woods, hedges, and meadows. — Perennial. *May, June.*

7. A. *Schœnoprasum* Linn. Chives. E. B. 34. 2441.
Stalk naked, round, the height of the foliage. Leaves cylindrical, somewhat tapering at the point. Stamens simple. *Smith.*
In meadows and pastures. — Perennial. *June.*

3. GAGEA *Salisb.*

Perianthium somewhat herbaceous, 6-parted, persistent, converging at the base, spreading at the apex. *Stamens* 6; *filaments* not dilated at the base. *Stigma* gaping. *Capsule* 3-cornered. — Flowers *yellow, corymbose, with leafy* bracteæ.

1. G. *lutea* Ker. E. B. 1. 21.
Radical leaves 1 or 2, linear, much longer than the angular leafless scape. Pedicels solitary, shorter than the taper-pointed bracteæ. Segments of the perianth lanceolate, obtuse.
Ornithogalum luteum *Linn.*
In groves and pastures. — Perennial. *April.*

4. ORNITHOGALUM *Linn.*

Perianthium petaloid, 6-parted, withering upon the stalk, converging at the base, diverging at the apex. *Stamens* 6, the 3 outer with dilated filaments. *Stigma* small, capitate. *Ovary* bluntly 3-cornered. — Flowers *racemose, white, green, or yellow, with membranous* bracteæ.

1. O. *pyrenaicum* Linn. E. B. 7. 499.
Cluster very long. Filaments all dilated. Flower-stalks equal, spreading; growing erect, and close-pressed, as the fruit ripens. *Smith.*
In pastures. — Perennial. *June, July.*

2. O. *umbellatum* Linn. Star of Bethlehem. E. B. 2. 130.
Flowers corymbose; their partial stalks overtopping the main one. Filaments dilated, tapering, entire. *Smith.*
In meadows, pastures, and groves. — Perennial. *April, May.*

3. O. *nutans* Linn. E. B. 28. 1997.
Flowers pendulous, unilateral. Filaments dilated, cloven, converging; 3 of them longer, their lobes nearly equal to the anther. *Smith.*
In fields and orchards. — Perennial. *April, May.*

5. ANTHERICUM *Linn.*

Perianthium 6-parted, petaloid, equal, spreading, deciduous. *Stamens* 6. *Filaments* bearded. *Anthers* versatile. *Ovary* polyspermous. *Style* filiform. *Capsule* roundish, 3-celled, with 3 valves bearing the dissepiments in the middle. *Seeds* few, angular, naked at the hilum. *R. Br.*

1. A. *serotinum* Linn. Mountain Spiderwort. E. B. 12. 793.
Leaves semicylindrical; those on the stem dilated at their base. Flower mostly solitary. *Smith.*
On the loftiest Welsh mountains. — Perennial. *June.*

6. SCILLA *Linn.*

Perianthium 6-parted, generally spreading and deciduous. *Filaments* filiform, smooth, inserted into the base of the perianthium. *Seeds* roundish. — Root *bulbous.* Duby.

1. S. *verna* Huds. E. B. 1. 23.
Bulb coated. Corymb hemispherical, of few flowers. Bracteas lanceolate, obtuse. Leaves linear, channelled. *Smith.*
On maritime rocks and cliffs. — Perennial. *April.*

2. S. *bifolia* Linn. E. B. 1. 24.
Bulb coated. Cluster slightly corymbose, without bracteas. Flowers nearly erect. Leaves lanceolate, generally 2. *Smith.*
In groves in the west of England. — Perennial. *March, April.*

3. S. *autumnalis* Linn. E. B. 2. 78.
Leaves linear, numerous. Cluster somewhat corymbose. Flower-stalks ascending, the length of the flowers, without bracteas. *Smith.*
In dry pastures or on rocks. — Perennial. *September.*

7. MUSCARI *Tourn.*

Perianthium ovate, inflated, 6-toothed. *Capsule* 3-cornered, with prominent angles. *Cells* 2-seeded.

1. M. *racemosum* Mill. Starch Hyacinth. E. B. 27. 1931.
Flowers ovate, with 6 furrows; the upper ones sessile and abortive. Leaves linear, channelled, flaccid. *Smith.*
Hyacinthus racemosus *Linn.*
In grassy fields, or among ruins. — Perennial. *May.*

8. HYACINTHUS *Linn.*

Perianthium 6-cleft, tubular; segments spreading at the apex. *Stamens* inserted about the middle of the perianthium. *Capsule* obtusely 3-cornered; *cells* many-seeded.

1. H. *non scriptus* Linn. Harebells. E. B. 6. 377.
Leaves linear. Bracteæ in pairs.
Scilla nutans *Smith.*
S. non scripta *Redouté.*
In thickets. — Perennial. *May.*

Order 98. Smilaceæ *R. Br.*

Flowers hermaphrodite or diœcious.
Perianthium inferior, petaloid, 6-parted.
Stamens 6, inserted into the segments near their base; seldom hypogynous.
Ovarium 3 celled; the cells 1- or many-seeded; *style* usually trifid; *stigmas* 3.
Fruit a roundish berry.
Seeds with a membranous testa (not black or brittle); *albumen* between fleshy and cartilaginous; *embryo* usually distant from the hilum. *R. Br.*
Herbaceous plants or *under-shrubs,* often with a tendency to climb. *Leaves* with parallel veins.

N. B. Tamus has the perianthium superior.

1. CONVALLARIA *Linn.*

Sepals and *petals* united in a perianthium, which is either globose or cylindrical, and 6-toothed. *Stamens* 6. *Berry* round, before maturity spotted, 3-celled, with 1-seeded cells.

1. C. *majalis* Linn. Lily of the Valley. E. B. 1035.
Flower-stalk radical, naked, semicylindrical. Cluster simple. Flowers drooping, cup-shaped, with rather distinct segments. *Smith.*
In groves or rocky woods. — Perennial. *May.*

2. C. *verticillata* Linn. E. B. 2. 128.
Leaves linear-lanceolate, whorled. *Smith.*
In woods at the bases of the Scottish mountains. — Perennial. *June.*

3. C. *Polygonatum* Linn. E. B. 4. 280.
Leaves alternate, clasping the angular stem. Stalks axillary, mostly single-flowered. Stamens smooth. *Smith.*
In rocky mountainous woods. — Perennial. *May, June.*

4. C. *multiflora* Linn. Common Solomon's Seal. E. B. 4. 279.
Leaves alternate, clasping the round stem. Stalks axillary, many-flowered. Stamens downy. *Smith.*

In woods and thickets. — Perennial. *May, June.*

2. RUSCUS *Linn.*

Sepals 3. *Petals* 3, spreading. *Filaments* united in a tube, which is either with or without anthers. *Style* 1. *Stigma* 1. *Berry* globose, 3-celled, with 2-seeded cells.

1. R. *aculeatus* Linn. Butcher's Broom. E. B. 8. 560.
Leaves ovate, sharp-pointed, flowering on the upper side without a leaflet. *Smith.*

β. laxus.
Ruscus laxus *Smith.*

On bushy heaths, and in woods. — Perennial. *March, April.*

3. PARIS *Linn.*

Sepals 4. *Petals* 4. *Stamens* 8. *Anthers* attached to the middle of the filaments. *Stigmas* 4. *Berry* 4-celled; cells 6 or 8-seeded.

1. P. *quadrifolia* Linn. Herb Paris. E. B. 1. 7.
Leaves ovate, about 4. *Smith.*

In groves and shady places. — Perennial. *May, June.*

4. TAMUS *Linn.*

Flowers diœcious. *Perianthium* campanulate, 6-parted; in the males, which are hexandrous, spreading; in the females superior and contracted at the neck. *Style* 1. *Stigmas* 3. *Berry* 3-celled.

1. T. *communis* Linn. Black Bryony. E. B. 2. 91.
Leaves heart-shaped, undivided, acute. *Smith.*

In tall hedges, thickets, and woods. — Perennial. *June.*

Order 99. Butomeæ *Richard.*

Sepals 3, herbaceous.

Petals 3, coloured, petaloid.

Stamens definite or indefinite, hypogynous.

Ovaries superior, 3, 6, or more, either distinct or united into a single mass; *stigmas* the same number as the ovaries, simple.

Follicles many-seeded, either distinct and rostrate, or united in a single mass.

Seeds minute, very numerous, attached to the whole of the inner surface of the fruit; *albumen* none; *embryo* with the same direction as the seed.

Aquatic plants. *Leaves* very vascular, often yielding a milky juice, with parallel veins. *Flowers* in umbels, conspicuous, purple, or yellow.

1. BUTOMUS *Linn.*

Stamens 9, of which 3 are internal and petaloid. *Ovaries* 6, with long beaks. *Fruit* capsular, dehiscing at the inner edge. *Seeds* linear-oblong, straight, with longitudinal streaks.

1. B. *umbellatus* Linn. Flowering-rush. E. B. 10. 651.
In ditches, and the margins of rivers. — Perennial. *June, July.*

Order 100. Restiaceæ *R. Br.*

Perianthium inferior, 2-6-parted, seldom wanting.
Stamens definite, 1-6; when they are from 2 to 3 in number, and attached to a perianthium of 4 or 6 divisions, they are then opposite the inner segments (*petals*).
Ovarium 1- or more celled; cells monospermous; ovules pendulous
Fruit capsular, or nucamentaceous.
Seeds inverted; *albumen* of the same figure as the seed; *embryo* lenticular, on the outside of the albumen, at that end of the seed which is most remote from the hilum.
Herbaceous plants or *under-shrubs*. *Leaves* simple, narrow, or none. *Culms* naked, or more usually protected by sheaths, which are slit, and have equitant margins. *Flowers* generally aggregate, in spikes or heads, separated by bracteæ, and most frequently bisexual. *R. Br.*

1. ERIOCAULON *Linn.*

Head androgynous. *Scales* 1-flowered, the exterior generally empty, and forming an involucrum. *Petals* 2 or 3. *Sepals* 2 or 3. *Males* in the disk. *Petals* cohering at the base, or inserted rather higher up than the sepals. *Stamens* 4 or 6. *Anthers* 2-celled. *Females* in the circumference. *Petals* distinct. *Style* 1. *Stigmas* 2 or 3. *Capsule* 2- or 3-celled, 2- or 3-lobed, dehiscing by the salient angles. *Seeds* solitary.

1. E. *septangulare* With. Pipewort. E. B. 11. 733.
Capsule of 2 cells. Stem with about 7 angles, many times taller than the channelled, taper-pointed leaves. Head convex. Outer scales empty, rounded, smooth.
E. decangulare *Lightf.*
Nasmythia articulata *Hudson.*
In lakes, in the isle of Skye, and on the west coast of Ireland. — Perennial. *September.*

Order 101. JUNCEÆ *Dec*

Flowers hermaphrodite or bisexual.

Calyx and *corolla* forming an inferior, 6-parted, somewhat glumaceous *perianthium*.

Stamens 6, inserted into the base of the segments; sometimes 3, and then opposite the sepals. *Anthers* 2-celled.

Ovarium 1- or 3-celled, 1- or many-seeded, or 1-celled and 3-seeded. *Style* 1. *Stigmas* generally 3, sometimes only 1.

Fruit capsular, with 3 valves, which have the dissepiment in their middle, sometimes destitute of valves, and 1-seeded by abortion.

Seeds with a testa, which is neither black nor crustaceous; *albumen* firm, fleshy, or cartilaginous; *embryo* within it. *R. Br.*

Herbaceous plants with fascicled or fibrous roots. *Leaves* fistular, or flat and channelled with parallel veins. *Inflorescence* often more or less capitate. *Flowers* generally brown.

1. JUNCUS *Linn.* Rush.

Perianthium 6-parted, glumaceous. *Stamens* 3 or 6. *Capsule* 3-celled, 3-valved, with a loculicidal dehiscence. *Seeds* numerous, attached to the dissepiment. — Stems *taper, usually leafless, with internal phragmata.*

* *Leaves none.*

1. J. *acutus* Linn. E. B. 23. 1614.
   Stem naked, sharp-pointed. Panicle aggregate, near the summit. Bractea spinous. Capsule twice as long as the calyx, roundish, with a blunt point.
   On the sea-coast, in deep sand. — Perennial. *July.*

2. J. *maritimus* Smith. E. B. 24. 1725.
   Stem naked, sharp-pointed. Panicle proliferous, near the erect summit. Bractea spinous. Capsule oblong, the length of the calyx.
   In marshes near the sea, along with the preceding, but much more plentiful. — Perennial. *August.*

3. J. *glaucus* Sibth. E. B. 10. 665.
   Stem naked, straight, glaucous. Panicle upright, far below the summit. Capsule elliptical, pointed, rather shorter than the calyx. *Smith.*
   J. inflexus *Relhan.*
   In wet pastures, or moist waste ground. — Perennial. *July.*

4. J. *conglomeratus* Linn. Common Rush. E. B. 12. 835.
   Stem naked, straight. Panicle dense, globular, far below the summit. Capsule abrupt. Stamens 3. *Smith.*
   n pastures, and by road sides. — Perennial. *July.*

5. J *effusus* Linn. E. B. 12. 836.
   Stem naked, straight. Panicle loose, repeatedly compound, very far below the summit. Capsule obtuse. *Smith.*
   In wet pastures, and boggy places by road sides. — Perennial. *July.*

6. J. *filiformis* Linn. E. B. 17. 1175.
Stem naked, thread-shaped, drooping. Panicle nearly simple, corymbose, of few flowers, very far below the summit. Bractea taper-pointed. Capsule almost globular. *Smith.*

About the margins of lakes in the north. — Perennial. *August.*

7. J. *arcticus* Willd.
Stem naked, straight, acute. Panicle towards the summit, dense, capitate of few flowers. Bractea shorter than the panicle. Capsule oblong, bluntish. *Smith.*

On the sands of Barry, near Dundee. *Mr. Drummond.*—Perennial. *July, August.*

** *Herb leafy.*

8. J. *trifidus* Linn. E. B. 21. 1482.
Stem naked. Radical leaves very few. Bracteas 3, leafy, channelled, with from 1 to 3 terminal flowers. *Smith.*
J. monanthus *Jacq.*

In alpine bogs in Scotland. — Perennial. *July.*

9. J. *squarrosus* Linn. Moss Rush. Goose Corn. E. B. 13. 933.
Stem naked. Leaves numerous, radical, channelled. Panicle terminal, compound, with cymose branches. *Smith.*
J. Sprengelii *Willd.*

In boggy spots, on the most barren sandy heaths. — Perennial. *June, July.*

10. J. *compressus* Jacq. E. B. 13. 934.
Stem simple, compressed; leafy below. Leaves linear, incurved at the edges. Panicle cymose, terminal, shorter than the bractea. Capsule roundish-obovate, longer than the obtuse calyx. *Smith.*
J. bulbosus *Linn.*

In moist pastures. — Perennial. *July, August.*

11. J. *cœnosus* Bicheno.
Stem simple, leafy. Leaves linear, channelled. Panicle cymose, terminal, longer than the bractea. Capsule obovate, the length of the rather obtuse calyx. *Smith.*

In salt marshes, and muddy places towards the sea, abundantly. — Perennial. *July, August.*

12. J. *Gesneri* Smith. E. B. 31. 2174.
Stem simple, naked. Leaves slightly channelled. Panicle forked, racemose, shorter than the bractea. Calyx-leaves lanceolate, taper-pointed, 3-ribbed, longer than the oval capsule. *Smith.*
J. gracilis *Smith.*
J. tenuis *Hooker.*

By a rivulet in marshy ground, among the mountains of Clova, Angusshire, very rare. *Mr. G. Don.* — Perennial. *July.*

13. J. *bufonius* Linn. E. B. 12. 802.
Stem leafy. Leaves angular, channelled. Panicle forked, racemose, longer than the bracteas. Calyx-leaves lanceolate, taper-pointed, membranous, 2-ribbed, longer than the oblong capsule. *Smith.*

In marshy ground. — Annual. *July, August.*

14. J. *uliginosus* Sibth. E. B. 12. 801.
Stem leafy, bulbous at the base. Leaves bristle-shaped, channelled. Heads lateral and terminal, about 3-flowered. Capsule obtuse, rather longer than the calyx. *Smith.*
J. bulbosus *Linn.*
J. supinus *Don.*
On moist, sandy, or turfy heaths. — Perennial. *June, July.*

15. J. *subverticillatus* Wulfen.
Stem leafy, trailing. Leaves bristle-shaped, channelled, very slightly jointed. Panicle forked. Heads lateral and terminal, about 5-flowered, somewhat whorled. Capsule obtuse, rather longer than the calyx. *Smith.*
J. setifolius *Ehr.*
In boggy and watery places. — Perennial. *July, August.*

16. J. *capitatus* Weigel.
Stem erect, unbranched; leafy at the base. Leaves bristle-shaped, channelled. Heads 1 or 2, lateral and terminal. Stamens 3. Calyx keeled, bristle-pointed, twice as long as the capsule. *Smith.*
J. gracilis *Roth.*
J. supinus *Bicheno.*
Schœnus minimus *Forst.*
In sandy ground. — Annual? *May—July.*

17. J. *biglumis* Linn. E. B. 13. 898.
Stem erect, unbranched; leafy at the base. Leaves flat. Head solitary, of 2 unilateral flowers, surmounted by a leafy bractea. *Smith.*
About mountain rills, in the Highlands of Scotland. — Perennial. *August.*

18. J. *triglumis* Linn. E. B. 13. 899.
Stem erect, unbranched; leafy in the lower part. Leaves flat. Head solitary, terminal, of about 3 upright flowers, with elliptical bracteas. *Smith.*
In alpine rivulets. — Perennial. *July.*

19. J. *castaneus* Smith. E. B. 13. 900.
Stem unbranched, leafy. Leaves keeled, flat; sheathing at the base. Heads terminal, mostly in pairs, many-flowered, with leafy bracteas. Capsule twice the length of the calyx. *Smith.*
n the Highlands of Scotland, also in the north of England. — Perennial. *July.*

20. J. *acutiflorus* Ehr. E. B. 4. 238.
Leaves apparently jointed, slightly compressed. Panicle repeatedly compound, forked. Petals and sepals all bristle-pointed, shorter than the taper beak of the capsule. *Smith.*
J. sylvaticus *Willd.*
J. articulatus *E. Bot.*
J. nemorosus *Sibth.*
In woods and watery places. — Perennial. *June, July*

21. J. *lampocarpus* Ehr. E. B. 30. 2143.
Leaves apparently jointed, compressed. Panicle erect, compound,

forked. Petals bordered. Capsule ovate, coloured, highly polished, longer than the calyx.

J. compressus *Relh.*

In meadows and watery places. — Perennial. *July, August.*

22. J. *obtusiflorus* Ehr. E. B. 30. 2144.

Leaves apparently jointed, cylindrical. Stem with internal partitions. Panicle repeatedly compound; branches divaricated and reflexed. Sepals obtuse, as long as the capsule. *Smith.*

In marshes. — Perennial. *August.*

23. J. *polycephalus* D. Don.

Leaves apparently jointed, awl-shaped, cylindrical. Panicle erect, twice or thrice forked, with nearly simple branches. Heads many-flowered. Sepals lanceolate, acute, rather shorter than the elliptic-ovate bluntish capsule.

In the Highlands of Scotland. — Perennial.

2. LUZULA *Dec.*

Perianthium 6-parted, glumaceous. *Stamens* 6. *Capsule* 1-celled, 3-seeded, 3-valved. — Leaves *flat, usually hairy.*

1. L. *pilosa* Willd. E. B. 11. 736.

Panicle cymose, widely spreading and reflexed. Flowers solitary. Capsule pointless. Crest of the seeds hooked. *Smith.*

Juncus pilosus *Linn.*

J. vernalis *Ehr.*

In shady groves, and on banks. — Perennial. *March, April.*

2. L. *Forsteri* Dec. E. B. 18. 1293.

Panicle cymose, erect. Flowers solitary. Capsule pointed. Crest of the seeds straight and obtuse. *Smith.*

Juncus Forsteri *Smith.*

In groves and thickets. — Perennial. *May.*

3. L. *sylvatica* Bicheno. E. B. 11. 737.

Panicle cymose, doubly compound. Flowers and bracteas aggregate. Capsule pointed. Crest of the seeds obsolete. *Smith.*

Juncus sylvaticus *Hudson.*

Luzula maxima *Willd.*

Juncus maximus *Ehr.*

J. latifolius *Jacq.*

In woods, and hilly, open, or bushy ground. — Perennial. *May, June.*

4. L. *campestris* Willd. E. B. 10. 672.

Panicle of 3 or 4 ovate, dense, partly stalked, clusters. Capsule obovate, obtuse, with a small point, shorter than the calyx. Seeds stalked, without a crest. Leaves flat. *Smith.*

Juncus campestris *Linn.*

In dry barren pastures. — Perennial. *April, May.*

5. L. *congesta* Dec.

Panicle of numerous, roundish-ovate, dense, partly stalked, clusters. Capsule obovate, obtuse, with a small point, as long as the calyx. Seeds stalked, without a crest. Leaves flat, rough-edged. *Smith.*

L. campestris β. *Bicheno.*

Juncus liniger *Purton.*

In marshy turfy ground. — Perennial. *June.*

6. L. *spicata* Bicheno. E. B. 17. 1176.

Panicle dense, compound, oblong, lobed, drooping. Capsule elliptical, with a small point. Crest of the seeds obsolete. Stem-leaves channelled. *Smith.*

Juncus spicatus *Linn.*

On the loftiest mountains of Scotland and Westmoreland. — Perennial. *July, August.*

7. L. *arcuata* Smith.

Panicle somewhat umbellate, partly compound, with drooping branches. Heads globose, of few flowers. Bracteas membranous, fringed. Capsule elliptical. Leaves channelled. *Smith.*

On the most stony and barren summits of Cairngorum, and others of the Grampian mountains. *Professor Hooker.* — Perennial. *July.*

3. NARTHECIUM *Huds.*

Perianthium 6-parted. *Stamens* 6. *Filaments* woolly, persistent. *Ovary* pyramidal. *Style* short. *Capsule* 3-celled, 3-valved. *Seeds* numerous, with an appendage at each end.

1. N. *ossifragum* Huds. Lancashire Bog-asphodel. E. B. 8. 535.

Cluster uninterrupted. One bractea at the base, the other above the middle, of each partial stalk. *Smith*

Anthericum ossifragum *Linn.*

Abama ossifraga *Dec.*

In black turfy bogs. — Perennial. *July, August.*

Division II. GLUMACEÆ.

ANALYSIS OF THE ORDERS.

Bracteæ to each flower 1; or 3, of which 2 are connate. Embryo undivided, included within the albumen. Stem angular.. Leaves with their sheaths entire - - -	102. CYPERACEÆ.
Bracteæ to each flower several, imbricated. Embryo with a naked plumula on the outside of the albumen. Stem cylindrical. Leaves with their sheaths split - - -	103. GRAMINEÆ.

Order 102. CYPERACEÆ *Juss.*

Flowers hermaphrodite, or bisexual, consisting of imbricated solitary bracteæ, very rarely enclosing other bracteæ called *glumes.*

Stamens hypogynous, definite, 1- 2- 3- 4- 6-12; *anthers* fixed by their base, entire, 2-celled.

Ovary 1-seeded, often surrounded by bristles called hypogynous setæ, probably constituting the rudiments of a perianthium; *ovulum* erect; *style* single, trifid, or bifid; *stigmas* undivided, occasionally bifid.

Nut crustaceous or bony.

Albumen of the same figure as the seed; *embryo* lenticular, undivided, enclosed within the base of the albumen.

Roots fibrous. *Stems* very often without joints, 3-cornered, or taper. *Leaves* with their sheaths entire. The lowermost bracteæ often sterile.

ANALYSIS OF THE GENERA.

Florets hermaphrodite
  Bracteæ distichous - - - - - - - - - - 1. CYPERUS.
  Bracteæ regularly imbricated
    Style persistent
      Bracteæ all fertile - - - - - - - 5. HELIOCHARIS.
      Lower bracteæ empty - - - - - - 2. RHYNCHOSPORA.
    Style deciduous
      Hypogynous setæ present
        longer than the bracteæ - - - - - 7. ERIOPHORUM.
        shorter than the bracteæ
          Bracteæ all fertile - - - - - 6. SCIRPUS.
          Lower bracteæ empty - - - - 3. BLYSMUS.

Hypogynous setæ wanting
Spikes 1- or 2-flowered, panicled - - - 8. Cladium.
Spikes many-flowered
panicled and round - - - - - 9. Holoschœnus.
terminal and oval
Leaves with parallel veins
Bracteæ all fertile - - - 10. Isolepis.
Lower bracteæ empty - - 4. Schœnus.
Lateral veins of leaves straggling - 11. Heliogiton.
Florets bisexual
Female florets naked - - - - - - - - - 12. Kobresia.
Female florets enclosed in 2 united glumes - - - 13. Carex.

CYPERUS *Linn.*

Spikes terminal, with involucral leaves. *Bracteæ* distichous, in pairs; the outer usually smallest at the base and empty; the inner adhering to the rachis. *Hypogynous setæ* wanting. *Fruit* 2-edged, or 3-cornered. *Style* filiform, deciduous.

1. C. *longus* Linn. English Galingale. E. B. 19. 1309.
Stem triangular. Umbel leafy, twice compounded, with naked stalks. Spikes alternate. *Smith.*

By a rivulet between St. David's town and St. David's head; at Walton in Gordan, Somersetshire. — Perennial. *July.*

2. C. *fuscus* Linn.
Stem triangular. Umbel compound, with 3 unequal leaves beneath. Spikes crowded, spreading every way. Stigmas 3. *Smith.*

Found by Mr. Haworth in a low marshy meadow, half a mile from Little Chelsea. — Annual. *September.*

2. RHYNCHOSPORA *Vahl.*

Spikes terminal and axillary, with a many-leaved involucrum. *Rachis* nearly straight, or somewhat flexuose. Lower *bracteæ* smaller than the others and empty. *Hypogynous setæ* from 6 to 10. *Fruit* compressed, somewhat 3-cornered. *Style* dilated at the base, and persistent upon the fruit.

1. R. *alba* Vahl. E. B. 14 985.
Heads abrupt. Stamens 2. Leaves tapering. Hypogynous setæ many.
Schœnus albus *Linn.*

On turfy bogs. — Perennial. *July, August.*

2. R. *fusca* Smith. E. B. 22. 1575.
Heads ovate-oblong. Stamens 3. Leaves thread-shaped. Hypogynous setæ 3.
Schœnus fuscus *Linn.*

In bogs, rare. — Perennial. *July, August.*

3. BLYSMUS *Panzer.*

Spike compound, terminal. *Spiculæ* furnished with an involucral leaf. *Rachis* nearly straight. *Bracteæ* gradually diminishing in size; the lowest empty. *Hypogynous setæ* several or none. *Fruit* somewhat compressed, tapering, with a filiform, deciduous *style.*

1. B. *compressus* Panzer. E. B. 11. 791.
Stem roundish, leafy at the bottom. Spikes aggregate, 2-ranked, many-flowered. Leaves flat, with rough edges and keel. Hypogynous setæ 6.
Schœnus compressus *Linn.*
Scirpus caricinus *Schrad.*
Carex uliginosa *Linn.*
Scirpus compressus *Pers.*
In boggy meadows. — Perennial. *July.*

4. SCHŒNUS *Linn.*

Spikes terminal. *Involucrum* 2- or many-leaved. *Rachis* nearly straight. Lower *bracteæ* smaller than the rest and empty. *Hypogynous setæ* 0. *Fruit* 3-cornered, with a very short point. *Style* filiform, deciduous.

1. S. *nigricans* Linn. E. B. 16. 1121.
Stem round, naked. Head roundish, abrupt, overtopped by one of the two involucral leaves.
Cyperus nigricans *With.*
On turfy bogs. — Perennial. *June.*

2. S. *rufus* Hudson. E. B. 15. 1010.
Stem round, leafy at the bottom. Spikes aggregate, distichous. Leaves channelled, smooth, without a keel
Scirpus rufus *Schrader.*
Blysmus rufus *Link.*
In marshes towards the sea-coast. — Perennial. *June, July.*

5. HELIOCHARIS.

Lamina of the leaf absent, or nearly so. *Spike* terminal. *Rachis* nearly straight. *Bracteæ* gradually diminishing in size. *Hypogynous setæ* from 2 to 10. *Fruit* 2-edged, seldom 3-cornered. *Style* 2- or 3-parted, persistent.

1. H. *palustris.* E. B. 2. 131.
Stem round. Root creeping. Stigmas 2. Fruit lenticular, most convex at one side.
Scirpus palustris *Linn.*
Eleocharis palustris *R. Br.*
In ditches, rivulets, and boggy ground. — Perennial. *June, July.*

2. H. *multicaulis.* E. B. 17. 1187.
Stem round. Root fibrous. Stigmas 3. Fruit acutely triangular, as well as the permanent base of the style.
Scirpus multicaulis *Smith.*
Eleocharis multicaulis *Smith.*
On turfy bogs and wet commons. — Perennial. *July.*

3. H. *acicularis.* E. B. 11. 749.
Stem quadrangular. Stigmas 3. Fruit numerously furrowed. Filaments permanent.
Scirpus acicularis *Linn.*
Eleocharis acicularis *R. & S.*
In damp spots upon heaths. — Perennial. *August.*

4. H. *cæspitosa.* E. B. 15. 1029.
Stem round, striated; sheathed and invested with numerous scales at the base. Spike terminal. Outer bracteæ largest, with leafy points.
Scirpus cæspitosus *Linn.*
Eleocharis cæspitosa *Link.*
On turfy barren heaths. — Perennial. *July.*

5. H. *pauciflora.* E. B. 16. 1122.
Stem round, with a tight leafless sheath at the base. Spike terminal, of few flowers, longer than its blunt membranous-tipped outer bracteæ.
Scirpus pauciflorus *Lightf.*
Sc. Bæothryon *Ehr.*
Sc. campestris *Roth.*
Eleocharis pauciflora *Link.*
On moors and mountains. — Perennial. *August.*

6. SCIRPUS *Linn.*

Spikes lateral or terminal. *Rachis* nearly straight. *Bracteæ* gradually diminishing in size. *Hypogynous setæ* shorter than the bracteæ, or nearly of the same length. *Style* filiform, 2- or 3-parted, deciduous. *Fruit* 2-edged or 3-cornered, mucronate, usually plano-convex.

1. S. *triqueter* Linn. E. B. 24. 1694.
Stem acutely triangular, straight, naked, sharp-pointed. Spikes lateral; sessile or stalked. Stigmas 2. Fruit smo
Scirpus pungens *Vahl.*
About the muddy banks of rivers. — Perennial. *August.*

2. S. *carinatus* Smith. E. B. 28. 1983.
Stem bluntly triangular upwards, naked; round at the base. Panicle cymose, terminal. Involucral leaf pungent, channelled, erect. Stigmas 2.
About the banks of large rivers. — Perennial. *August.*

3. S. *lacustris* Linn. Bull-rush. E. B. 10. 666.
Stem round, naked. Panicle cymose, twice compound, terminal. Spikes ovate. Involucral leaves generally much shorter than the panicle.
In clear ditches, ponds, and the borders of rivers. — Perennial. *July, August.*

4. S. *glaucus* Smith. E. B. 33. 2321.
Stem round, naked, glaucous. Panicle cymose, not higher than the bractea. Spikes ovate, conglomerate. Stigmas 2. *Smith.*
In salt marshes. — Perennial. *August.*

5. S. *maritimus* Linn. E. B. 8. 542.
Stem triangular. Panicle terminal, leafy. Spikes conglomerate. Bracteæ torn, with an intermediate point. Stigmas 3.
Sc. tuberosus *Desf.*
In salt marshes. — Perennial. *July, August.*

6. S. *sylvaticus* Linn. E. B. 13. 919.
Stem triangular, leafy throughout. Panicle terminal, leafy, cy-

mose, repeatedly compound. Flower-stalks sheathed at the base. Spikes aggregate. *Smith*.

In moist shady woods. — Perennial. *June, July.*

7. ERIOPHORUM *Linn.* Cotton-grass.

Spike terminal. *Rachis* nearly straight. *Bracteæ* gradually diminishing in size. *Hypogynous setæ* much longer than the bracteæ, persistent. *Style* 2- or 3-parted, filiform, deciduous. *Fruit* 3-cornered, pointed.

* *Spike solitary.*

1. E. *vaginatum* Linn. Hare's-tail Cotton-grass. E. B. 13. 873.
Stem triangular above; round below, with a swelling sheath. Spike ovate. Bracteæ membranous.
E. cæspitosum *Host.*

On barren mountainous moors. — Perennial. *March, April.*

2. E. *capitatum* Host. E. B 34. 2387.
Stem entirely round, with a swelling sheath. Spike roundish. Bracteæ membranous.
E. Scheuchzeri *Roth.*

On a sand bank by an alpine rivulet on Ben Lawers, Scotland, near the limits of perpetual snow. — Perennial. *August.*

3. E. *alpinum* Linn. E. B. 5. 311.
Stem triangular, naked above the leaves, which are shorter than their sheaths. Spike oblong-ovate. Glumes firm, strongly keeled.

On turfy alpine bogs in Scotland. — Perennial. *June, July.*

** *Spikes several.*

4. E. *polystachion* Linn. E. B. 8. 563.
Stem round. Leaves flat, lanceolate, with a triangular point. Stalks of the spikes smooth. Setæ thrice the length of the spike.
E. latifolium *Schrad.*

In boggy meadows. — Perennial. *April.*

5 E. *pubescens* Smith.
Stem angular upwards. Leaves flat, lanceolate, with a triangular point. Stalks of the spikes downy. Setæ twice the length of the spike.
E. angustifolium *Poit. & Turp.*

In bogs and marshes. — Perennial. *April, May?*

6. E. *angustifolium* Roth. E. B. 8. 564.
Stem nearly round. Leaves linear, triangular; channelled towards the base. Stalks of the spikes smooth. Setæ four times the length of the spike.
E. Vaillantii *Poit.*

In turfy, boggy, and muddy meadows. — Perennial. *April.*

7. E. *gracile* Roth. E. B. 34. 2402.
Stem round, with 3 slight angles. Leaves triangular; channelled towards the base. Spikes longer than the bractea. Setæ twice the length of the spike.

E. triquetrum *Schrad.*

On Ben Lawers and the Clova mountains, in a micaceous soil. — Perennial *July.*

8. CLADIUM *Schrad.*

Inflorescence terminal and axillary, panicled; the heads formed of 1- or 2-flowered *spikelets*. *Lower bracteæ* smaller and empty. *Hypogynous setæ* wanting. *Stamens* 2. *Style* 3-parted, filiform, deciduous. *Fruit* covered with a brittle pericarpium.

1. C. *Mariscus* R. Br. E. B. 14. 950.
Panicle repeatedly compound, leafy. Spikes capitate. Stem round, smooth, leafy. Leaves prickly at the margin and keel. *Smith*.
Schœnus Mariscus *Linn.*
Cladium germanicum *Schrad.*
In fens and boggy places. — Perennial. *July, August.*

9. HOLOSCHŒNUS *Link.*

Spikes terminal, clustered. *Rachis* nearly straight. *Bracteæ* gradually diminishing in size. *Hypogynous setæ* wanting. *Fruit* 3-cornered, with a short point. *Style* 2- or 3-parted, filiform, deciduous.

1. H. *vulgaris* Link. E. B. 23. 1612.
Stem round, naked. Involucrum many-leaved. Leaves channelled.
Scirpus Holoschœnus *Linn.*
Sc. australis *Linn.*
Sc. romanus *Linn.*
Isolepis Holoschœnus *R. & S.*
On sandy sea-shores. — Perennial. *September.*

10. ISOLEPIS *R. Br.*

Spikes terminal or lateral. *Rachis* nearly straight. *Bracteæ* gradually diminishing in size. *Hypogynous setæ* wanting. *Fruit* 2-edged, rarely 3-cornered, with a short point. *Style* 2- or 3-parted, filiform, deciduous.

1. I. *setacea* R. Br. E. B. 24. 1693.
Stem bristle-shaped, leafy at the base. Spikes about 2, sessile, surmounted by a leafy bractea. Fruit furrowed.
Scirpus setaceus *Linn.*
In watery places. — Annual. *July, August.*

11. HELIOGITON.

Leaves with short straggling lateral veins. *Spikes* terminal. *Rachis* nearly straight. *Bracteæ* gradually diminishing in size. *Hypogynous setæ* 0. *Fruit* 2-edged, seldom 3-cornered, with a short point. *Style* 2- or 3-parted, filiform, deciduous.

1. H. *fluitans.* E. B. 3. 216.
Stem branched, leafy, pliant, and floating. Flower-stalks alternate, naked. Spikes terminal of few flowers. *Smith.*
Scirpus fluitans *Linn.*

Isolepis fluitans *R. Br.*
Eleogiton fluitans *Link.*
In ditches and ponds. — Perennial. *June, July.*

12. KOBRESIA *Willd.*

Spikes terminal, compound, the lateral clustered. *Bracteæ* male above, female below. *Hypogynous setæ* wanting. *Style* 3-parted. *Fruit* somewhat 3-cornered.

1. K. *caricina* Willd. E. B. 20. 1410.
Spikes aggregate, crowded, alternate.
Carex hybrida *Schk.*
Schœnus monoicus *Smith.*
Carex mirabilis *Host.*
ountains, in muddy spots. — Perennial. *August.*

3. CAREX *Linn.* Sedge.

Spikes bisexual or unisexual (diœcious or androgynous). *Bracteæ* single. *Glumes* of the male florets wanting, of the female 2, united at the margins, ribbed, becoming hard and enclosing a nut. *Style* 2- or 3-parted. *Hypogynous setæ* wanting.

* *Spike solitary, simple.*

1. C. *dioica* Linn. E. B. 8. 543.
Spikes simple, diœcious. Fruit ovate, ribbed, ascending, finely serrated at the edges. Root creeping. *Smith.*
C. capitata *Hudson.*
In spongy bogs. — Perennial. *May, June.*

2. C. *Davalliana* Smith. E. B. 30. 2123.
Spikes simple, diœcious. Fruit lanceolate, triangular, ribbed, deflexed; its angles rough towards the summit. Root tufted. *Smith.*
C. dioica *Hudson.*
In boggy mountainous situations. — Perennial. *May, June.*

3. C. *pulicaris* Linn. E. B. 15. 1051.
Spike simple; bracteæ in the upper half barren; in the lower fertile. Fruit spreading, deflexed, polished, tapering at each end. Stigmas 2.
C. Psyllophora *Ehr.*
In spongy or muddy bogs. — Perennial. *June.*

4. C. *pauciflora* Lightf. E. B. 29. 2041.
Spike simple, lax, of few bracteæ; the uppermost barren. Fruit spreading, deflexed, awl-shaped, smooth. Stigmas 3.
C. patula *Hudson.*
C. Leucoglochin *Ehr.*
In alpine bogs. — Perennial. *June.*

** *Spikes aggregate, each composed of barren and fertile bracteæ. Stigmas 2.*

5. C. *stellulata* Goodenough. E. B. 12. 806.
Spikes 3 or 4, roundish, slightly distant. Barren bracteæ inferior. Fruit spreading, with a tapering undivided beak.

C. muricata *Hudson.*
C. echinata *Sibth.*

In boggy meadows. — Perennial. *May, June.*

6. C. *curta* Goodenough. E. B. 20. 1386.
Spikes about 6, elliptical, slightly distant, with scarcely any involucral leaf. Bracteæ ovate, membranous, about as long as the ovate, tumid, smooth fruit.
C. brizoides *Huds.*
C. canescens *Lightf.*
C. elongata *Leers.*
C. tenella *Ehr.*

In watery meadows. — Perennial. *June.*

7. C. *elongata* Linn.
Spikes numerous, oblong, rather distant, without involucral leaves. Fruit ovate-oblong, tapering, cloven, many-ribbed, recurved, longer than the bracteæ.

At Aldwark, near the river Don, below Sheffield, Yorkshire. — Perennial. *June.*

8. C. *ovalis* Goodenough. E. B. 5. 306.
Spikes about 6, oval, crowded, alternate, with an involucral leaf under the lowermost. Fruit lanceolate, rough-edged, striated, nearly entire, the length of the lanceolate acute bracteæ.
C. leporina *Hudson.*

In marshes and watery meadows. — Perennial. *June.*

9. C. *tenella* Schk.
Spikes 3, distant, minute, of about 3 bracteæ, with involucral leaves. Fruit elliptical, convex at each side, very smooth and even, with a blunt, entire beak. Stamens 2.

In a wood by the river Esk, Angusshire, very rare. — Perennial. *June.*

10. C. *remota* Linn. E. B. 12. 832.
Spikes several, solitary, simple, remote, nearly sessile. Involucral leaves very long, overtopping the stem. Fruit ovate, with a slightly cloven beak.
C. axillaris *Linn.*

In moist shady places. — Perennial. *May, June.*

11. C. *axillaris* Goodenough. E. B. 14. 993.
Spikes several, remote, sessile; the lower ones compound, with very long involucral leaves. Fruit ovate; its beak deeply cloven.

In marshes, and the neighbourhood of wet ditches. — Perennial. *June.*

12. C. *incurva* Lightf. E. B. 13. 927.
Spikes crowded into a dense head. Lower bracteæ fertile. Involucral leaves membranous. Stem roundish, smooth. Leaves channelled.
C. juncifolia *All.*

About the mouths of alpine rivers. — Perennial. *July, August.*

13. C. *arenaria* Linn. E. B. 13. 928.
Spikes numerous, crowded into an oblong head; upper ones chiefl

of barren, lower of fertile, bracteæ. Involucral leaves membranous; lower ones leafy. Stem triangular Leaves flat. Fruit winged.

C. repens *Bellardi.*

On the sea-shore. — Perennial. *June.*

14. C. *intermedia* Goodenough. E. B. 29. 2042.

Spikes numerous, crowded into an oblong dense head; the lowermost and terminal ones fertile; intermediate ones barren. Stem upright, triangular.

C. disticha *Huds.*

C. arenaria *Leers.*

In marshy watery meadows. — Perennial. *May, June.*

15. C. *divisa* Hudson. E. B. 16. 1096.

Head dense, once or twice compounded. Spikes of barren and fertile bracteæ, the latter inferior, most numerous. Involucral leaf leafy, erect. Fruit not spreading. Root creeping.

In marshes. — Perennial. *May, June.*

16. C. *muricata* Linn. E. B. 16. 1097.

Head oblong, dense, prickly with the broad, rough-edged, cloven, spreading beaks of the fruit. Spikes roundish, mostly simple. Root fibrous.

C. spicata *Hudson.*

In moist pastures and shady places. — Perennial. *May,*

17. C. *divulsa* Goodenough.

Head elongated, lax. Spikes of its lower half finally very distant, mostly single. Fruit erect, smooth-edged; roughish at the cloven point of the beak. Root fibrous.

C. canescens *Hudson.*

C. divisa *Don.*

In moist shady pastures. — Perennial. *May.*

18. C. *vulpina* Linn. E. B. 5. 307.

Head thrice compound, dense, obtuse. Fruit spreading, with a notched rough-edged beak. Bracteæ pointed. Angles of the stem compressed, very sharp.

In watery places. — Perennial. *May.*

19. C. *teretiuscula* Goodenough. E. B. 15. 1065.

Head twice or thrice compound, dense. Fruit spreading, tumid at one side, with a tapering, serrated beak. Stem triangular, with convex interstices.

In boggy meadows. — Perennial. *May.*

20. C. *paniculata* Linn. E. B. 15. 1064.

Head thrice compound, loosely panicled, interrupted, acute. Fruit spreading, with an abrupt, serrated beak. Stem sharply triangular, with flat interstices.

In wet pastures. — Perennial. *June.*

*** *Barren and fertile bracteæ in separate spikes; the barren spike solitary. Involucral leaves membranous. Stigmas 3.*

21. C. *digitata* Linn. E. B. 9. 615.
nvolucral leaves membranous, sheathing, scarcely leafy. Spikes linear, lax, erect; the barren one shortest; fertile 2 or 3. Leaves flat.
In thickets, among limestone rocks. — Perennial. *May.*

22. C. *clandestina* Goodenough. E. B. 30. 2124.
Involucral leaves membranous, scarcely leafy. Fertile spikes remote, of few bracteæ, inclosed in the sheathing involucral leaves. Leaves channelled.
C. humilis *Leys.*
C. prostrata *Allioni.*
On St. Vincent's rocks, Bristol, just below the hot wells. — Perennial. *May.*

**** *Barren and fertile bracteæ in separate spikes; the barren spike solitary, very rarely or occasionally more than one. Involucral leaves leafy, often sheathing.*

23. C. *pendula* Hudson. E. B. 33. 2315.
Sheaths nearly as long as the flower-stalks. Fertile spikes cylindrical, very long, drooping. Fruit densely crowded, ovate, beaked. *Smith.*
C. Agastachys *Ehr.*
C. maxima *Scop.*
In moist woods and hedges. — Perennial. *May, June.*

24. C. *strigosa* Hudson. E. B. 14. 994.
Sheaths nearly equal to the flower-stalks. Spikes slender, loose, slightly drooping. Fruit lanceolate, triangular, ribbed. *Smith.*
C. leptostachys *Ehr.*
In groves and thickets. — Perennial. *April, May.*

25. C. *sylvatica* Hudson. E. B. 14. 995.
Sheaths not half the length of the flower-stalks. Spikes slender, rather loose, drooping. Fruit ovate, triangular, beaked, without ribs. *Smith.*
C. Drymeia *Ehr.*
C. patula *Scop.*
C. capillaris *Leers.*
In woods. — Perennial. *May, June.*

26. C. *depauperata* Goodenough. E. B. 16. 1098.
Sheaths much shorter than the flower-stalks. Fertile spikes distant, erect, of about 3 bracteæ. Fruit inflated, ribbed, with a notched beak.
C. ventricosa *Curtis.*
C. triflora *Willd.*
In dry woods. — Perennial. *May, June.*

27. C. *Mielichoferi* Willd. E. B. 32. 2293.
Sheaths not half the length of the flower-stalks. Fertile spikes 3,

distant, erect, lax. Fruit ovate, tumid, triangular, rough-edged; its beak cloven, membranous at the summit. *Smith.*

C. alpina *Hoppe.*

Upon the rocky ledges of Craig Challoch, Breadalbane. — Perennial. *August.*

28. C. *speirostachya* Swartz.

Sheaths shorter than the flower-stalks. Fertile spikes about 3, distant, erect, ovate, dense, many-flowered. Fruit ovate, triangular, ribbed, smooth, with a deeply-cloven beak, membranous at the orifice. *Smith.*

About Mugdoch castle, 9 miles north of Glasgow; also on the hills of Lanarkshire and Perthshire. — Perennial. *July, August.*

29. C. *phæostachya* Smith.

Sheaths shorter than the flower-stalks. Fertile spikes 2, distant, erect, ovate. Fruit ovate, triangular, smooth, with a cloven beak. Bracteæ of the barren spike pointed; of the fertile ones obtuse. *Smith.*

Upon rocks on the high mountains of Cairngorum, Inverness-shire; also on the Clova mountains; and on Ben Macdowie, near the head of the river Dee. — Perennial. *June.*

30. C. *capillaris* Linn. E. B. 29. 2069.

Common sheath much shorter than the 2 or 3 capillary drooping flower-stalks. Fertile spikes ovate, rather loose, pendulous. Fruit ovate, triangular, pointed, without ribs; membranous at the tip. Root fibrous. *Smith.*

On the Highland mountains of Scotland. — Perennial. *July, August.*

31. C. *rariflora* Smith. E. B. 35. 2516.

Sheaths scarcely any. Fertile spikes lax, drooping, of few bracteæ. Fruit obovate, triangular, slightly pointed, without ribs. Root creeping.

C. limosa β. *Wahl.*

At the head of a glen, called the Dell, among the mountains of Clova, Angusshire, near the limits of perpetual snow. — Perennial. *July.*

32. C. *Pseudo-cyperus* Linn. E. B. 4. 242.

Sheaths scarcely any. Fertile spikes dense, cylindrical, drooping, many-flowered. Bracteæ awl-shaped. Fruit spreading, lanceolate, furrowed, rough-edged, with a deeply-cloven beak. *Smith.*

In wet shady places. — Perennial. *June.*

33. C. *limosa* Linn. E. B. 29. 2043.

Sheaths scarcely any. Fertile spikes ovate, dense, drooping, many-flowered. Fruit elliptical, compressed, ribbed, smooth-edged, without a beak. Root creeping. *Smith.*

C. elegans *Willd.*

In deep rotten bogs. — Perennial. *July.*

34. C. *ustulata* Willd. E. B. 34. 2404.

Sheaths very short. Fertile spikes ovate, dense, pendulous. Fruit elliptical, compressed, rough-edged, with a cloven beak. Root tufted, somewhat creeping. *Smith.*

C. nigra *All.*

C. atro-fusca *Schk.*

On Ben Lawers. — Perennial. *July.*

35. C. *atrata* Linn. E. B. 29. 2044.
Sheaths scarcely any. Spikes uniform, stalked, ovate, drooping; the terminal one with many barren bracteæ below. Fruit elliptical, compressed, smooth, with a notched beak. Stamens 2 or 3. *Smith.*

In alpine pastures. — Perennial. *June, July.*

36. C. *pulla* Goodenough. E. B. 29. 2045.
Sheaths none. Fertile spikes ovate; the lower one stalked. Fruit elliptical, slightly inflated, with a short notched beak. Stigmas 2. *Smith.*
C. fusca *Schk.*

On the Highland mountains of Scotland. — Perennial. *July.*

37. C. *pallescens* Linn. E. B. 31. 2185.
Sheaths very short. Fertile spikes cylindrical, stalked; at length pendulous. Fruit obovate, triangular, inflated, smooth, obtuse, with a minute abrupt beak. *Smith.*

In pastures and shady places. — Perennial. *May, June.*

38. C. *flava* Linn E. B. 18. 1294.
Sheaths short, nearly equal to the flower-stalks. Fertile spikes roundish-ovate. Fruit triangular, smooth, with a cloven beak curved downward. Stem nearly smooth. *Smith.*

In boggy meadows. — Perennial. *May, June.*

39. C. *Oederi* Ehr. E. B. 25. 1773.
Sheaths and flower-stalks all very short. Fertile spikes roundish-ovate. Fruit globular, triangular, direct, smooth, with a straight cloven beak. Stem smooth. *Smith.*
C. flava β. *Hooker.*

On wet commons. — Perennial. *July.*

40. C. *fulva* Goodenough. E. B. 18. 1295.
Sheaths tubular, elongated, shorter than the flower-stalks. Fertile spikes ovate, erect. Bracteæ pointless. Fruit ovate, triangular, direct, smooth, with a straight, cloven, rough-edged beak. Stem rough. *Smith.*

In boggy meadows. — Perennial. *June, July.*

41. C. *extensa* Goodenough. E. B. 12. 833.
Sheaths and flower-stalks very short. Involucral leaves very long and spreading. Fertile spikes elliptic-oblong, near together. Bracteæ somewhat awned. Fruit ovate, triangular, with a short, smooth, cloven beak. Stem very smooth.

On the sea-coast. — Perennial. *June.*

42. C. *distans* Linn. E. B. 18. 1234.
Sheaths tubular, elongated, nearly equal to the flower-stalks. Fertile spikes elliptic-oblong, widely distant. Bracteæ pointed. Stem smooth.

In muddy marshes. — Perennial. *June.*

43. C. *binervis* Smith. E. B. 18. 1235.
Sheaths tubular, elongated, shorter than the flower-stalks. Fertile

spikes cylindrical, distant; partly compound. Bracteæ pointed. Stem smooth. Fruit with 2 principal ribs.

On dry heaths. — Perennial. *June.*

44. C. *præcox* Jacq. E. B. 16. 1099.

Sheaths about equal to the very short flower-stalks. Spikes all elliptical, rather crowded. Bracteæ of the fertile ones pointed. Fruit pear-shaped, downy, with an abrupt entire point.

C. saxatilis *Hudson.*

C. montana *Lightf.*

C. filiformis *Leers.*

C. stolonifera *Ehr.*

On dry heaths and hillocks. — Perennial. *April.*

45. C. *pilulifera* Linn. E. B. 13. 885.

Sheaths none. Fertile spikes 2 or 3, sessile, crowded, almost globular, with pointed bracteæ. Fruit triangular, roundish, downy, with a short cloven beak.

C. montana *Linn.*

C. filiformis *Fl. Dan.*

C. decumbens *Ehr.*

On heaths. — Perennial. *April, May.*

46. C. *tomentosa* Linn. E. B. 29. 2046.

Sheaths extremely short. Fertile spikes 1 or 2, nearly sessile, cylindrical, obtuse, with acute bracteæ. Fruit globose, slightly triangular, densely downy, with a short cloven beak.

C. sphærocarpa *Ehr.*

In meadows in the south of England. — Perennial. *June.*

47. C. *panicea* Linn. E. B. 21. 1505.

Sheaths elongated, about half the length of the flower-stalks. Fertile spikes 1 or 2, distant; lower one rather lax. Fruit tumid, smooth, cloven at the summit. Stem smooth, obtusely triangular. *Smith.*

In meadows. — Perennial. *May, June.*

48. C. *recurva* Huds. E. B. 21. 1506.

Sheaths short. Fertile spikes 2 or 3, cylindrical, dense, drooping, on very long recurved stalks. Fruit elliptical, triangular, roughish, obtuse, slightly notched. *Smith.*

C. glauca *Scop.*

C. flacca *Schreb.*

C. pendula *Schreb.*

β *Micheliana.* Fruit smooth. E. B. 32. 2236.

C. Micheliana *Smith.*

C. ambleocarpa *Willd.*

In moist meadows. — Perennial. *May, June.*

49. C. *rigida* Goodenough. E. B. 29. 2047.

Stigmas 2. Sheaths none. Fertile spikes ovate; the lowermost stalked. Involucral leaves lanceolate, recurved, as well as the leaves. Fruit triangular somewhat compressed, with a short abrupt beak.

C. saxatilis *Fl. Dan.*
C. fusca *All.*
C. mucronata *Jacq.*
On the tops of mountains. — Perennial. *June, July.*

50. C. *cæspitosa* Linn. E. B. 21. 1507.
Stigmas 2. Sheaths none. Fertile spikes cylindrical, obtuse, erect; the lowermost rarely stalked. Leaves, and auricled involucral leaves, linear, erect. Fruit permanent, elliptical, flat, many-ribbed, with a very short abrupt beak.
In wet places. — Perennial. *May, June.*

51. C. *stricta* Goodenough. E. B. 13. 914.
Stigmas 2. Sheaths none. Fertile spikes nearly sessile, erect, cylindrical, elongated, acute; often barren-flowered at the top. Bases of the leaves reticulated. Fruit elliptical, flat, with a short cloven beak, deciduous. *Smith.*
C. cæspitosa *Hudson.*
In marshes.— Perennial. *April.*

***** *Barren and fertile bracteæ in separate spikes. Barren spikes 2 or more.*

52. C. *acuta* Linn. E. B. 9. 580.
Stigmas 2. Spikes cylindrical, slender; drooping in flower; afterwards erect. Fruit elliptical, with a blunt undivided beak. *Smith.*
C. gracilis *Curtis.*
In watery meadows. — Perennial. *May.*

53. C. *paludosa* Goodenough. E. B. 12. 807.
Stigmas 3. Spikes cylindrical, bluntish, erect; the fertile ones with taper-pointed bracteæ. Fruit ovate, triangular, compressed, with a notched beak. *Smith.*
C. acutiformis *Ehr.*
C. acuta *Curt.*
In meadows. — Perennial. *May.*

54. C. *riparia* Curtis. E. B. 9. 579.
Stigmas 3. Spikes erect, with taper-pointed bracteæ. Fruit ovate, tumid, with a deeply-cloven beak. *Smith.*
C. acuta *Hudson.*
C. crassa *Ehr.*
In watery places. — Perennial. *April, May.*

55. C. *lævigata* Smith. E. B. 20. 1387.
Spikes cylindrical; fertile ones stalked. Bracteæ all pointed. Sheaths very long. Fruit triangular, with a cloven beak. *Smith.*
C. patula *Schk.*
C. æthiopica *Schk.*
In marshes. — Perennial. *June.*

56. C. *vesicaria* Linn. E. B. 11. 779.
Fertile spikes cylindrical, short, abrupt, on short stalks. Bracteæ all

lanceolate, acute. Sheaths none. Fruit ovate, inflated, with an elongated cloven beak. *Smith.*

C. inflata *Hudson.*

In marshes. — Perennial. *May.*

57. C. *ampullacea* Goodenough. E. B. 11. 780.

Fertile spikes cylindrical, elongated, nearly sessile. Bracteæ all lanceolate, acute. Sheaths none. Fruit inflated, globose, with a linear cloven beak. *Smith.*

C. vesicaria *Hudson.*

C. rostrata *Sibth.*

C. obtusangula *Ehr.*

In marshes. — Perennial. *May.*

58. C. *hirta* Linn. E. B. 10. 685.

Herbage hairy. Fertile spikes ovate-cylindrical, remote. Bracteæ awned. Sheaths nearly as long as the flower-stalks. Fruit hairy, tumid, with a deeply-cloven beak. Stem rough-edged. *Smith.*

C. anonyma *Fl. Dan.*

In wet meadows. — Perennial. *May, June.*

59. C. *secalina* Willd.

Fertile spikes ovate-cylindrical; the lower one very remote. Bracteæ acute. Sheaths as long as the flower-stalks. Fruit ovate, rough-edged, compressed; concave at the inner side; with an elongated, linear, cloven beak. Stem smooth. *Smith.*

In a den near Panmure, about 9 miles south-east of Forfar. — Perennial. *June, July?*

60. C. *stictocarpa* Smith.

Fertile spikes 2, ovate, stalked. Bracteæ pointed. Sheaths scarcely any. Fruit obovate, obtuse, pointless, finely dotted. *Smith.*

On the lofty mountains of Clova, Angusshire. — Perennial. *June, July?*

1. C. *angustifolia* Smith.

Fertile spikes 1 or 2, ovate, stalked. Bracteæ obtuse. Sheaths none. Fruit ovate, compressed, smooth, with a short abrupt beak. Leaves linear, channelled. *Smith.*

In a marsh in Angusshire. — Perennial. *June?*

62. C. *filiformis* Linn. E. B. 13. 904.

Fertile spikes ovate. Bracteæ pointed. Sheaths nearly equal to the short flower-stalks. Fruit ovate, hairy, with a deeply-cloven beak. Leaves linear, channelled, smooth. *Smith.*

C. tomentosa *Lightf.*

C. hirta *Fl. Dan.*

C. lasiocarpa *Ehr.*

C. splendida *Willd.*

In boggy meadows. — Perennial. *June.*

Order 103. GRAMINEÆ *Juss.*

Flowers usually hermâphrodite, sometimes monœcious or polygamous; consisting of imbricated bracteæ, of which the most exterior are called *glumes*, the interior immediately enclosing the stamens *paleæ*, and the innermost at the base of the ovarium *scales*.

Glumes usually 2, alternate; sometimes single, most commonly unequal.

Paleæ 2, alternate; the lower or exterior simple, without keel; the upper or interior composed of two united by one margin, and usually with 2 keels.

Scales 2, sometimes wanting, collateral, alternate with the paleæ, and opposite the lower of them; either distinct, or united.

Stamens hypogynous, 1, 2, 3, 4, 6, or more; *anthers* versatile.

Ovarium single; *styles* 2, very rarely 1 or 3; *stigmas* feathery or hairy.

Pericarpium usually undistinguishable from the seed, membranous.

Albumen farinaceous; *embryo* lying on one side of the albumen at the base, lenticular, with a broad cotyledon, and a developed plumula; and occasionally, but very rarely, with a second cotyledon on the outside of the plumula, and alternate with the usual cotyledon.

Roots fibrous or bulbous. *Stems* cylindrical, fistular, closed at the joints, covered with a coat of silex. *Leaves* alternate, with a split sheath. *Flowers* in spikes, racemes, or panicles.

ANALYSIS OF THE GENERA.

- Inflorescence in imbricated or 2-sided spikes
  - Spikelets imbedded in the rachis - - - - - 1. OPHIURUS.
  - Spikelets not imbedded
    - A bractea at the base of the spiculæ - - 2. LOLIUM.
    - Bractea wanting
      - Glumes none - - - - - - - 3. NARDUS.
      - Glumes twin
        - Spikelets 1-flowered - - - - - 4. HORDEUM.
        - Spikelets 2- or many-flowered
          - in pairs, threes, or fours - - 5. ELYMUS.
          - solitary
            - Rachis flat - - - - - 6. CATOPODIUM.
            - Rachis nearly round or angular
              - Florets equal-sized - - 7. BRACHYPODIUM.
              - Florets smallest upwards - 8. AGROPYRUM.
- Inflorescence in 1-sided spikes
  - Spikelets 1-flowered. Glumes keeled
    - Scales present - - - - - - - 9. CYNODON.
    - Scales wanting - - - - - - 10. SPARTINA.
  - Spikelets 2-flowered. Glumes flat - - - - 11. DIGITARIA.
- Inflorescence panicled. Spikelets 1-flowered
  - Florets enveloped in hair - - - - - - 12. LAGURUS.
  - Florets naked

Panicle spiked, or densely contracted
  Glumes awnless
    Palea single, with a basal awn - - 13. ALOPECURUS.
    Paleæ 2, awnless
      Glumes navicular - - - - 14. PHALARIS.
      Glumes not navicular - - - 15. ACHNODON.
  Glumes awned - - - - - - 16. PHLEUM.
Panicle loose
  Paleæ cartilaginous, plane - - - 17. MILIUM.
  Paleæ membranous
    awnless
      Glumes with a tuft of hairs in the inside at the base - - } 18. DIGRAPHIS.
      Glumes naked at the base - - 19. CHAMAGROSTIS.
    awned
      Awns twisted - - - - - 20. STIPA.
      Awns straight
        Glumes awned - - - 21. POLYPOGON.
        Glumes awnless
          ventricose - - - 22. GASTRIDIUM.
          keeled - - - - - 23. AMMOPHILA.
          convex
            Palea 1 - - - 24. TRICHODIUM.
            Paleæ 2
              No rudiment of a second floret
                Paleæ naked at the base } 25. AGROSTIS.
                Paleæ surrounded by hairs - - } 26. CALAMAGROSTIS.
              A rudiment of a second floret - } 27. ANEMAGROSTIS

Inflorescence panicled. Spikelets 2- or 3-flowered
  Florets of 2 sexes
Florets 2
  upper hermaphrodite, lower male or neuter
    Glumes nearly equal - - - - - 28. ARRHENATHERUM.
    Lower glume much smaller - - - - 29. ECHINOCHLOA.
  upper male, lower hermaphrodite - - - 30. HOLCUS.
Florets 3
  Lateral florets, neuter - - - - - - 31. ANTHOXANTHUM.
  Lateral florets, male - - - - - - 32. HIEROCHLOE.
  Florets all of the same sex
    Lower glume largest
      Spikelets with a pinnated bractea - - 33. CYNOSURUS.
      Spikelets destitute of bracteæ
        Glumes truncate - - - - - 34. CATABROSA.
        Glumes acute
          Paleæ awnless
            Panicle loose
              Glumes much shorter than the florets - - } 35. MOLINIA.
              Glumes the length of the florets - - - } 36. MELICA.
            Panicle contracted - - - 37. AIROCHLOA.
          Paleæ awned
            Awn articulated clavate - 38. CORYNEPHORUS.
            Awn twisted
              Paleæ emarginate - - 39. AIRA.
              Paleæ deeply cleft - - 40. TRISETUM.
            Awn straight - - - - - 41. DESCHAMPSIA.
    Lower glume smallest - - - - - - 42. SETARIA.

Inflorescence panicled. Spikelets many-flowered
  Panicle spiked - - - - - - - - 43. SESLERIA.
  Panicle loose
    Florets of different sexes
      lower male, upper hermaphrodite - - 44. ARUNDO.
      lower hermaphrodite, upper neuter - 45. AVENA.
    Florets all hermaphrodite
      Paleæ awned
        Spikelets clustered at the end of the ramifications - - - - - } 46. DACTYLIS.
        Spikelets regularly panicled or racemose

Lower palea with 3 nearly equal teeth - - - - - 47. Triodia.
Lower palea awned under the apex 48. Bromus.
Lower palea awned at the apex
Glumes nearly equal, or not very unequal
Pedicels of the florets with 3 tufts of hair at the end - - - 49. Schedonorus.
Pedicels of the florets naked at the end - - 50. Festuca.
Glumes very unequal, sometimes single - - - 51. Vulpia.
Paleæ awnless
Scale single - - - - - - 52. Glyceria.
Scales 2
Spikelets cordate - - - 53. Briza.
Spikelets not cordate
Pedicels articulated with the spikelets - - - - 54. Sclerochloa.
Pedicels not articulated with the spikelets
Scales truncate - - 55. Hydrochloa.
Scales acute - - - 56. Poa.

I. *Inflorescence in imbricated or 2-sided spikes.*

1. OPHIURUS *Beauv.*

Spikelets 1-flowered, seldom 2-flowered. *Glumes* 2, seldom single, lateral, longer than the floret. *Paleæ* 2, awnless, about as long as the glumes. *Scales* lanceolate, acute. *Styles* pencil-shaped.

1. O. *incurvatus* Beauv. Sea Hard-grass. E. B. 11. 760.
Spikes cylindrical, tapering. Glumes combined below. Floret solitary. Paleæ awnless.
Rottböllia incurvata *Linn.*
Ægilops incurvata *Linn.*
On the sea-coast. — Annual. *August.*

2. LOLIUM *Linn.*

Spikelets many-flowered, at right angles with the rachis. A *bractea* at the base of the spikelet. *Glumes* 2, lateral, often deficient. *Paleæ* 2, nearly equal; the outer often awned under the apex. *Scales* oval, gibbous, nearly acute. *Styles* feathery.

1. L. *perenne* Linn. Perennial Darnel. Rye-grass. E. B. 5. 315.
Paleæ very slightly awned. Spikelets longer than the glumes. Florets lanceolate.
Lolium tenere *Linn.*, a variety.
In meadows and waste ground. — Perennial. *June.*

2. L. *temulentum* Linn. E. B. 16. 1124.
Awns longer than the paleæ. Spikelets shorter than the glumes. Florets elliptical. Stem rough in the upper part.
In fields, among wheat, barley, or flax. — Annual. *July.*

3. L. *arvense* With. Annual Darnel. E. B. 16. 1125.
Paleæ slightly awned. Spikelets as long as the glumes. Florets elliptical. Stem very smooth.
In fields.— Annual. *July.*

O 4

3. NARDUS *Linn.*

Glumes none. *Paleæ* 2; the outer wrapping up the inner, which is flat. *Style* hairy.

1. N. *stricta* Linn. Mat-grass. E. B. 5. 290.
Spike bristle-shaped, straight. Leaves thrice the length of their sheaths. *Smith.*
On barren heaths and moors. — Perennial. *July.*

4. HORDEUM *Linn.*

Spikelets in pairs or threes, 1-flowered. *Glumes* 2. *Paleæ* 2; the lower awned at the apex. *Scales* obtuse, fringed. *Styles* feathery. *Ovarium* villous at the end.

1. H. *murinum* Linn. Wall Barley. Mouse Barley. Way Bennet. E. B. 28. 1971.
Lateral flowers barren. Glumes of the intermediate one lanceolate, fringed. *Smith.*
In waste ground. — Annual. *June—August.*

2. H. *pratense* Huds. E. B. 6. 409.
Lateral flowers imperfect, with shorter awns. All the glumes bristle-shaped and rough. *Smith.*
H. nodosum *Linn.*
H. secalinum *Willd.*
In meadows and pastures. — Perennial. *June.*

3. H. *maritimum* With. Sea Barley. Squirrel-tail-grass. E. B. 17. 1205.
Lateral flowers imperfect, with shorter awns; the inner glume half-ovate. *Smith.*
H. marinum *Huds.*
H. geniculatum *Allion.*
H. rigidum *Roth.*
In pastures and sandy ground. — Annual. *June, July.*

5. ELYMUS *Linn.*

Spikelets in pairs, threes, or fours, many-flowered or somewhat 2-flowered. *Glumes* 2, as long as the spikelet, or shorter. *Paleæ* 2; the lower often awned at the apex. *Scales* obtuse, generally fringed. *Styles* 2, feathery.

1. E. *arenarius* Linn. Sea Lyme-grass. E. B. 24. 1672.
Spike upright, close; main stalk not winged. Glumes lanceolate, the length of the spikelets. Leaves spinous-pointed. *Smith.*
On the sea-shore. — Perennial. *July*, but rarely.

2. E. *geniculatus* Curtis. E. B. 23. 1586.
Spike bent perpendicularly downward, lax; main stalk winged. Glumes awl-shaped, longer than the spikelets. Leaves spinous-pointed. *Smith.*
In marshes near the coast. — Perennial. *July.*

3. E. *europæus* Linn. E. B. 19. 1317
Spike erect. Florets about 2, rough, awned, as well as the glumes. Leaves flat, pliant. *Smith.*
Hordeum sylvaticum *Huds.*
In woods and hedges. — Perennial. *June.*

6. CATOPODIUM *Link.*

Inflorescence spiked. *Rachis* flat, flexuose. *Spikelets* alternate, 1-sided, many-flowered. *Glumes* 2, unequal, shorter than the lower florets. *Paleæ* 2; the lower awnless. *Scales* 2, lanceolate. *Styles* somewhat pencil-shaped.

1. C. *loliaceum* Link. E. B. 4. 221.
Glumes obtuse, awnless. Florets numerous, awnless, elliptical, ribbed. Spike unilateral. Stem branched. Root fibrous. *Smith.*
Triticum loliaceum *Smith.*
T. unilaterale *H. Kew.*
Poa loliacea *Huds.*
On the sea-coast. — Annual. *June, July.*

7. BRACHYPODIUM *Beauv.*

Inflorescence somewhat racemose. *Spikelets* many-flowered, with a short pedicel. *Glumes* 2, unequal, shorter than the lower florets. *Paleæ* 2; the lower awned under the point, or awnless. *Scales* lanceolate, hairy above, occasionally bifid. *Styles* feathery. *Ovarium* villous at the end.

1. B. *sylvaticum* Beauv. E. B. 11. 729.
Spike simple, drooping. Spikelets nearly cylindrical, turned to one side. Awns longer than their glumes. Leaves hairy. Root fibrous. *Smith.*
Festuca sylvatica *Huds.*
F. gracilis *Mœnch.*
Bromus sylvaticus *Pollich.*
B. gracilis *Weig.*
In dry copses and hedges. — Perennial. *July.*

2. B. *pinnatum* Beauv. E. B. 11. 730.
Spike simple, erect, 2-ranked. Spikelets nearly cylindrical. Awns shorter than their glumes. Leaves nearly smooth. Root somewhat creeping. *Smith.*
Festuca pinnata *Huds.*
Bromus pinnatus *Linn.*
In open fields and heaths. — Perennial. *July.*

8. AGROPYRUM *Beauv.*

Inflorescence spiked, and somewhat racemose. *Spikelets* many-flowered. *Florets* all fertile, diminishing upwards. *Glumes* 2, shorter than the lower florets. *Paleæ* 2; the lower generally awned. *Scales* 2, acute, seldom partially bifid. *Styles* feathery. *Ovarium* hairy at the point.

1. A. *junceum* Beauv. E. B. 12. 814.
Glumes blunt, many-ribbed. Florets about 5, awnless. Main stalk smooth. Leaves involute, sharp-pointed. Root creeping. *Smith.*
Triticum junceum *Linn.*
On the sea-coast. — Perennial. *July.*

2. A. *repens* Beauv. Couch-grass. E. B. 13. 909.
Glumes pointed or awned, lanceolate, many-ribbed. Florets about 5, sharp-pointed or awned. Leaves flat. Root creeping. *Smith.*
Triticum repens *Linn.*
In waste, as well as cultivated, land. — Perennial. *July.*

3. A. *caninum* Beauv. E. B. 20. 1372.
Glumes somewhat awned, with 3 or 5 ribs. Florets 4, awned. Leaves flat. Root fibrous. *Smith.*
Elymus caninus *Linn.*
Triticum caninum *Huds.*
In woods and shady hedges. — Perennial. *July.*

4. A. *cristatum* Beauv. E. B. 32. 2267.
Glumes elliptical, awned, keeled, obscurely ribbed. Florets awned. Spikelets closely imbricated, depressed, straight. Stems simple. *Smith.*
Triticum cristatum *Linn.*
Bromus cristatus *Linn.*
On steep banks and rocks by the sea-side, between Arbroath and Montrose. — Perennial. *July*, but rarely.

II. *Inflorescence in 1-sided spikes.*

9. CYNODON *Rich.*

Spikelets 1-sided, in 2 or more rows, 1-flowered, attached to a flat rachis. *Glumes* 2, keeled, nearly equal, shorter than the paleæ. *Paleæ* 2, keeled, the upper enwrapped by the lower. *Styles* pencil-shaped. The rudiment of an abortive floret.

1. C. *Dactylon* Rich. Dog's-tooth-grass. E. B. 12. 850.
Spikes 4 or 5, crowded together. Paleæ smooth.
Panicum Dactylon *Linn.*
Digitaria stolonifera *Schrad.*
Agrostis linearis *Retz.*
On the sandy shores of Cornwall. — Perennial. *July, August.*

10. SPARTINA *Willd.*

Spikelets 1-flowered, 1-sided, in 2 rows, pressed close to the rachis. *Glumes* 2, unequal, often awned; the lower smaller and narrower, the upper about the same length as the paleæ. *Paleæ* 2, nearly equal, both keeled. *Scales* wanting.

1. S. *stricta* Smith. E. B. 6. 380.
Spikes 2 or 3, erect, with very smooth stalks. Glumes downy; outer smallest. *Smith.*

Dactylis stricta *H. Kew.*
D. cynosuroides *Hudson.*
In muddy salt marshes, on the eastern coast. — Perennial. *August.*

11. DIGITARIA *Scopoli.*

Spikes somewhat fascicled. *Spikelets* 2, about 2-flowered, their back turned to the rachis; one on a longer stalk than the other. *Glumes* 2, the lower very small, sometimes wanting. *Paleæ* of the neuter floret single. *Paleæ* of the hermaphrodite floret 2, nearly equal, cartilaginous, awnless.

1. D. *sanguinalis* Scop. Cock's-foot Finger-grass. E. B. 12. 849.
Leaves and their sheaths somewhat hairy. Flowers in pairs. Glumes rough at the edges of the largest only.
Panicum sanguinale *Linn.*
Syntherisma vulgare *Schreb.*
In sandy cultivated fields. — Annual. *July, August.*

III. *Inflorescence panicled. Spikelets 1-flowered.*

12. LAGURUS *Linn.*

Panicle contracted. *Spikelets* 1-flowered. *Glumes* 2, equal, terminating in a long plumed awn. *Paleæ* 2, the lower with a dorsal awn, membranous and bifid at the apex, with long, sharp, awn-like lobes.

1. L. *ovatus* Linn. Hare's-tail-grass. E. B. 19. 1334.
In Guernsey. — Annual. *June.*

13. ALOPECURUS *Linn.*

Panicle contracted. *Glumes* 2, equal, keeled, often connate at the base, about as long as the palea. *Palea* single, with an awn arising from its base, ribbed. *Style* single or double, hairy.

1. A. *pratensis* Linn. Meadow Fox-tail-grass. E. B. 11. 759.
Stem erect, smooth. Spike somewhat panicled. Glumes acute, hairy, combined at the base, shorter than the awn of the palea.
In meadows and pastures. — Perennial. *May.*

2. A. *alpinus* Smith. E. B. 16. 1126.
Stem erect, smooth. Spike ovate, somewhat panicled. Glumes woolly, obliquely abrupt, nearly as long as the awn of the palea.
? A. ovatus *Knapp.*
On the mountains about Loch-ne-gare, Aberdeenshire. — Perennial. *July.*

3. A. *agrestis* Linn. E. B. 12. 848.
Stem erect, roughish. Spike racemose, nearly simple, tapering. Glumes almost naked, combined at the base, dilated at the keel. *Smith.*
In cultivated fields, and by way sides. — Annual. *July.*

4. A. *bulbosus* Linn. E. B. 18. 1249.
Stem erect. Spike tapering, perfectly simple. Glumes distinct, linear, pointed, downy. Root bulbous. *Smith.*
In wet salt marshes. — Perennial. *July.*

5. A. *geniculatus* Linn. E. B. 18. 1250.
Stem ascending, bent at the joints. Spike cylindrical, slightly panicled. Glumes combined at the base, abrupt, fringed. Palea notched, its awn twice the length of the glumes.
A. paniceus *Fl. Dan.*
Floats in ponds and slow streams. — Perennial. *July.*

6. A. *fulvus* Smith. E. B. 21. 1467.
Stem ascending, bent at the joints. Spike cylindrical, panicled. Glumes combined at the base, obtuse, fringed. Awn the length of the palea. Anthers roundish.
Floats in ponds and ditches. — Perennial. *July.*

14. PHALARIS *Linn.*

Panicle contracted. *Glumes* 2, navicular, equal, longer than the paleæ, each with a scale at its base inside, representing an abortive floret. *Paleæ* 2, awnless, equal, forming a hard shining coat to the grain.

1. Ph. *canariensis* Linn. Common Canary-grass. E. B. 19. 1310.
Panicle ovate, resembling a spike. Glumes entire at the summit.
In cultivated and waste ground. — Annual. *June—August.*

15. ACHNODON *Trinius.*

Panicle contracted. *Glumes* 2, equal, acute, converging, much longer than the paleæ. *Paleæ* 2, equal, awnless, convex, membranous, ribbed. A *rudiment* of a second floret at the back of the upper palea.

1. A. *arenarius* Trinius. E. B. 4. 222.
Spike slightly panicled, ovate-lanceolate, obtuse. Glumes lanceolate, fringed, thrice the length of the abrupt, notched paleæ.
Phleum arenarium *Linn.*
Phalaris arenaria *Willd.*
Chilochloa arenaria *Trinius.*
On the sandy sea-coast. — Annual. *May.*

16. PHLEUM *Linn.*

Panicle contracted *Glumes* 2, keeled, equal, longer than the paleæ, with an awn proceeding from the midrib. *Paleæ* 2, equal, awnless, membranous, usually convex, and ribless. *Styles* half pencil-shaped.

1. Ph. *pratense* Linn. Cat's-tail-grass. Timothy-grass. E. B. 15. 1076.
Cluster spiked, cylindrical. Glumes abrupt, fringed at the keel, longer than its awns.
P. nodosum *Linn.*
In moist meadows and pastures. — Perennial. *June—October.*

2. Ph. *alpinum* Linn. E. B. 8. 519.
Cluster spiked, ovate-oblong. Glumes abrupt, fringed at the keel, as long as their awns. Root tuberous.
On the mountains of Scotland. — Perennial. *July.*

3. Ph. *asperum* Jacq. E. B. 15. 1077.
Panicle spiked, cylindrical. Glumes wedge-shaped, swelling upward, pointed, rough; keel naked. Stem branched.
P. paniculatum *Hudson.*
P. viride *Allioni.*
Phalaris aspera *Retz.*
Phalaris paniculata *H. Kew.*
In dry open fields. — Annual. *July.*

4. Ph. *Boehmeri* Schrad. E. B. 7. 459.
Panicle spiked, nearly cylindrical. Glumes linear-lanceolate, slightly pointed, nearly smooth, abrupt at the inner margin. Stem simple. *Smith.*
Phalaris phleoides *Linn.*
In sandy or chalky fields. — Perennial. *July.*

5. Ph. *Michelii* Allioni. E. B. 32. 2265.
Panicle spiked, nearly cylindrical. Glumes lanceolate, taper-pointed, hairy, fringed. Paleæ oblong, firm, hairy; upper cloven.
Phalaris alpina *Hænke.*
On the rocky parts of the mountains of Clova, Angusshire. — Perennial. *July.*

17. MILIUM *Linn.*

Panicle loose. *Glumes* 2, flattish, equal, ribbed, as long as the paleæ, or a little longer. *Paleæ* 2, equal, ribless, very smooth, awnless; the upper flat. *Styles* feathery.

1. M. *effusum* Linn. E. B. 16. 1106.
Flowers in a loose spreading panicle.
In moist shady places. — Perennial. *June, July.*

18. DIGRAPHIS *Trinius.*

Panicle open. *Glumes* 2, keeled, equal, longer than the paleæ; each with a tuft of hairs at its base, representing an abortive floret. *Paleæ* 2, equal, awnless, coating the grain.

1. D. *arundinacea* Trinius. E. B. 6. 402.
Panicle upright, with spreading branches. Flowers crowded, unilateral.
Phalaris arundinacea *Linn.*
Baldingera arundinacea *Dumort.*
Arundo colorata *H. Kew.*
By the side of water. — Perennial. *July.*

19. CHAMAGROSTIS *Borkh.*

Raceme simple. *Glumes* 2, equal, awnless, longer than the paleæ. *Paleæ* 2, equal, convex, awnless. *Styles* feathery.

1. C. *minima* Schrader. E. B. 16. 1127.
Agrostis minima *Linn.*
Sturmia minima *Hoppe.*
Knappia agrostidea *Smith.*
Mibora verna *Beauv.*
In sandy pastures upon the sea-coast. — Annual. *March, April.*

20. STIPA *Linn.*

Glumes 2, equal, or nearly so, and longer than the paleæ. *Paleæ* 2; the lower convolute, bearing from its apex a twisted articulated awn. *Anthers* naked.

1. S. *pennata* Linn. Feather-grass. E. B. 19. 1356.
Awns feathery. *Smith.*

Found by Dr. Richardson, in company with Thomas Lawson, on the lime-stone rocks hanging over a little valley, called Long Sleadale, about six miles north of Kendall, Westmoreland. Nobody has been able to meet with it since. *Smith.* — Perennial. *June.*

21. POLYPOGON *Desf.*

Panicle contracted. *Glumes* 2, equal, much longer than the paleæ; the lower awned under the apex, the upper from the apex. *Paleæ* 2, equal, convex; the lower awned beneath the apex. *Styles* feathery.

1. P. *monspeliensis* Desf. E. B. 24. 1704.
Awns straight, thrice as long as the glumes. Root fibrous.
Alopecurus monspeliensis *Linn.*
Alopecurus aristatus *Hudson.*
Phleum crinitum *Schreb.*
Agrostis panicea *H. Kew.*
Agrostis triaristata *Knapp.*
Cynosurus paniceus *Linn.*
In moist pastures near the sea. — Annual. *July, August.*

2. P. *littoralis* Smith. E. B. 18. 1251.
Awns straight, about the length of the glumes. Root creeping.
Agrostis littoralis *Smith.*
Polypogon Lagascæ *Trin.*
In muddy salt marshes. — Perennial. *July, August.*

22. GASTRIDIUM *Beauv.*

Panicle contracted into the form of a spike. *Glumes* 2, awnless, nearly equal, ventricose, much longer than the paleæ. *Paleæ* 2, equal, membranous; the lower usually awned under the apex. Sometimes a rudiment of a second floret at the back of the upper palea. *Styles* pencil-shaped.

1. G. *lendigerum* Link. E. B. 16. 1107.
Flowers in a dense spiked panicle.
Agrostis australis *Linn.*
Milium lendigerum *Linn.*
Agrostis rubra *Hudson.*
Agrostis ventricosa *Gouan.*
Alopecurus ventricosus *Hudson.*
Gastridium australe *Beauv.*
In fields where water has stagnated. — Annual. *August.*

23. AMMOPHILA *Host.*

Glumes 2, keeled, unequal, the lower smaller, longer than the paleæ. *Paleæ* 2; the lower awned under the apex, with hairs at the base. *Scales* longer than the ovarium.

1. A. *arenaria* Host. Sea Reed. Marram. Sea Mat-weed. E. B. 8. 520.

Panicle spiked. Flowers erect, slightly awned. Leaves involute, sharp-pointed.

Arundo arenaria *Linn.*

Calamagrostis arenaria *Roth.*

Frequent on sea-coast. — Perennial. *July.*

24. TRICHODIUM *Schrad.*

Panicle loose. *Glumes* 2, nearly equal; the lower larger, longer than the paleæ. *Paleæ* single, usually with a dorsal awn. *Styles* feathery.

1. T. *caninum* Schrad. E. B. 26. 1856.

Awn incurved, from below the middle of the palea. Glumes ovate, coloured Stems decumbent, with prostrate shoots.

Agrostis canina *Linn.*

A. vinealis *With.*

A. stricta *Sinclair.*

Agragulus caninus *Beauv.*

Agrostis tenuifolia *Curtis.* } a variety according to *Smith.*

A. fascicularis *Sinclair.* }

In meadows and pastures. — Perennial. *June, July.*

2. T. *setaceum* R. & S. E. B. 17. 1188.

Awn incurved, from near the base of the palea. Glumes lanceolate, tapering, rough. Radical leaves bristle-shaped. Stem nearly erect. Panicle close, oblong.

Agrostis setacea *Curtis.*

A. alpina *Withering.*

On dry turfy heaths. — Perennial. *July, August*

25. AGROSTIS *Linn.*

Panicle loose. *Glumes* 2, nearly equal, the lower larger, longer than the paleæ. *Paleæ* 2, unequal, the lower larger, sometimes with a dorsal awn. *Styles* feathery.

1. A. *vulgaris* Withering. Bent-grass. E. B. 24. 1671.

Panicle spreading; with divaricated, capillary branches. Glumes nearly equal. Stem erect. Ligula abrupt, very short.

A. hispida *Willd.*

A. tenuis *Sibth.*

A. capillaris *Abbott.*

A. polymorpha *Hudson.*

A. canina *Withering*; a variety.

A. pumila *Linn.*; a variety.

Everywhere. Perennial. *July, August.*

2. A. *alba* Linn. Fiorin-grass. E. B. 17. 1189.

Panicle condensed at the base of the main divisions; stalks rough. Glumes lanceolate, bristly at the keel. Stem spreading, creeping. Ligula oblong, ribbed.

A. mutabilis *Knapp.*

A. palustris *Sinclair.*

A. stolonifera *Linn.*; a variety.
A. sylvatica *Hudson*; a variety.
In moist meadows and fields. — Perennial. *July, August.*

26. CALAMAGROSTIS *Roth.*

Glumes 2, convex, equal or unequal, larger than the paleæ. *Paleæ* 2, unequal, membranous, ribbed, surrounded with hairs at the base; the lower awned. *Styles* 2, feathery.

1. C. *epigejos* Roth. E. B. 6. 403.
Panicle erect, close. Flowers crowded, unilateral. Paleæ with a dorsal awn about as long as the hairs and glumes. Leaves lanceolate.
Arundo epigejos *Linn.*
A. Calamagrostis *Hudson.*
Calamagrostis lanceolata *With.*
In shady ditches, and moist woods. — Perennial. *July.*

2. C. *lanceolata* Roth. E. B. 30. 2159.
Panicle erect, loose. Flowers scattered, spreading every way. Awn terminal, short. Hairs longer than the paleæ. Leaves linear.
Arundo Calamagrostis *Linn.*
Calamagrostis epigejos *With.*
In moist woods and fens. — Perennial. *June, July.*

3. C. *stricta.* E. B. 30. 2160.
Panicle erect, close. Flowers scattered, spreading every way, with a dorsal awn. Hairs shorter than the paleæ. Ligula very short.
Arundo stricta *Schrad.*
A. neglecta *Ehr.*
In the White Mire, a mile from Forfar. — Perennial. *June.*

27. ANEMAGROSTIS *Trinius.*

Panicle loose. *Glumes* 2, unequal; the lower smaller, the upper a little longer than the paleæ. *Paleæ* 2; the lower awned under the apex, the upper enwrapped in it. The *rudiment* of a second floret at the back of the upper palea. *Styles* feathery.

1. A. *Spica venti* Trinius. E. B. 14. 951.
Awn straight, rigid, many times longer than the paleæ. Panicle loosely spreading.
Agrostis Spica venti *Linn.*
Apera Spica venti *Beauv.*
In sandy corn-fields, occasionally overflowed. — Annual. *June, July.*

IV. *Inflorescence panicled. Spikelets 2- or 3-flowered.*

28. ARRHENATHERUM *Beauv.*

Panicle loose. *Spikelets* 2-flowered; upper floret hermaphrodite, lower male. *Glumes* 2, nearly equal, shorter than the paleæ. *Paleæ* 2; the lower emarginate, of the male floret with a twisted awn at the base, of the hermaphrodite floret with a straight bristle under the apex.

1. A. *avenaceum* Beauv. E. B. 12. 813.
Root knotty. Knots of the stem smooth. Lower palea villous. *Dumortier*.
Holcus avenaceus *Scop.*
Avena elatior *Linn.*
pastures, and by road-sides. — Perennial. *June, July.*

2. A *bulbosum* Dumortier.
Root bulbous. Knots of the stem villous. Lower palea smooth. *Dumortier.*
Avena nodosa *Cullum.*
In waste places. — Perennial. *June, July.*

29. ECHINOCHLOA *Beauv.*

Spike compound. *Glumes* 2; the lower smaller, mucronate, the upper as long as the fertile palea, mucronate or awned. *Paleæ* of the fertile floret 2, equal; one convex, the other plane. *Paleæ* of the neuter or male floret 1- or 2-valved; the lower mucronate or with a long awn, the upper smaller, membranous, and often wanting.

1. E. *Crus-galli* Beauv. E. B. 13. 876.
Panicle erect, branched, bristly. Flowers awned, unilateral. Leaves lanceolate, harsh, naked, without ligulæ.
Panicum Crus-galli *Linn.*
Oplismenus Crus-galli *Dumort.*
In moist arable land. — Annual. *July.*

30. HOLCUS *Linn.* Soft-grass.

Panicle loose. *Spikelets* 2-flowered; lower floret awnless and hermaphrodite; upper awned, and male. *Glumes* 2, nearly equal, rather longer than the florets. *Paleæ* 2; the lower awnless, or awned under the apex.

H. *lanatus* Linn. E. B. 17. 1169.
Glumes woolly. Lower floret perfect, awnless; upper with an arched awn. Leaves downy on both sides. Root fibrous. *Smith.*
Abundant in meadows and pastures. — Perennial. *June, July.*

2. H. *mollis* Linn. E. B. 17. 1170.
Glumes partly naked. Lower floret perfect, awnless; upper with a sharply bent prominent awn. Leaves slightly downy. Root creeping. *Smith.*
In pastures and hedges. — Perennial. *July.*

31. ANTHOXANTHUM *Linn.*

Glumes 2; the lower smaller, the upper enfolding the paleæ, and longer than them. *Florets* 3; the two lateral neuter, the middle hermaphrodite. *Paleæ* of the neuter florets single; one with a dorsal awn, the other with an awn from the base. *Paleæ* of the hermaphrodite floret 2, nearly equal, awnless. *Stamens* 2.

1. A. *odoratum* Linn. Sweet-scented Vernal-grass. E. B. 9. 647.
Panicle spiked, ovate-oblong. Florets longer than their awns, on short partial stalks. *Smith.*
In meadows and pastures. — Perennial. *May, June.*

32. HIEROCHLOE *Gmelin.*

Panicle loose. *Spikelets* 3-flowered: lateral florets male, triandrous; terminal, hermaphrodite, diandrous. *Glumes* 2, nearly equal. *Paleæ* 2, awnless, or awned.

1. H. *borealis* R. & S. Northern Holy-grass.
Panicle somewhat unilateral, with smooth flower-stalks. Perfect floret awnless; barren ones slightly awned. Scales unequal, linear. Leaves flat.
Holcus borealis *Schrad.*
Holcus odoratus *Linn.*
In a narrow mountain valley called Kella, Angusshire. —Perennial. *May, June.*

33. CYNOSURUS *Linn.*

Panicle contracted. *Spikelets* 2- or many-flowered, resting upon pinnate bracteæ. *Glumes* 2, about the same length as the florets. *Paleæ* 2; the lower awned from the apex or mucronate. *Scales* lanceolate, acute. *Styles* feathery or hairy.

1. C. *cristatus* Linn. Crested Dog's-tail-grass. E. B. 5. 316.
Spike simple, linear. Neuter spikelets without awns. *Smith.*
In dry pasture, parks, and lawns. — Perennial. *June, July.*

2. C. *echinatus* Linn. Rough Dog's-tail-grass. E. B. 19. 1333.
Spike compound, ovate. Neuter spikelets awned. Awns of the paleæ full as long as the glume.
Chrysurus echinatus *Beauv.*
Phalona echinata *Dumort.*
On sandy ground in the south of England. — Annual. *July.*

34. CATABROSA *Beauv.*

Panicle loose. *Spikelets* 2-flowered. *Glumes* 2, truncate, unequal, much shorter than the florets. *Paleæ* 2, truncate, nearly equal, awnless; the upper free from the lower.

1. C. *aquatica* Beauv. Water Hair-grass. E. B. 22. 1557.
Panicle spreading. Florets awnless, even, obtuse, longer than the glumes. Leaves flat. Ligula oblong.
Aira aquatica *Linn.*
In ditches, and the margins of rivers. — Perennial. *May, June.*

35. MOLINIA *Mœnch.*

Joints of the stem not prominent. *Panicle* loose. *Spikelets* 2-, 3-, or many-flowered, sometimes with a rudimentary floret. *Glumes* 2, unequal, much shorter than the florets. *Paleæ* 2, nearly equal, awnless. *Scales* short, truncate. *Styles* pencil-shaped.

1. M. *cærulea* Mœnch. E. B. 11. 750.
Leaves much shorter than the panicle. Panicle dense, coloured. Glumes nearly equal, 3- or 4-flowered. Lower palea obtuse, 3-ribbed.
Melica cærulea *Linn.*
Aira cærulea *Linn.*
Enodium cæruleum *Dumort.*
n barren, sandy, boggy ground. — Perennial. *August.*

2. M. *depauperata*
Leaves much longer than the panicle. Panicle thin, few-flowered, colourless. Glumes very unequal, 1-flowered. Lower palea acuminate, obtuse, 5-ribbed.
On the mountains of Clova. *Mr. Donald Munro.* — Perennial. *August.*
N. B. This very distinct species is most nearly related to Molinia atrovirens (*Enodium cæruleum var atrovirens* Dumort.); but that plant has the glumes constantly 2-flowered, and its lower paleæ have 3 ribs.

36. MELICA *Linn.* Melic-grass.

Panicle loose. *Spikelets* 1- or 2-flowered; the apex of the spikelet involute. *Glumes* 2, nearly equal, about as long as the florets. *Paleæ* 2, unequal, awnless. *Scales* 2, gibbous, cuspidate.

1. M. *uniflora* Retz. E. B. 15. 1058.
Paleæ beardless. Panicle branched, drooping toward one side. Spikelets erect, with only one perfect floret.
M. Lobelii *Villars.*
M. nutans *Hudson.*
In groves and thickets. — Perennial. *May, June.*

2. M. *nutans* Linn. E. B. 15. 1059.
Paleæ beardless. Panicle close, drooping, nearly simple. Spikelets pendulous, with 2 perfect florets.
M. montana *Hudson.*
In mountainous woods. — Perennial. *June, July.*

37. AIROCHLOA *Link.*

Panicle contracted. *Spikelets* compressed, 2- or 3-flowered. *Glumes* 2. *Paleæ* 2; the lower awnless or mucronate. *Scales* usually bifid.

1. A. *cristata* Link. E. B. 9. 648.
Panicle spiked. Glumes longer than their flower-stalk, shorter than the florets, all pointed.
Aira cristata *Linn.*
Poa cristata *Willd.*
Köleria cristata *Pers.*
Köleria glauca *Dec.* }
Aira glauca *Schrad.* } a variety.
In dry, elevated, or calcareous, pastures, or on walls. — Perennial. *July, August.*

38. CORYNEPHORUS *Beauv.*

Panicle loose. *Spikelets* 2-flowered; florets equally perfect. *Glumes* 2, equal, longer than the paleæ. *Paleæ* 2; the lower entire, bearing a twisted, articulated, clavate awn at its base.

1. C. *canescens* Beauv. E. B. 17. 1190.

Panicle rather dense. Florets shorter than the glumes. Awn club-shaped, not longer; hairy at the joint. Leaves bristle-shaped. *Smith.*

Aira canescens *Linn.*

On the sea-coasts of Norfolk and Suffolk. — Perennial. *July.*

39. AIRA *Linn.*

Panicle loose. *Spikelets* 2-flowered; florets equally perfect. *Glumes* 2, nearly equal. *Paleæ* 2; the lower bifid, with a dorsal slightly-twisted awn.

1. A. *flexuosa* Linn. E. B. 22. 1519.

Panicle spreading, triple-forked, with wavy branches. Florets about the length of the glumes, acute. Awn from the middle of the outer paleæ, longer than the glumes, twisted. Leaves bristle-shaped.

Avena flexuosa *Link.*

Aira montana *Huds.* }
A. scabrosetacea *Knapp.* } a variety.
A. setacea *Hudson.* }

In heathy, sandy places. — Perennial. *July.*

2. A. *præcox* Linn. E. B. 18. 1296.

Panicle close, erect. Florets the length of the glume, both sessile. Awn nearly twice as long, from the base of the palea. Leaves bristle-shaped, with angular sheaths.

Avena præcox *Link.*

Common on dry gravelly ground. — Annual. *May, June.*

3. A. *caryophyllea* Linn. E. B. 12. 812.

Panicle spreading, triple-forked. Florets not longer than the glumes, both sessile. Awn twice as long, from above the middle of the palea. Leaves bristle-shaped, with ribbed close sheaths.

Avena caryophyllea *Link.*

On barren sandy heaths or hillocks. — Annual. *June, July.*

4. A. *alpina* Linn. E. B. 30. 2102.

Panicle rather close and upright. Florets the length of the glumes, acute; one of them on a smooth stalk. Awn short, from near the top of the outer palea. Leaves involute-awl-shaped, with smooth sheaths. *Smith.*

A. lævigata *Smith.*

On the highland mountains. — Perennial. *June, July.*

40. TRISETUM *Pers.*

Panicle cylindrical, dense. *Spikelets* 2-flowered; florets equally perfect. *Glumes* 2, longer than the paleæ. *Paleæ* 2; the lower deeply cleft, with a twisted dorsal awn.

1. T. *pubescens* Pers. E. B. 23. 1640.

Panicle erect, nearly simple. Florets longer than the glumes. Partial stalk bearded. Leaves flat, downy. Root somewhat creeping.

Avena pubescens *Linn.*
A. sesquitertia *Linn.*
In pastures on a chalky, or limestone soil. — Perennial. *June.*

2. T. *flavescens* Beauv. E. B. 14. 952.
Panicle much branched, spreading, erect. Florets longer than the very unequal glumes. Leaves flat, a little downy. Root somewhat creeping.
Avena flavescens *Linn.*
In meadows and by road-sides. — Perennial. *July.*

41. DESCHAMPSIA *Beauv.*

Panicle loose. *Spikelets* 2-flowered; florets equally perfect. *Glumes* 2, nearly equal. *Paleæ* 2; the lower toothed at the point, with a straight awn proceeding from its base.

1. D. *cæspitosa* Beauv. E. B. 21. 1453.
Panicle spreading. Florets about the length of the glumes, abrupt, hairy at the base; one of them on a hairy stalk. Awn short, from the bottom of the outer palea. Leaves flat.
Aira cæspitosa *Linn.*
Campella cæspitosa *Link.*
In moist shady groves. — Perennial. *June, July.*

42. SETARIA *Beauv.*

Spike often compound. *Involucrum* of many bristles surrounding 2 spikelets. *Glumes* 2; the lower smaller. *Paleæ* of the fertile floret 2, equal, cartilaginous. *Paleæ* of the male or neuter floret 1 or 2; the upper smaller and membranous, often wanting.

1. S. *verticillata* Beauv. E. B. 13. 874.
Panicle spiked, cylindrical, lobed, with whorled branches. Paleæ of the perfect floret slightly uneven.
Panicum verticillatum *Linn.*
In moist cultivated fields. — Annual. *July, August.*

2. S. *viridis* Beauv. E. B. 13. 875.
Panicle spiked, cylindrical, continuous. Paleæ of the perfect floret slightly uneven.
Panicum viride *Linn.*
In sandy fields. — Annual. *July, August.*

V. *Inflorescence panicled. Spikelets many-flowered.*

43. SESLERIA *Arduin.* MOOR-GRASS.

Panicle contracted. *Spikelets* many-flowered, with a sheathing or glume-like bractea. *Glumes* 2, nearly equal, somewhat awned at the apex. *Paleæ* 2, nearly equal; the lower awned beneath the apex. *Styles* 1 or 2, long.

1. S. *cærulea* Scop. E. B. 23. 1613
Spikes ovate-oblong, imbricated. Bracteæ alternate. Outer palea with 3 teeth.
Cynosurus cæruleus *Linn.*
On moist, alpine, limestone rocks. — Perennial. *April—June.*

44. ARUNDO *Linn.*

Panicle loose. *Spikelets* many-flowered; the lower floret male and naked, the upper hermaphrodite and surrounded by hairs. *Glumes* 2; the lower smaller, the upper about as long as the florets. *Paleæ* 2, unequal.

1. A. *Phragmites* Linn. Common Reed. E. B. 6. 401.
Florets about 5, awnless, longer than the glumes. Panicle loose. *Smith.*
Phragmites communis *Trin.*
In marshes, ditches, and about the banks of pools and rivers.—Perennial. *July.*

45. AVENA *Linn.*

Panicle loose. *Spikelets* many-flowered; upper florets sterile and imperfect. *Glumes* 2, nearly equal, as long as the paleæ. *Paleæ* 2; the lower bifid, with a twisted dorsal awn.

1. A. *fatua* Linn. Wild Oat, or Haver. E. B. 31. 2221.
Panicle erect, compound. Spikelets pendulous. Florets about 3, shorter than the glumes, bristly at the base, with an oblique scar, all awned. *Smith.*
In corn-fields.—Annual. *June, July.*

2. A. *strigosa* Schreb. E. B. 18. 1266.
Panicle oblong, turned to one side. Florets 2. Outer palea tipped with a double straight bristle.
In corn-fields.—Annual. *June, July.*

3. A. *pratensis* Linn. E. B. 17. 1204.
Panicle erect, with very short simple branches. Florets about 5, longer than the glumes. Partial stalk all over hairy. Leaves involute, finely serrated, naked; sheaths smooth. *Smith.*
A. bromoides *Linn.*
In dry chalky, or limestone, pastures, and heathy spots.—Perennial. *July.*

4. A. *alpina* Smith. E. B. 30. 2141.
Panicle erect, slightly branched. Florets about 5, longer than the glumes. Partial stalk bearded under each. Leaves flat, minutely serrated, naked; sheaths rough. Root fibrous. *Smith.*
A. planiculmis *Smith.*
Found upon the rocky summits of the highest mountains of Clova, Angusshire.—Perennial. *July.*

46. DACTYLIS *Linn.*

Panicle loose or contracted; branches solitary; terminal ramifications always very short. *Spikelets* clustered, many-flowered, horizontal. *Glumes* 2, unequal-sided.* *Paleæ* 2; the lower awned under the apex, the upper of nearly the same size. *Scales* toothed.

1. D. *glomerata* Linn. Cock's-foot-grass. E. B. 5. 335
Panicle distantly branched. Flowers in dense globular tufts, unilateral. Paleæ somewhat awned, 5-ribbed, taper-pointed.
In meadows and shady places.—Perennial. *June—August.*

47. TRIODIA *R. Br*

Panicle somewhat racemose. *Spikelets* many-flowered. *Glumes* 2, nearly equal. *Paleæ* 2; the lower with 3 nearly equal teeth, the middle one of which is stiff. *Grain* 3-toothed.

1. T. *decumbens* Beauv. E. B. 11. 792.
Panicle nearly simple, close, erect. Florets 4; their middle tooth shortest. Glumes smooth. Ligula hairy.
Festuca decumbens *Linn.*
Poa decumbens *Withering.*
Melica decumbens *Weber.*
In spongy bogs. — Perennial. *July.*

48. BROMUS *Linn.*

Panicle loose. *Spikelets* more than 4-flowered. *Glumes* 2, unequal, shorter than the lower florets. *Paleæ* 2; the lower awned under the apex, very seldom awnless. *Scales* lanceolate, entire.

1. B. *secalinus* Linn. Rye Brome-grass. E. B. 17. 1171.
Panicle spreading; slightly subdivided below. Spikelets ovate, of about ten, distinct, somewhat cylindrical, smooth florets. Awns wavy, shorter than the glumes. Leaves slightly hairy. *Smith.*
B. vitiosus *Weigel.*
In corn-fields. — Annual. *July—September.*

2. B. *velutinus* Schrad. E. B. 27. 1884.
Panicle spreading; scarcely subdivided. Spikelets ovate-oblong, of from 10 to 15 crowded, elliptical, downy florets. Awns as long as the glumes. Leaves slightly hairy. *Smith.*
B. multiflorus *Smith.*
In corn-fields. — Annual. *July.*

B. *mollis* Linn. Brome-grass. E. B. 15. 1078.
Panicle erect, rather close, compound. Spikelets ovate, downy. Florets imbricated, depressed, ribbed. Awns as long as the glumes. Leaves and sheaths very soft and downy. *Smith.*
B. polymorphus *Hudson.*
B. hordeaceus *Linn.*
In fields and pastures. — Biennial. *June.*

4. B. *racemosus* Linn. E. B. 15. 1079.
Panicle nearly erect, spreading, slightly branched. Spikelets ovate-oblong, naked. Florets imbricated, depressed, ribbed. Awns as long as the glumes. Leaves somewhat downy. *Smith.*
B. pratensis *Ehr.*
B. arvensis *Knapp.*
In meadows and pastures. — Annual or biennial. *June.*

5. B. *squarrosus* Linn. E. B. 27. 1885.
Panicle drooping, scarcely branched. Spikelets ovate-oblong. Florets about 12, imbricated, depressed, ribbed. Awns widely spreading. Leaves downy. *Smith.*
In corn-fields. — Annual. *July.*

6. B. *arvensis* Linn. E. B. 28. 1984.
Panicle spreading, drooping, compound, half-whorled. Spikelets lanceolate, acute. Florets about 8, imbricated, smoothish, with 2 close ribs at each side. Leaves hairy. *Smith.*
B. spiculitenuata *Knapp.*
B. versicolor *Pollich.*
B. verticillatus *Cav.*
In corn-fields. — Annual. *July.*

7. B. *erectus* Huds. E. B. 7. 471.
Panicle erect, slightly branched. Spikelets linear-lanceolate. Florets about 8, loosely imbricated, lanceolate, compressed. Awn shorter than the glumes, straight. Radical leaves very narrow, fringed with scattered hairs. *Smith.*
B. agrestis *Allioni.*
B. perennis *Villars.*
In fields and by road-sides. — Perennial. *July.*

8. B. *asper* Linn. E. B. 17. 1172.
Panicle drooping, branched. Spikelets linear-oblong. Florets about 8, rather distant, lanceolate, compressed, downy. Awns shorter than the glumes. Leaves uniform; lower ones hairy. *Smith.*
B. ramosus *Linn.*
B. nemoralis *Huds.*
B. nemorosus *Vill.*
B. hirsutus *Curtis.*
B. montanus *Pollich.*
In woods and hedges. — Annual or biennial. *July, August.*

9. B. *sterilis* Linn. E. B. 15. 1030.
Panicle drooping, mostly simple. Spikelets linear-lanceolate. Florets about 7, lanceolate, compressed, 7-ribbed, furrowed. Awns longer than the glumes. Leaves downy. *Smith.*
In fields and on walls. — Annual. *June, July.*

10. B. *diandrus* Curtis. E. B. 14. 1006.
Panicle upright, a little spreading, scarcely subdivided. Florets lanceolate, with 2 close marginal ribs, and only 2 stamens. *Smith.*
B. muralis *Huds.*
B. ciliatus *Huds.*
In sandy ground and on walls. — Annual. *June.*

49. SCHEDONORUS *Beauv.*

Panicle loose. *Spikelets* many-flowered. *Glumes* 2, nearly equal, shorter than the lower paleæ. *Paleæ* 2; the lower awned at the apex. *Pedicels* of the florets with 3 tufts of hairs at the end. *Scales* oblong, 2-toothed.

1. S. *pratensis* Beauv. E. B. 23. 1592.
Panicle nearly upright, branched, spreading, turned to one side. Spikelets linear, compressed. Florets numerous, cylindrical, obscurely ribbed. Root fibrous. *Smith.*

Festuca pratensis *Huds.*
F. elatior *Host.*
In pastures and meadows. — Perennial. *June, July.*

2. S. *elatior.* E. B. 23. 1593.
Panicle somewhat drooping, much branched, spreading loosely every way. Spikelets ovate-lanceolate. Florets numerous, cylindrical, somewhat awned, obscurely ribbed. Root creeping. *Smith.*
Festuca arundinacea *Schreb.*
Festuca elatior *Smith.*
Schedonorus radicans *Dumort.*
Bromus littoreus *Willd.*
In meadows, and the borders of ditches. — Perennial. *June, July.*

3. S. *sylvaticus* Beauv. E. B. 14. 1005.
Panicle repeatedly compound, spreading, erect. Florets from 2 to 5, oblong, cylindrical, keeled, angular, pointed; inner paleæ folded in the middle. *Smith.*
Festuca calamaria *Smith.*
F. sylvatica *Villars.*
Poa sylvatica *Pollich.*
P. trinervata *Ehr.*
Festuca decidua *Smith.* E. B. 32. 2266. a variety.
In mountainous woods of Scotland, Ireland, and the north-west part of England. — Perennial. *June, July.*

4. S. *loliaceus* Dumort. E. B. 26. 1821.
Spike 2-ranked, drooping. Spikelets nearly sessile, linear-oblong. Florets cylindrical, awnless, pointed, with 5 slight ribs at the top. *Smith.*
Festuca loliacea *Huds.*
F. elongata *Ehr.*
In rich moist pastures and meadows. — Perennial. *June, July.*

50. FESTUCA *Linn.*

Panicle loose. *Spikelets* many-flowered; the florets deciduous. *Glumes* 2, unequal, or nearly equal, acute. *Paleæ* 2; the lower mucronate or awned at the point. *Scales* 2, usually toothed.

1. F. *ovina* Linn. Sheep's Fescue-grass. E. B. 9. 585.
Panicle unilateral, rather close. Florets cylindrical, pointed or awned; smooth at the base, and at the edges of the inner paleæ. Stem square. Leaves folded, bristle-shaped. Ligula short and obtuse.
Festuca rubra *Withering*; a variety.
F. cæsia *E. B.* 27. 1917.; a variety.
F. tenuifolia *Sibth.*; a variety.
In dry open pastures. — Perennial. *June.*

2. F. *vivipara* Smith. E. B. 19. 1355.
Panicle unilateral, rather close. Florets compressed, keeled, awnless, somewhat downy, as well as the edges of their inner

paleæ and glumes. Stem square. Leaves folded, bristle-shaped, smooth.

On the tops of the loftiest mountains. — Perennial. *July.*

3. F. *duriuscula* Linn. Hard Fescue-grass. E. B. 7. 470.
Panicle unilateral, spreading. Florets longer than their awns. Stem round. Upper leaves flat. Root fibrous. *Smith.*
F. heterophylla *Hænke.*
F. nemorum *Leyss.*
F. dumetorum *Linn.*; a variety.

In pastures, waste ground, and thickets. — Perennial. *June, July.*

4. F. *rubra* Linn. Creeping Fescue-grass. E. B. 29. 2056.
Panicle unilateral, spreading. Florets longer than their awns. Leaves downy on the upper side, more or less involute. Root extensively creeping. *Smith.*
Festuca cambrica *Huds.*; a variety.
F. glabra *Lightf.*; a variety.
F. glauca *Winch.*; a variety.

In mountainous pastures. — Perennial. *July.*

5. F. *gigantea* Villars. Tall Fescue-grass. E. B. 26. 1820.
Panicle drooping, twice compound, spreading. Florets from 3 to 6, ovate-lanceolate, shorter than their awns. Ligula abrupt, auricled, clasping the stem.
Bromus giganteus *Linn.*
Festuca triflora *E. B.* 27. 1918. } a variety.
Bromus triflorus *Linn.* }

In woods and hedges. — Perennial. *July, August.*

51. VULPIA *Gmelin.*

Panicle racemose or contracted, with very thick pedicels. *Spikelets* many-flowered; the florets deciduous. *Glumes* 2; the lower much smaller than the other, or wholly wanting. *Paleæ* 2; the lower ending in a long awn. *Scales* oval, acute.

1. V. *Myurus* Gmel. E. B. 20. 1412.
Panicle drooping, elongated, rather close. Florets tapering, shorter than their awns, rough at the top. Leaves awl-shaped. Stem leafy to the very summit. *Smith.*
Festuca Myurus *Linn.*
Mygalurus caudatus *Link.*

On walls, and barren sandy ground. — Annual. *June, July*

2. V. *uniglumis* Dumort.
Panicle erect, nearly simple. Florets tapering, compressed, awned. One glume very short.
Festuca uniglumis *H. Kew.*
Stipa membranacea *Linn.*
Mygalurus uniglumis *Link.*
Vulpia membranacea *Link.*
Lolium bromoides *Hudson.*

On the sandy sea-coast. — Biennial. *June.*

3. V. *bromoides* Dumort. E. B. 20. 1411.
Panicle nearly erect, racemose. Florets tapering, shorter than their awns, rough at the top. Leaves tapering, shorter than their sheaths. Upper half of the stem naked. *Smith.*
Festuca bromoides *Linn.*
Mygalurus bromoides *Link.*

On walls and barren sandy ground. — Annual. *June.*

52. GLYCERIA *R. Br.*

Panicle loose. Spikelets cylindrical, many-flowered; florets articulated with their pedicels. *Glumes* 2, nearly equal, obtuse, 1-ribbed, shorter than the lower florets. *Paleæ* 2, many-ribbed, nearly equal, obtuse, awnless. *Scales* cohering in 1, truncate. *Styles* dichotomous, feathery.

1. G. *fluitans* R. Br. Manna Grass. E. B. 22. 1520.
Panicle oblong, branched, divaricating. Spikelets close-pressed. Florets numerous, obtuse, 7-ribbed, with short intermediate ribs at the base.
Festuca fluitans *Linn.*
Poa fluitans *Scop.*

In stagnant waters and slow streams. — Perennial. *June—August.*

53. BRIZA *Linn.* Quaking-Grass.

Panicle loose. *Spikelets* many-flowered, cordate. *Glumes* 2, equal, convex, about as long as the lower florets. *Paleæ* 2, convex, awnless; their margins not involute. *Scales* acuminate, gibbous at the base. *Styles* feathery almost to the base.

1. B. *minor* Linn. E. B. 19. 1316.
Spikelets triangular, 7-flowered. Glumes longer than the florets. Ligula lanceolate, elongated.
B. aspera *Knapp.*

In cultivated fields in the south of England. — Annual. *July.*

2. B. *media* Linn. Maiden's Hair. E. B. 5. 340.
Spikelets ovate, about 7-flowered. Glumes shorter than the florets. Ligula very short and blunt.

In pastures. — Perennial. *May, June.*

54. SCLEROCHLOA *Beauv.*

Panicle contracted. *Pedicels* articulated with the spikelets. *Spikelets* many-flowered, cylindrical, and compressed. *Glumes* 2, shorter than the lower florets. *Paleæ* 2, awnless, or mucronate.

1. S. *maritima.* E. B. 16. 1140.
Panicle branched, rather close; erect after flowering. Florets about 5, somewhat pointed, slightly 5-ribbed. Root creeping. *Smith.*
? Sclerochloa dichotoma *Link.*
Poa maritima *Hudson.*

Glyceria maritima *Smith*.

In salt marshes on the coast. — Perennial. *July—October.*

2. S. *procumbens* Beauv. E. B. 8. 532.

Panicle lanceolate, unilateral, 2-ranked, close, with rough stalks; the main one cylindrical. Florets about 5, bluntish, 5-ribbed. *Smith.*

Poa procumbens *Curtis.*

Poa rupestris *Withering.*

Glyceria procumbens *Smith.*

In waste ground near the sea. — Annual. *July, August.*

3. S. *rigida* Panzer. E. B. 20. 1371.

Panicle lanceolate, unilateral, 2-ranked, close, with smooth stalks; the main one bordered. Florets about 7, acute, scarcely ribbed. *Smith.*

Megastachya rigida *R. & S.*

Poa rigida *Linn.*

Glyceria rigida *Smith.*

On walls and dry gravelly banks. — Annual. *June.*

55. HYDROCHLOA *Hartman.*

Panicle loose. *Spikelets* many-flowered, compressed. *Glumes* 2, nearly equal, very obtuse, 1-ribbed, membranous, shorter than the lowest floret. *Paleæ* 2, nearly equal, awnless, many-ribbed. *Scales* 2, truncate. *Styles* simple, pencil-shaped.

1. H. *aquatica* Hartman. E. B. 19. 1315.

Panicle erect, repeatedly branched, spreading. Florets numerous, obtuse, with 7 ribs.

Poa aquatica *Linn.*

Glyceria aquatica *Smith.*

G. spectabilis *Mert. & Koch.*

In ditches and the margins of streams. — Perennial. *July.*

56. POA *Linn.*

Panicle loose, seldom contracted. *Spikelets* 3- or many-flowered, or even 2-flowered, with the pedicels of a greater number of florets; florets articulated with their rachis. *Paleæ* 2, nearly equal, awnless. *Scales* oval, acute, gibbous at the base.

1. P. *compressa* Linn. E. B. 6. 365.

Panicle unilateral, rather dense. Stem compressed. Root creeping. Spikelets ovate oblong. Florets connected by a web. *Smith.*

On walls and in dry ground. — Perennial. *June—September.*

2. P. *alpina* Linn. E. B. 14. 1003.

Panicle loosely spreading. Spikelets heart-shaped, 4- or 5-flowered. Florets rather sickle-shaped, hairy at the base without a web. Lower ligulæ very short; upper oblong, acute. *Smith.*

On lofty mountains. — Perennial. *July, August.*

3. P. *laxa* Hænke. E. B. 16. 1123.

Panicle drooping, loosely spreading, zigzag. Spikelets ovate, 3-

flowered. Florets connected by a web. Ligulæ all lanceolate. *Smith.*

P. flexuosa *Smith.*

In the Highlands of Scotland. — Perennial. *July.*

4. P. *bulbosa* Linn. E. B. 15. 1071.

Panicle close, slightly zigzag. Spikelets 4-flowered. Florets hairy at the keel, connected by a web. Leaves finely serrated. Stem bulbous at the base. *Smith.*

On the sandy sea-shore. — Perennial. *April, May.*

5. P. *trivialis* Linn. E. B. 15. 1072.

Panicle spreading. Spikelets 3-flowered. Florets lanceolate, 5-ribbed, connected by a web. Ligula oblong. Stem and leaves roughish. Root fibrous. *Smith.*

P. dubia *Linn.*

P. scabra *Ehr.*

P. setacea *Hudson;* a variety

In meadows and pastures. — Perennial. *June—October.*

6. P. *pratensis* Linn. Meadow-grass. E. B. 15. 1073.

Panicle spreading. Spikelets 4-flowered. Florets lanceolate, 5-ribbed, connected by a web. Ligula short and obtuse. Stem and leaves smooth. Root creeping. *Smith.*

P. glabra *Ehr.*

P. angustifolia *Linn.*; a variety.

P. subcærulea *Smith;* }
P. humilis *Ehr.*; } a variety
P. cærulea *Knapp;* }

In meadows and pastures. — Perennial. *May, June.*

7. P. *annua* Linn. E. B. 16. 1141.

Panicle widely spreading. Spikelets ovate, 5-flowered. Florets a little remote, 5-ribbed, without a web. Stems oblique, compressed. *Smith.*

In meadows and cultivated ground everywhere. — Annual. *April—November.*

8. P. *glauca* Fl. Dan. E. B. 24. 1720.

Panicle spreading. Spikelets ovate. Florets from 2 to 5, obscurely 5-ribbed, bluntish; silky at the keel and lateral ribs; hairy at the base, without a web. Ligulæ of the lower leaves very short and blunt.

P. cæsia *E. Bot.*; a variety.

On mountains. — Perennial. *June, July,*

9. P. *nemoralis* Linn. E. B. 18. 1265.

Panicle spreading, capillary. Glumes lanceolate, taper-pointed, each 3-ribbed. Spikelets lanceolate. Florets about 3, 5-ribbed, acute; silky at the keel and lateral ribs; hairy at the base, without a web. Ligulæ very short, notched. *Smith.*

P. angustifolia *Hudson.*

In groves and woods. — Perennial. *June, July.*

10. P. *distans* Linn. E. B. 14. 986.

Panicle branched, spreading; the branches finally reflexed. Florets about 5, obtuse, obscurely 5-ribbed, polished. Root fibrous. *Smith.*

Glyceria distans *Smith.*
P. retroflexa *Curtis.*
P. salina *Pollich.*

On sandy waste ground. — Perennial. *July, August.*

APPENDIX.

The genus Bryonia having been accidentally omitted in its place, the reader will be so good as observe, that the natural order Cucurbitaceæ, to which it belongs, should be inserted between *Loranthcæ* and *Vaccinieæ*, with the following characters : —

Order 41.* Cucurbitaceæ *Juss.*

Flowers usually bisexual, sometimes hermaphrodite.

Calyx 5-toothed.

Corolla 5-parted, scarcely distinguishable from the calyx, very vascular, with strongly marked, reticulated veins.

Stamens 5, either distinct, or cohering in 3 parcels; *anthers* 2-celled, very sinuous.

Ovarium inferior, 1-celled, with 3 parietal placentæ; *style* short; *stigmas* very thick, velvety or fringed.

Fruit fleshy, more or less succulent, crowned by the scar of the calyx, 1-celled, with 3 parietal placentæ.

Seeds flat, ovate, enveloped in an arillus, which is either juicy, or dry and membranous; *testa* coriaceous, often thick at the margin; *embryo* flat, with no *albumen*; *cotyledons* foliaceous veined; *radicle* next the hilum.

Roots annual or perennial, fibrous or tuberous. *Stem* succulent, climbing by means of tendrils formed by abortive leaves. *Leaves* palmated, very succulent, covered with numerous asperities. *Flowers* white, red, or yellow.

1. BRYONIA *Linn.*

Flowers monœcious or diœcious. *Petals* scarcely cohering at the base. *Males*. *Calyx* 5-toothed. *Stamens* in 3 parcels. *Females*. *Styles* 3-fid. *Fruit* succulent, with small, ovate, compressed seeds, which are more or less bordered. — Tendrils *simple*.

1. B. *dioica* Jacq. Bryony. E. B. 7. 439.

Leaves palmate, rough on both sides with callous points. Barren and fertile flowers on separate plants. *Smith*.

B. alba *Hudson*.

In hedges and thickets. — Perennial. *May—September*.

INDEX

OF

THE ORDERS, GENERA, AND SPECIES.

Note. — The Orders are in Roman small capitals, the Genera and Species in small Roman characters. Those names which are in Italics are either synonyms or merely subjects of reference.

Abama ossifraga Dec. 277.
Aceras *R. Br.* 262.
  anthropophora *R. Br.* 262.
ACERINEÆ *Juss.* 55.
Acer *Linn.* 55.
  campestre *Linn.* 55.
  Pseudo-platanus *Linn.* 55.
Achillea *Linn.* 150.
  Millefolium *Linn.* 151.
  Ptarmica *Linn.* 151.
  serrata *Retz,* 151.
  tomentosa *Linn.* 151.
ACHLAMYDEÆ, 5, 228.
Achnodon *Trin.* 300.
  arenarius *Trin.* 300.
Aconitum *Linn.* 13.
  Napellus Smith, 13.
  vulgare *Dec.* 13.
Acorus *Linn.* 246.
  Calamus *Linn.* 246.
Acotyledoneæ Juss. 3.
Actæa *Linn.* 13.
  spicata *Linn.* 14.
Adonis *Linn.* 9.
  æstivalis With. 9.
  autumnalis *Linn.* 9.
Adoxa *Linn.* 67.
  Moschatellina *Linn.* 67.
Ægilops incurvata Linn. 295.
Ægopodium *Linn.* 123.
  Podagraria *Linn.* 123.
Æthusa *Linn.* 119.
  Cynapium *Linn.* 119.
  Meum Linn. 118.
Agraulus caninus Beauv. 303.
Agrimonia *Linn.* 99.
  Eupatoria *Linn.* 99.
Agropyrum *Beauv.* 297.
  caninum *Beauv.* 298.
  cristatum *Beauv.* 298.
  junceum *Beauv.* 298.
  repens *Beauv.* 298.
Agrostemma *Linn.* 47.
  Githago *Linn.* 47.
Agrostis *Linn.* 303.
  alba *Linn.* 303.
  alpina With. 303.
  australis Linn. 302.
  canina Linn. 303.
  canina With. 303.
  capillaris Abbot, 303.
  fascicularis Sincl. 303.
  hispida Willd. 303.
  linearis Retz, 298.
  littoralis Smith, 302.
  minima Linn. 301.
  mutabilis Knapp, 303.
  palustris Sincl. 303.
  inicea H. Kew. 302.
  polymorpha Huds. 303.
  pumila Linn. 303.
  rubra Huds. 302.
  setacea Curt. 303.
  Spica Venti Linn. 304.
  stolonifera Linn. 304.
  stricta Sincl. 303.
  sylvatica Huds. 304.
  tenuis Sibth. 303.

tenuifolia Curt. 303.
triaristata Knapp, 302.
ventricosa Gouan, 303.
vinealis With. 303.
vulgaris *With.* 303.
Aira *Linn.* 308.
alpina *Linn.* 308.
aquatica Linn. 306.
cærulea Linn. 307.
cæspitosa Linn. 309.
canescens Linn. 308.
caryophyllea *Linn.* 308.
cristata Linn. 307.
flexuosa *Linn.* 308.
glauca Schrad. 307.
lævigata Smith, 308.
montana Huds. 308.
præcox *Linn.* 308.
scabrosetacea Knapp, 308.
setacea Huds. 308.
Airochloa *Link.* 307.
cristata *Link.* 307.
Ajuga *Linn.* 197.
alpina *Linn.* 198.
Chamæpitys *Smith,* 198.
genevensis With. 198.
pyramidalis Huds. 198.
pyramidalis *Linn.* 198.
reptans *Linn.* 198.
Alchemilla *Linn.* 103.
alpina *Linn.* 103.
aphanes Leers, 103.
arvensis *Smith,* 103.
minor Hudson, 103.
vulgaris *Linn.* 103.
ALISMACEÆ *Juss.* 253.
Alisma *Linn.* 253.
Damasonium *Linn.* 253.
lanceolata With. 253.
natans *Linn.* 253.
Plantago *Linn.* 253.
ranunculoides *Linn.* 253.
repens *Cav.* 253.
Alliaria *Adans.* 29.
officinalis *Dec.* 29.
Allium *Linn.* 267.
Ampeloprasum *Linn.* 267.
arenarium *Linn.* 267.
carinatum *Linn.* 268.
oleraceum *Linn.* 268.
Schœnoprasum *Linn.* 268.
ursinum *Linn.* 268.
vineale *Linn.* 268.
Alnus *Tourn.* 229.
glutinosa *Gærtn.* 229.
Alopecurus *Linn.* 299.
agrestis *Linn.* 299.
alpinus *Smith,* 299.
aristatus Huds. 302.
bulbosus *Linn.* 299.
fulvus *Smith,* 300.
geniculatus *Linn.* 300.
monspeliensis Linn. 302.
ovatus Knapp, 299.
paniceus Fl. Dan. 300.
pratensis *Linn.* 299.
ventricosus Huds. 302.
ALSINEÆ *Dec.* 47.
Alsine media Linn. 52.
rubella Wahl. 50.
Althæa *Linn.* 41.
officinalis *Linn.* 41.
ALYSSINEÆ *Dec.* 25.
Alyssum halimifolium Bot. Mag. 26.
maritimum Willd. 26.
minimum Linn. 26.
sativum Smith, 30.
AMARANTACEÆ *Juss.* 213.
Amaranthus *Linn.* 213.
Blitum *Linn.* 213.
AMARYLLIDEÆ *R.Br.* 264.
AMENTACEÆ *Juss.* 228.
AMMINEÆ *Koch.* 120.
Ammophila *Host.* 302.
arenaria *Host.* 303.
Amygdaleæ Juss. 89.
Anacamptis *Rich.* 261.
pyramidalis *Rich.* 261.
Anagallis *Linn.* 185.
arvensis *Linn.* 185.
cærulea *Schreb.* 185.
tenella *Linn.* 185.
Anchusa *Linn.* 165.
arvensis Lehm. 165.
officinalis *Linn.* 165.
sempervirens *Linn.* 165.
Andromeda *Linn.* 173.
cærulea Linn. 173.
Daboecia Linn. 173.
polifolia *Linn.* 173.
taxifolia Pall. 173.

Androsæmum *All.* 43.
officinalis *All.* 43.
Anemagrostis *Trin.* 304.
Spica Venti *Trin.* 304.
Anemone *Linn.* 9.
apennina *Linn.* 9.
nemorosa *Linn.* 9.
pratensis Sibth. 9.
Pulsatilla *Linn.* 9.
ranunculoides *Linn.* 10.
Anethum Fœniculum Linn. 119.
ANGELICEÆ *Koch.* 117.
Angelica *Linn.* 117.
archangelica Linn. 117.
sylvestris *Linn.* 117.
Antennaria *Gærtn.* 144.
dioica *Gærtn.* 144.
margaritacea *Gærtn.* 144.
Anthemis *Linn.* 150.
arvensis *Linn.* 150.
cotula Linn. 150.
maritima *Linn.* 150.
nobilis *Linn.* 150.
tinctoria *Linn.* 150.
Anthericum *Linn.* 269.
serotinum *Linn.* 269.
Anthericum calyculatum Linn. 264.
Anthericum ossifragum Linn. 277.
Anthoxanthum *Linn.* 305.
odoratum *Linn.* 306.
Anthriscus *Spreng.* 124.
cerefolium *Hoffm.* 124.
sylvestris *Hoffm.* 124.
vulgaris *Spreng.* 124.
Anthyllis *Linn.* 77.
Dillenii Schultes, 78.
vulneraria *Linn.* 78.
Antirrhinum *Linn.* 192.
Cymbalaria Linn. 191.
Elatine Linn. 191.
Linaria Linn. 191.
majus *Linn.* 192.
minus Linn. 192.
monspessulanum Linn. 191.
Orontium *Linn.* 192.
epens Linn. 191.
spurium Linn. 191.
Apargia *Schreb.* 162.
autumnalis *Willd.* 162.
hirta Hoffm. 162.
hispida *Willd.* 162.
Taraxaci *Willd.* 162.
Apera Spica Venti Beauv. 304.
Aphanes arvensis Linn. 103.
Apium *Linn.* 123.
graveolens *Linn.* 123.
APOCYNEÆ *Juss.* 176.
Aquilegia *Linn.* 13.
alpina Huds. 13.
vulgaris *Linn.* 13.
ARABIDEÆ *Dec.* 22.
Arabis *Linn.* 24.
ciliata *R. Br.* 24.
Crantziana Willd. 24.
hirsuta *R. Br.* 24.
hispida *Linn.* 24.
petræa *Lam.* 24.
stricta *Huds.* 24.
thaliana *Linn.* 24.
Turrita *Linn.* 24.
umbrosa Crantz, 24.
Arbutus *Linn.* 174.
alpina Linn. 174.
Unedo *Linn.* 174.
Uva ursi Linn. 174.
Archangelica *Hoffm.* 117.
officinalis *Hoffm.* 117.
Arctium Bardana Willd. 154.
Lappa Linn. 154.
Arctostaphylos *Kunth.* 174.
alpina *Spreng.* 174.
Uva ursi *Spreng.* 174.
Arenaria *Linn.* 49.
cæspitosa Ehr. 50.
campestris Linn. 50.
ciliata *Linn.* 40.
fasciculata *Jacq.* 50.
fastigiata Smith, 50.
hirta *Wormsk.* 49.
juniperina With. 50.
laricifolia With. 50.
marina Fl. Dan. 50.
media *Linn.* 50.
multicaulis Linn. 50.
peploides *Linn.* 49.
rubella Smith, 50.
rubra *Linn.* 50.
saxatilis Huds. 50.
serpyllifolia *Linn.* 49.
tenuifolia *Linn.* 49.
trinervis *Linn.* 49.
verna *Linn.* 50.

Arethuseæ *Lindl.* 258.
Arhizæ Rich. 3.
Aristolochiæ *Juss.* 224.
Aristolochia *Linn.* 225.
Clematitis *Linn.* 225.
Armeria *Dec.* 170.
maritima *Willd.* 170.
Arnoseris pusilla Gærtn. 157.
Aroideæ *Juss.* 246.
Arrhenatherum *Beauv.* 304.
avenaceum *Beauv.* 305.
bulbosum *Dumort,* 305.
Artemisia *Linn.* 149.
Absinthium *Linn.* 149.
cærulescens *Linn.* 149.
campestris *Linn.* 149.
gallica *Willd.* 149.
maritima *Linn.* 149.
vulgaris *Linn.* 149.
Arum *Linn.* 246.
maculatum *Linn.* 246.
Arundo *Linn.* 310.
arenaria Linn. 303.
Calamagrostis Huds. 304.
Calamagrostis Linn. 304.
colorata H. Kew. 301.
epigejos Linn. 304.
neglecta Ehr. 304.
Phragmites *Linn.* 310.
stricta Schrad. 304.
Asarum *Linn.* 224.
europæum *Linn.* 225.
Asparagus *Linn.* 267.
officinalis *Linn.* 267.
Asperula *Linn.* 130.
cynanchica *Linn.* 130.
odorata *Linn.* 130.
Asperugo *Linn.* 165.
procumbens *Linn.* 165.
Asphodeleæ *R. Br.* 266.
Aster *Linn.* 143.
Tripolium *Linn.* 143.
Ast agalus *Linn.* 78.
arenarius Huds. 78.
campestris Linn. 79.
danicus Retz, 78.
epiglottis Dicks. 78.
glycyphyllos *Linn.* 78.
hypoglottis *Linn.* 78.
sordidus Willd. 79.
uralensis Linn. 79.
Athamanta Meum Linn. 118.
Libanotis Linn. 119.
Athanasia maritima Linn. 150
Atriplices Juss. 213.
Atriplex *Linn.* 216.
angustifolia *Smith,* 217.
erecta *Huds.* 217.
laciniata *Linn.* 217.
littoralis *Linn.* 217.
marina *Linn.* 217.
patula *Linn.* 217.
pedunculata *Linn.* 217.
portulacoides *Linn.* 217.
serrata Huds. 217.
Atropa *Linn.* 182.
Belladonna *Linn.* 182.
Avena *Linn.* 310.
alpina *Sm.* 310.
bromoides Linn. 310.
caryophyllea Link. 308.
elatior Linn. 305.
fatua *Linn.* 310.
flavescens Linn. 309.
flexuosa Link. 308.
nodosa Cullum, 305.
planiculmis Sm. 310.
præcox Linn. 308.
pratensis *Linn.* 310.
pubescens Linn. 309.
sesquitertia Linn. 309.
strigosa *Schreb.* 310.
Azalea *Linn.* 172.
procumbens *Linn.* 172.
Baldingera arundinacea Dum. 301.
Ballota *Linn.* 201.
nigra *Linn.* 201.
Balsamineæ *A. Rich.* 59.
Barbarea *R. Br.* 23.
præcox *R. Br.* 23.
vulgaris *R. Br.* 23.
Bartsia *Linn.* 190.
alpina *Linn.* 190.
Odontites *Linn.* 191.
viscosa *Linn.* 190.
Bellis *Linn.* 148.
perennis *Linn.* 148.
Berberideæ *Vent.* 14.
Berberis *Linn.* 14.
vulgaris *Linn.* 14.
Beta *Linn.* 216.
maritima *Linn.* 216.

Betonica *Linn.* 202.
officinalis *Linn.* 202.
BETULINEÆ *Rich.* 228.
Betula *Linn.* 228.
alba *Linn.* 229.
Alnus Linn. 229.
margaritacea of some, 229.
nana *Linn.* 229.
pendula *Roth.* 229.
verrucosa *Ehr.* 229.
Bidens *Linn.* 151.
cernua *Linn.* 151.
minima Linn. 151.
tripartita *Linn.* 151.
Blysmus *Panz.* 279.
compressus *Panz.* 280.
rufus Link. 280.
BORAGINEÆ *Juss.* 163.
Borago *Linn.* 164.
officinalis *Linn.* 164.
Brachypodium *Beauv.* 297.
pinnatum *Beauv.* 297.
sylvaticum *Beauv.* 297.
BRASSIÇEÆ *Dec.* 32.
Brassica *Linn.* 32.
campestris *Linn.* 32.
monensis *Huds.* 32.
Napus *Linn.* 32.
oleracea *Linn.* 32.
orientalis Linn. 30.
Rapa *Linn.* 32.
Briza *Linn.* 315.
aspera Knapp, 315.
media *Linn.* 315.
minor *Linn.* 315.
Bromus *Linn.* 311.
agrestis All. 312.
arvensis *Linn.* 312.
arvensis Knapp, 311.
asper *Linn.* 312.
ciliatus Huds. 312.
cristatus Linn. 298.
diandrus *Curt.* 312.
erectus *Huds.* 312.
giganteus Linn. 314.
gracilis Weig. 297.
hirsutus Curt. 312.
hordeaceus Linn. 311.
littoreus Willd. 313.
mollis *Linn.* 311.
montanus Poll. 312.
multiflorus Smith, 311.
muralis Huds. 312.
nemoralis Linn. 312.
nemorosus Vill. 312.
perennis Vill. 312.
pinnatus Linn. 297.
polymorphus Huds. 311.
pratensis Ehr. 311.
racemosus *Linn.* 311.
ramosus Linn. 312.
secalinus *Linn.* 311.
spiculitenuatus Knapp, 312.
squarrosus *Linn.* 311.
sterilis *Linn.* 312.
sylvaticus, Poll. 297.
triflorus Linn. 314.
velutinus *Schrad.* 311
versicolor Poll. 312.
verticillatus Cav. 312.
vitiosus Weig. 311.
Bryonia *Linn.* 319.
alba Huds. 319.
dioica *Jacq.* 319.
Buffonia *Linn.* 47.
annua *Dec.* 47.
tenuifolia Linn. 48.
Bunias Kakile Linn. 28.
Bunium flexuosum With. 121.
Bupleurum *Linn.* 120.
Odontites *Linn.* 120.
rotundifolium *Linn.* 120.
tenuissimum *Linn.* 120.
BUTOMEÆ *Rich.* 271.
Butomus *Linn.* 272.
umbellatus *Linn.* 272.
Buxus *Linn.* 223.
sempervirens *Linn.* 223.
Cakile *Tournef.* 28.
maritima *Willd.* 28.
Calamagrostis *Roth.* 304.
arenaria Roth. 303.
epigejos *Roth.* 304.
epigejos With. 304.
lanceolata *Roth.* 304.
lanceolata With. 304.
stricta 304.
CALLITRICHINEÆ *Link.* 242.
Callitriche *Link.* 243.
autumnalis *Linn.* 243.
verna *Linn.* 243.

Calluna *Salisb.* 173.
vulgaris *Salisb.* 173.
Caltha *Linn.* 12.
palustris *Linn.* 12.
radicans *Forst.* 12.
Calystegia *R. Br.* 167.
sepium *R. Br.* 167.
Soldanella *R. Br.* 167.
Camelina *Crantz,* 30.
sativa *Crantz,* 30.
CAMELINEÆ *Dec.* 30.
CAMPANULACEÆ *Juss.* 135.
Campanula *Linn.* 135.
glomerata *Linn.* 136.
hederacea *Linn.* 136.
hybrida Linn. 136
latifolia *Linn.* 136.
patula *Linn.* 136.
persicifolia *Linn.* 136.
Rapunculus *Linn.* 136.
rapunculoides *Linn.* 136.
rotundifolia *Linn.* 136.
Trachelium *Linn.* 136.
Campella cæspitosa Link. 309
CAPRIFOLIACEÆ *Juss.* 131.
Caprifolium *Tourn.* 131.
perfoliatum 131.
periclymenum 131.
Capsella *Dec.* 31.
bursa pastoris *Dec.* 31.
Cardamine *Linn.* 25.
amara *Linn.* 25.
bellidifolia *Linn.* 25.
flexuosa With. 25.
hastulata E. Bot. 24.
hirsuta *Linn.* 25.
impatiens *Linn.* 25.
parviflora Lightf. 25.
petræa Huds. 24.
pratensis *Linn.* 25.
Carduus *Linn.* 155.
acanthoides *Linn.* 155.
acanthoides Huds. 156.
acaulis Linn 154.
arvensis Curtis, 153.
crispus Huds. 156.
dissectus Vill. 153.
eriophorus Linn. 153.
helenioides Huds. 153.
heterophyllus Linn. 153.
lanceolatus Linn. 152.
marianus Linn. 154.
nutans *Linn.* 155.
palustris Linn. 152.
pratensis Huds. 153.
polyacanthos Curtis, 156.
tenuiflorus *Curtis,* 156.
tuberosus Linn. 153.
Carex *Linn.* 284.
acuta *Linn.* 291.
acuta Curt. 291.
acuta Huds. 291.
acutiformis Ehr. 291.
æthiopica Schk. 291.
Agastachys Ehr. 287.
alpina Hoppe, 288.
ambleocarpa Willd. 290.
ampullacea *Gooden.* 292.
angustifolia *Sm.* 292.
anonyma Fl. Dan. 292.
arenaria *Linn.* 285.
arenaria Leers, 286.
atrata *Linn.* 289.
atro-fusca *Schk.* 288.
axillaris *Gooden.* 285.
axillaris Linn. 285.
binervis *Sm.* 289.
brizoides Huds. 285.
cæspitosa *Linn.* 291.
cæspitosa Huds. 291.
canescens Lightf. 285.
canescens Huds. 286.
capillaris *Linn.* 288.
capillaris Leers, 287.
capitata Huds. 284.
clandestina *Gooden.* 287.
crassa Ehr. 291.
curta *Gooden.* 285.
Davalliana *Sm.* 284.
decumbens Ehr. 290.
depauperata *Gooden.* 287.
digitata *Linn.* 287.
dioica Huds. 284.
dioica *Linn.* 284.
distans *Linn.* 289.
disticha Huds. 286.
divisa *Huds.* 286.
divisa Don, 286.
divulsa *Gooden.* 286.
Drymeia *Ehr.* 287.
echinata *Sibth.* 285.
elegans Willd. 288.

elongata *Linn.* 285.
elongata Leers, 285.
extensa *Gooden.* 289.
filiformis *Linn.* 292.
filiformis Fl. Dan. 290.
filiformis Leers, 290.
flacca Schreb. 290.
flava *Linn.* 289.
fulva *Gooden.* 289.
fusca All. 291.
fusca Schk. 289.
glauca Scop. 290.
gracilis Curt. 291.
hirta *Linn.* 292.
hirta Fl. Dan. 292.
humilis Leys. 287.
hybrida Schk. 284.
incurva *Lightf.* 285.
inflata Huds. 292.
intermedia *Gooden.* 286.
juncifolia All. 285.
lævigata *Sm.* 291.
lasiocarpa Ehr. 292.
leporina Huds. 285.
leptostachys Ehr. 287.
Leucoglochin Ehr. 284.
limosa *Linn.* 288.
maxima Scop. 287.
Micheliana Sm. 290.
Mielichoferi *Willd.* 287.
mirabilis Host. 284.
montana Linn. 290.
montana Lightf. 290.
mucronata Jacq. 291.
muricata *Linn.* 286.
muricata Huds. 285.
nigra All. 288.
obtusangula Ehr. 292.
Œderi *Ehr.* 289.
ovalis *Gooden.* 285.
pallescens *Linn.* 289.
paludosa *Gooden.* 291.
panicea *Linn.* 290.
paniculata *Linn.* 286.
patula Huds. 284.
patula Scop. 287.
patula Schk. 291.
pauciflora *Lightf.* 284.
pendula *Huds.* 287.
pendula Schreb. 290.
phæostachya *Sm.* 288.
pilulifera *Linn.* 290.
præcox *Jacq.* 290.
prostrata All. 287.
pseudo-cyperus *Linn.* 288.
psyllophora Ehr. 284.
pulicaris *Linn.* 284.
pulla *Gooden.* 289.
rariflora *Sm.* 288.
recurva *Huds.* 290.
remota *Linn.* 285.
repens Bell. 286.
rigida *Gooden.* 290.
riparia *Curt.* 291.
rostrata Sibth. 292.
saxatilis Huds. 290.
saxatilis Fl. Dan. 291.
secalina *Willd.* 292.
speirostachya Swz. 288.
sphærocarpa Ehr. 290.
spicata Huds. 286.
splendida Willd. 292.
stellulata *Gooden.* 284.
stictocarpa *Sm.* 292.
stolonifera Ehr. 290.
stricta *Gooden.* 291.
strigosa *Huds.* 287.
sylvatica *Huds.* 287.
tenella *Schk.* 285.
tenella Ehr. 285.
teretiuscula *Gooden.* 286.
tomentosa *Linn.* 290.
tomentosa Lightf. 292.
triflora Willd. 287.
uliginosa Linn. 280.
ustulata *Willd.* 288.
ventricosa Curt. 287.
vesicaria *Linn.* 291.
vesicaria Huds. 292.
vulpina *Linn.* 286.
Carlina *Linn.* 154.
vulgaris, *Linn.* 154.
Carpinus *Linn.* 240.
Betulus *Linn.* 240.
Carrichtera *Dec.* 33.
Vellæ *Dec.* 33.
CAROPHYLLEÆ *Juss.* 43.
Carum *Linn.* 122.
Carvi *Linn.* 122.
verticillatum *Koch.* 122.
Castanea *Gærtn.* 239.
vesca *Gærtn.* 239.
vulgaris Dec. 239.

Catabrosa *Beauv.* 306.
aquatica *Beauv.* 306.
Catopodium *Link.* 297.
loliaceum *Link.* 297.
Caucalineæ *Koch.* 113.
Caucalis *Linn.* 113.
Anthriscus, Huds. 114.
arvensis Huds. 114.
daucoides *Linn.* 114.
helvetica Jacq. 114.
infesta Curtis, 114.
latifolia *Linn.* 114.
leptophylla Huds. 114.
nodosa Huds. 114.
scandicina Wigg. 124.
Celastrineæ *R. Br.* 74.
Cellulares Dec. 3.
Cellularia Fries. 3.
Centaurea *Linn.* 154.
Calcitrapa *Linn.* 155.
Cyanus *Linn.* 155.
Isnardi *Linn.* 155.
Jacea *Linn.* 155.
nigra *Linn.* 155.
Scabiosa *Linn.* 155.
solstitialis *Linn.* 155.
Centranthus *Dec.* 138.
latifolius *Dufr.* 139.
Centunculus *Linn.* 183.
minimus *Linn.* 183.
Cerastium *Linn.* 50.
alpinum *Linn.* 51.
aquaticum *Linn.* 51.
arvense *Linn.* 51.
latifolium Smith, 51.
pumilum Curtis, 51.
semidecandrum *Linn.* 51.
tetrandrum Smith, 51.
viscosum *Linn.* 51.
vulgatum *Linn.* 51.
Cerasus *Tournef.* 90.
avium *Mœnch.* 90.
Padus *Dec.* 90.
Ceratophylleæ *Dec.* 225.
Ceratophyllum *Linn.* 225.
demersum *Linn.* 225.
submersum *Linn.* 225.
Chærophyllum *Linn.* 125.
aromaticum *Linn.* 125.
aureum *Linn.* 125.
odoratum Hooker, 125.
sativum Spreng. 124.
sylvestre Linn. 124.
temulum *Linn.* 125.
Chamagrostis *Borkh.* 301.
minima *Schrad.* 301.
Cheiranthus *Linn.* 22.
Cheiri *Linn.* 22.
erysimoides Huds. 30.
fruticulosus Linn. 23.
incanus Linn. 22.
sinuatus Linn. 22.
tricuspidatus Hudson, 22.
Chelidonium *Linn.* 18.
Glaucium Linn. 17.
hybridum Linn. 18.
laciniatum *Mill.* 18.
majus *Linn.* 18.
Chenopodeæ *Vent.* 213.
Chenopodium *Linn.* 214.
acutifolium *Sm.* 216.
album *Linn.* 215.
arrectum Desm. 216.
Bonus Henricus *Linn.* 215.
botryodes *Sm.* 215.
ficifolium *Smith,* 215.
fruticosum *Linn.* 216.
glaucum *Linn.* 215.
hybridum *Linn.* 215.
maritimum *Linn.* 216.
murale *Linn.* 215.
olidum *Curtis,* 216.
polyspermum *Linn.* 216.
polyspermum Curtis, 216.
rubrum *Linn.* 215.
serotinum Huds. 215.
urbicum *Linn.* 215.
viride Linn. 215.
Vulvaria Linn. 216.
Cherleria *Linn.* 48.
sedoides *Linn.* 48.
Chilochloa arenaria Trin. 300.
Chironia Centaurium Curtis, 178.
littoralis Turner, 178.
pulchella Willd. 178.
pulchella Don, 178.
ramosissima Ehr. 178.
Chlora *Linn.* 179.
perfoliata *Linn.* 179.
Chondrilla *Gærtn.* 157.
muralis *Lam.* 157.
Chrysanthemum *Linn.* 148.
inodorum *Linn.* 148.

Leucanthemum *Linn.* 148.
maritimum *Smith,* 148.
Parthenium *Smith,* 148.
segetum *Linn.* 148.
Chrysocoma *Linn.* 142.
Linosyris *Linn.* 142.
Chrysosplenium *Linn.* 66.
alternifolium *Linn.* 66.
oppositifolium *Linn.* 67.
Chrysurus echinatus Beauv. 306.
CICHORACEÆ *Juss.* 156.
Cichorium *Linn.* 162.
Intybus *Linn.* 162.
Cicuta *Linn.* 123.
virosa *Linn.* 123.
Cineraria *Linn.* 147.
alpina Huds. 148.
campestris *Retz,* 147.
integrifolia Jacq. 147.
palustris *Linn.* 147.
CIRCÆACEÆ, 109.
Circæa *Linn.* 109.
alpina *Linn.* 110.
lutetiana *Linn.* 110.
Cirsium arvense Lam. 153.
acaule All. 153.
bulbosum Dec. 153.
eriophorum Scop. 153.
heterophyllum Dec. 153.
lanceolatum Scop. 152.
palustre Scop. 152.
pratense Dec. 153.
CISTINEÆ *Juss.* 36.
Cistus anglicus Linn. 37.
guttatus Linn. 37.
Helianthemum Linn. 37.
hirsutus Huds. 37.
ledifolius Linn. 37.
parvifolius Linn. 37.
polifolius Linn. 37.
salicifolius Huds. 37.
surrejanus Linn. 37.
tomentosus Smith, 37.
Cladium *Schrad.* 283.
germanicum Schrad. 283.
Mariscus *R. Br.* 283.
Clematis *Linn.* 8.
Vitalba *Linn.* 8.
Clinopodium *Linn.* 205.
vulgare *Linn.* 206.
Cnicus *Linn.* 152.
acaulis *Willd.* 153.
arvensis *Smith,* 153.
eriophorus *Willd.* 153.
Forsteri *Smith,* 153.
heterophyllus *Willd.* 153.
lanceolatus *Willd.* 152.
palustris *Willd.* 152.
pratensis *Willd.* 153.
tuberosus *Willd.* 153.
Cnidium Silaus Spreng. 118.
Cochlearia *Linn.* 27.
anglica *Linn.* 27.
Armoracia *Linn.* 27.
Coronopus Linn. 30.
danica *Linn.* 27.
grœnlandica *Linn.* 27.
officinalis *Linn.* 27.
Colchicaceæ Dec. 264.
Colchicum *Linn.* 264.
autumnale *Linn.* 264.
Comarum palustre Smith, 97.
COMPOSITÆ *Juss.* 140.
Condylocarpus *Hoffm.* 115.
officinalis *Koch.* 115.
CONIFERÆ *Juss.* 240.
Conium *Linn.* 126.
maculatum *Linn.* 126.
Conopodium *Koch.* 121.
flexuosum *Koch.* 121.
Convallaria *Linn.* 270.
majalis *Linn.* 270.
multiflora *Linn.* 271.
Polygonatum *Linn.* 270.
verticillata *Linn.* 270.
CONVOLVULACEÆ *Juss.* 167.
Convolvulus *Linn.* 167.
arvensis *Linn.* 167.
sepium Linn. 167.
Soldanella *Linn.* 168.
Conyza *Linn.* 142.
squarrosa *Linn.* 142.
Corallorhiza *Haller,* 258.
innata *R. Br.* 258.
Coreopsis Bidens Linn. 151.
CORIANDRIEÆ *Koch.* 114.
Coriandrum *Linn.* 115.
sativum *Linn.* 115.
Cornus *Linn.* 132.
sanguinea *Linn.* 133.
suecica *Linn.* 133.

Coronopus *Gærtn.* 30.
didyma *Smith,* 31.
Ruellii *Gærtn.* 30.
Corrigiola *Linn.* 60.
littoralis *Linn.* 60.
Corvisartia Helenium Merat. 143.
Corydalis *Dec.* 19.
bulbosa *Dec.* 19.
capnoides β Dec. 19.
claviculata *Dec.* 19.
lutea *Dec.* 19.
Corylaceæ Mirb. 239.
Corylus *Linn.* 240.
Avellana *Linn.* 240.
Corymbiferæ *Juss.* 142.
Corynephorus *Beauv.* 307.
canescens *Beauv.* 308.
Cotoneaster *Lindley,* 104.
vulgaris *Lindley,* 104.
Cotyledon lutea Huds. 64.
umbilicus Huds. 64.
Cotyledoneæ Juss. 3.
Crambe *Linn.* 34.
maritima *Linn.* 34.
Crassulaceæ *Dec.* 63.
Cratægus *Linn.* 104.
Aria Linn. 105.
oxyacantha *Linn.* 104.
scandica Wahl. 105.
Crepis *Linn.* 158.
biennis *Linn.* 158.
fœtida Linn. 158.
pulchra Linn. 157.
tuberosa *Linn.* 158.
Crithmum *Linn.* 118.
maritimum *Linn.* 118.
Crocus *Linn.* 255.
autumnalis E. Bot. 255.
nudiflorus *Smith,* 255.
officinalis Hudson, 255.
sativus *Linn.* 255.
vernus *Willd.* 255.
Cruciferæ *Juss.* 20.
Cryptocotyledoneæ Ag. 4.
Cryptogamia Linn. 3.
Cucubalus acaulis Linn. 46.
Behen Linn. 45.
Otites Linn. 46.
viscosus Huds. 46.
Cucurbitaceæ *Juss.* 319.
Cupuliferæ *Rich.* 239.
Cuscuta *Linn.* 168.
epithymum *Linn.* 168.
europæa *Linn.* 168.
major Dec. 168.
minor Dec. 168.
Cyclamen *Linn.* 183.
europæum E. Bot. 183.
hederifolium *Willd.* 183.
Cymbidium corallorrhiza Swartz, 258.
Löselii Swartz, 263.
Cynarocephalæ *Juss.* 152.
Cynodon *Rich.* 298.
Dactylon *Rich.* 298.
Cynoglossum *Linn.* 166.
officinale *Linn.* 166.
sylvaticum *Hænke,* 166.
Cynosurus *Linn.* 306.
cæruleus Linn. 309.
cristatus *Linn.* 306.
echinatus *Linn.* 306.
paniceus Linn. 302.
Cyperaceæ *Juss.* 278.
Cyperus *Linn.* 279.
fuscus *Linn.* 279.
longus *Linn.* 279.
nigricans With. 280.
Cypripedieæ *Lindley,* 263.
Cypripedium *Linn.* 263.
Calceolus *Linn.* 263.
Cytisus *Linn.* 77.
scoparius *Link.* 77.
Dactylis *Linn.* 310.
cynosuroïdes Huds. 299.
glomerata *Linn.* 310.
stricta H. K. 299.
Daphne *Linn.* 209.
Laureola *Linn.* 209.
Mezereum *Linn.* 209.
Datura *Linn.* 180.
Stramonium *Linn.* 181.
Daucineæ *Koch.* 113.
Daucus *Linn.* 113.
Carota *Linn.* 113.
maritimus *Linn.* 113.
Delphinium *Linn.* 13.
Consolida *Linn.* 13.
Dentaria *Linn.* 25.
bulbifera *Linn.* 25.
Deschampsia *Beauv.* 309.
cæspitosa *Beauv.* 309.

Dichlamydeæ, 5.
Dicotyledones *Juss.* 4, 5.
Digitalis *Linn.* 192.
  purpurea *Linn.* 192.
Digitaria *Scop.* 299.
  sanguinalis *Scop.* 299.
  stolonifera *Schrad.* 298.
Digraphis *Trin.* 301.
  arundinacea *Trin.* 301.
Diotis *Desf.* 149.
  maritima *Desf.* 150.
Diplecolobeæ *Dec.* 34.
Diplotaxis *Dec.* 33.
  muralis *Dec.* 33.
  tenuifolia *Dec.* 33.
Dipsaceæ *Juss.* 139.
Dipsacus *Linn.* 139.
  fullonum *Linn.* 139.
  pilosus *Linn.* 139.
  sylvestris *Linn.* 139.
Doronicum *Linn.* 147.
  Pardalianches *Linn.* 147.
Draba *Linn.* 26.
  aizoides *Linn.* 26.
  contorta Dec. 26.
  hirta *Linn.* 26.
  incana *Linn.* 26.
  muralis *Linn.* 26.
  rupestris *R. Br.* 26.
  verna Linn. 26.
Droseraceæ *Dec.* 38.
Drosera *Linn.* 38.
  anglica *Huds.* 38.
  longifolia *Linn.* 38.
  rotundifolia *Linn.* 38.
Drupaceæ *Dec.* 89.
Dryadeæ Vent. 90.
Dryas *Linn.* 98.
  octopetala *Linn.* 99.
Echinochloa *Beauv.* 305.
  Crus-galli *Beauv.* 305.
Echinophora *Linn.* 126.
  spinosa *Linn.* 126.
Echium *Linn.* 163.
  vulgare *Linn.* 163.
Elatine *Linn.* 48.
  Hydropiper E. B. 48.
  ? *triandra* Hoffm. 48.
  tripetala *Sm.* 48.
Elæagneæ *Juss.* 208.
Eleocharis acicularis R. & S. 280.
  cæspitosa Link. 281.
  multicaulis Smith, 280.
  palustris R. Br. 280.
  pauciflora Link. 281.
Eleogiton fluitans Link. 284.
Elymus *Linn.* 296.
  arenarius *Linn.* 296.
  caninus Linn. 298.
  europæus *Linn.* 297.
  geniculatus *Curtis*, 296.
Embryonatæ Rich. 3.
Empetreæ *Nutt.* 224.
Empetrum *Linn.* 224.
  nigrum *Linn.* 224.
Endogenæ Dec. 4.
Endorhizeæ Rich. 4.
Enodium cæruleum Dumort, 307.
Epilobium *Linn.* 107.
  alpinum *Linn.* 108.
  alsinefolium *Vill.* 108.
  angustifolium *Linn.* 108.
  hirsutum *Linn.* 108.
  hirsutum Huds. 108.
  montanum *Linn.* 108.
  palustre *Linn.* 108.
  parviflorum *Schreb.* 108.
  pubescens Willd. 108.
  ramosum Huds. 108.
  roseum *Schreb.* 108.
  tetragonum *Linn.* 108.
  villosum Curtis, 108.
Epimedium *Linn.* 14.
  alpinum *Linn.* 15.
Epipactis *Swartz*, 258.
  ensifolia *Swartz*, 259.
  grandiflora *Smith*, 259.
  latifolia *Swartz*, 258.
  Nidus avis Swartz, 258.
  pallens Swartz, 259.
  palustris *Swartz*, 259.
  purpurata *Smith*, 259.
  rubra *Swartz*, 259.
  xiphophylla Swartz, 259.
Ericeæ *Juss.* 172.
Erica *Linn.* 173.
  cærulea Willd. 173.
  ciliaris *Linn.* 174.
  cinerea *Linn.* 174.
  Dabeoci Linn. 173.
  didyma With. 174.

multiflora Huds. 174.
Tetralix *Linn.* 174.
vagans *Linn.* 174.
vulgaris Linn. 173.
Erigeron *Linn.* 143.
acre *Linn.* 144.
alpinum *Linn.* 144.
canadense *Linn.* 144.
uniflorum Linn. 144.
Eriocaulon *Linn.* 272.
decangulare Lightf. 272.
septangulare *With.* 272.
Eriolepis lanceolata Cass. 152.
lanigera Cass. 153.
Eriophorum *Linn.* 282.
alpinum *Linn.* 282.
angustifolium *Roth.* 282.
angustifolium Poit. & Turp. 282.
cæspitosum Host. 282.
capitatum *Host.* 282.
gracile *Roth.* 282.
latifolium Schrad. 282.
polystachyon *Linn.* 282.
pubescens *Smith,* 282.
Scheuchzeri Roth. 282.
triquetrum Schrad. 283.
vaginatum *Linn.* 282.
Vaillantii Poit. 282.
Erodium *L'Herit.* 58.
cicutarium *Smith,* 58.
maritimum *Smith,* 58.
moschatum *Smith,* 58.
Erophila *Dec.* 26.
vulgaris *Dec.* 26.
Eryngium *Linn.* 127.
campestre *Linn.* 127.
maritimum *Linn.* 127.
Erysimum *Linn.* 30.
Alliaria Linn. 29.
Barbarea Linn. 23.
Cheiranthoides *Linn.* 30.
orientale *R. Br.* 30.
perfoliatum Dec. 30.
præcox Smith, 25.
Erythræa *Renealm,* 177.
Centaurium *Pers.* 178.
latifolia *Smith,* 178.
littoralis *Hook.* 178.
pulchella *Hook.* 178.
Ervum *Linn.* 83.
hirsutum *Linn.* 83.
soloniense, Linn. 84.
tetraspermum *Linn.* 83.
Euonymus *Linn.* 74.
europæus *Linn.* 74.
Eupatorium *Linn.* 142.
cannabinum *Linn.* 142.
EUPHORBIACEÆ *Juss.* 220.
Euphorbia *Linn.* 220.
amygdaloides *Linn.* 223.
Characias *Linn.* 223.
Coderiana Fl. Fr. 221.
Cyparissias *Linn.* 221.
Esula *Linn.* 221.
exigua *Linn.* 222.
Helioscopia *Linn.* 221.
hiberna *Linn.* 221.
Lathyris *Linn.* 222.
paralias *Linn.* 222.
Peplis *Linn.* 220.
peploides Gouan, 222.
Peplus *Linn.* 222.
platyphylla *Linn.* 221.
Portlandica Linn. 222.
segetalis *Linn.* 222.
stricta Linn. 221.
Euphrasia *Linn.* 191.
officinalis *Linn.* 191.
Exacum *Linn.* 177.
filiforme *Smith,* 177.
Exembryonatæ Rich. 3.
Exogenæ Dec. 4.
Exorhizeæ Rich. 4.
Fagus *Linn.* 239.
Castanea Linn. 239.
sylvatica *Linn.* 239.
Fedia dentata Vahl. 138.
olitoria Vahl. 138.
Festuca *Linn.* 313.
arundinacea Schreb. 313
bromoides Linn. 315.
cæsia E. B. 315.
calamaria Smith, 313.
cambrica Huds. 314.
decidua Smith, 313.
decumbens Linn. 311.
dumetorum Linn. 314.
duriuscula *Linn.* 314.
elatior Host. 313.
elatior Smith, 312.

elongata Ehr. 313.
fluitans Linn. 315.
gigantea *Vill.* 314.
glabra Lightf. 314.
glauca Winch. 314.
gracilis Mœnch, 297.
heterophylla Hænke, 314.
loliacea Huds. 313.
Myurus Linn. 314.
nemorum Leyss. 314.
ovina *Linn.* 313.
pinnata Huds. 297.
rubra With. 313.
rubra *Linn.* 314.
sylvatica Huds. 297.
sylvatica Vill. 313.
tenuifolia Sibth. 313.
triflora E. B. 314.
uniglumis H. K. 314.
vivipara *Smith,* 313.
Ficaria ranunculoides Dec. 10.
Filago *Linn.* 145.
gallica *Linn.* 145.
germanica *Linn.* 145.
minima, 145.
FLUVIALES *Vent.* 248.
Fœniculum *Hoffm.* 119.
vulgare *Hoffm.* 119.
FRAGARIACEÆ *Rich.* 90.
Fragaria *Linn.* 95.
calycina *Lois.* 96.
elatior Smith, 96.
moschata *Duch.* 96.
sterilis Linn. 97.
vesca *Linn.* 95.
FRANKENIACEÆ *St. Hil.* 38.
Frankenia *Linn.* 39.
lævis *Linn.* 39.
pulverulenta *Linn.* 39.
Fraxinus *Linn.* 171.
excelsior *Linn.* 171.
heterophylla *Vahl.* 171.
simplicifolia Willd. 171.
Fritillaria *Linn.* 266.
Meleagris *Linn.* 266.
FUMARIACEÆ *Dec.* 18.
Fumaria *Tourn.* 19.
capreolata *Linn.* 19.
claviculata Smith, 19.
lutea Smith, 19.
media Loisel. 19.
officinalis *Linn.* 19
parviflora *Lam.* 19
solida Smith, 19.
Gagea *Salisb.* 268.
lutea *Ker.* 268.
Galanthus *Linn.* 265.
nivalis *Linn.* 265.
Galeobdolon *Huds.* 202.
luteum *Huds.* 202.
Galeopsis *Curtis,* 202.
Galeopsis *Linn.* 203.
angustifolia *Ehr.* 204.
cannabina Willd. 204.
Galeobdolon Linn. 202.
grandiflora Willd. 204.
Ladanum *Linn.* 204.
latifolia Ehr. 204.
Tetrahit *Linn.* 204.
versicolor *Curtis,* 204.
villosa *Huds.* 204.
Galium *Linn.* 128.
anglicum *Huds.* 130.
Aparine *Linn.* 130.
aristatum *Linn.* 129.
boreale *Linn.* 130.
cinereum *All.* 129.
cruciatum *Linn.* 128.
diffusum Hook. 129.
erectum Huds. 129.
hercynicum Weig. 129.
Mollugo *Linn.* 130.
montanum Huds. 128.
montanum With. 128.
obliquum Vill. 130.
palustre *Linn.* 128.
procumbens With. 129
pusillum *Linn.* 129.
saxatile *Linn.* 128.
scabrum Jacq. 130.
spurium Huds. 129.
spurium *Linn.* 129.
tricorne *With.* 129.
tricorne G. Don, 129.
uliginosum *Linn.* 129.
verrucosum *Smith,* 129.
verum *Linn.* 130.
Witheringii *Smith,* 128.
Gastridium *Beauv.* 302.
australe Beauv. 302.
lendigerum *Link.* 302.
Genista *Linn.* 77.

anglica *Linn.* 77.
pilosa *Linn.* 77.
tinctoria *Linn.* 77.
GENTIANEÆ *Juss.* 177.
Gentiana *Linn.* 178.
acaulis *Linn.* 178.
Amarella *Linn.* 178.
campestris *Linn.* 178.
Centaurium Linn. 178.
filiformis Linn. 177.
nivalis *Linn.* 178.
Pneumonanthe *Linn.* 178.
pulchella Swartz, 178.
verna *Linn.* 178.
GERANIACEÆ *Juss.* 56.
Geranium *Linn.* 56.
columbinum *Linn.* 58.
dissectum *Linn.* 57.
humile Cav. 57.
Lancastriense With. 58.
lucidum *Linn.* 57.
malvæfolium Scop. 57.
molle *Linn.* 57.
nodosum *Linn.* 56.
parviflorum Curtis, 57.
phæum *Linn.* 56.
pratense *Linn.* 56.
prostratum *Cav.* 58.
pusillum *Linn.* 57.
pyrenaicum *Linn.* 57.
Raii, 57.
Robertianum *Linn.* 57.
rotundifolium *Linn.* 57.
sanguineum *Linn.* 58.
sylvaticum *Linn.* 56.
Geum *Linn.* 98.
hybridum Wulf. 98.
intermedium Ehr. 98.
rivale *Linn.* 98.
urbanum *Linn.* 98.
Gifola vulgaris Cass. 146.
Githago segetum Desf. 47.
Glaucium *Linn.* 17.
corniculatum *Curtis,* 18.
flavum Crantz, 17.
luteum *Scop.* 17.
phœniceum Smith, 17.
violaceum Smith, 17.
Glaux *Linn.* 183.
maritima *Linn.* 183.
Glechoma *Linn.* 199.
hederacea *Linn.* 199
GLUMACEÆ, 245. 278
Glyce, 26.
maritima, 26.
Glyceria *R. Br.* 315.
aquatica Sm. 316.
distans Sm. 318.
fluitans *R. Br.* 315.
maritima Smith, 316.
procumbens Smith, 316.
rigida Smith, 316.
spectabilis M. & K. 316.
Gnaphalium *Linn.* 144.
alpinum Lightf. 145.
dioicum Linn. 144.
fuscum Scop. 145.
gallicum Huds. 145.
germanicum Huds. 146.
luteo-album *Linn.* 145.
margaritaceum Linn. 144.
minimum Smith, 145.
montanum Huds. 145.
rectum *Smith,* 145.
supinum *Linn.* 145.
sylvaticum *Linn.* 145.
uliginosum *Linn.* 145.
Goodyera *R. Br.* 257.
repens *R. Br.* 257.
GRAMINEÆ *Juss.* 293.
Graniferæ Ag. 4.
GROSSULACEÆ, 106.
Grossularieæ Dec. 106.
Gymnadenia *R. Br.* 261.
conopsea *R. Br.* 261.
Gymnopera Don, 70.
Habenaria albida R. Br. 261.
bifolia R. Br. 261.
viridis R. Br. 262.
HALORAGEÆ *R. Br.* 110.
HEDERACEÆ *Ach. Rich.* 132.
Hedera *Linn.* 133.
Helix *Linn* 133.
Hedypnois autumnalis Huds. 162.
biennis Huds. 158.
hieracioides Huds. 159.
hirta Fl. Brit. 162.
HEDYSAREÆ *Dec.* 87.
Hedysarum Onobrychis Linn. 88.
Helianthemum *Tournef.* 36.
apenninum *Dec.* 37.

canum *Dun.* 36.
guttatum *Miller,* 37.
ledifolium *Willd.* 37.
surrejanum *Mill.* 37.
vulgare *Gærtn.* 37.
Heleocharis, 280.
acicularis, 280.
cæspitosa, 281.
multicaulis, 280.
palustris, 280.
pauciflora, 281.
Heleogiton, 283.
fluitans, 283.
Helleborus *Linn.* 12.
fœtidus *Linn.* 13.
viridis *Linn.* 12.
Helminthia *Juss.* 158.
echioides *Gærtn.* 158.
Helonias borealis Willd. 264.
Helosciadium *Koch.* 121.
inundatum *Koch.* 122.
nodiflorum *Koch.* 122.
repens *Koch.* 122.
Hemerocallideæ R. Br. 266.
Heracleum *Linn.* 116.
angustifolium Sm. 116.
Sphondylium *Linn.* 116.
Herminium *R. Br.* 262.
Monorchis *R. Br.* 263.
Herniaria *Linn.* 61.
glabra *Linn.* 61.
hirsuta *Linn.* 61.
Hesperis *Linn.* 29.
inodora Linn. 29.
matronalis *Linn.* 29.
Hieracium *Linn.* 159.
alpinum *Linn.* 159.
aurantiacum *Linn.* 159.
Auricula *Linn.* 159.
cerinthoides *Linn.* 160.
denticulatum *Smith,* 161.
dubium *Linn.* 159.
Halleri *Vill.* 160.
Lawsoni *Vill.* 160.
maculatum *Smith,* 159.
molle *Jacq.* 160.
murorum *Linn.* 159.
paludosum *Linn.* 160.
Pilosella *Linn.* 159.
prenanthoides Fl. Br. 161.
prenanthoides *Vill.* 161.
pulmonarioides Vill. 160.
pulmonarium *Smith,* 160.
pumilum Willd. 160.
sabaudum *Linn.* 160.
spicatum All. 161.
sylvaticum *Smith,* 160.
Taraxaci Linn. 162.
umbellatum *Linn.* 161.
villosum *Linn.* 160.
Hierochloe *Gmel.* 306.
borealis *R. & S.* 306.
Hippocrepis *Linn.* 87.
comosa *Linn.* 88.
Hippophäe *Linn.* 208.
rhamnoides *Linn.* 208.
Hippuris *Linn.* 110.
vulgaris *Linn.* 110.
Hirculus *Haw.* 67.
ranunculoides *Haw.* 67.
Holcus *Linn.* 305.
avenaceus Scop. 305.
borealis Schrad. 306.
lanatus *Linn.* 305.
mollis *Linn.* 305.
odoratus Linn. 306.
Holoschœnus *Link.* 283.
vulgaris *Link.* 283.
Holosteum *Linn.* 50.
umbellatum *Linn.* 50.
Hordeum *Linn.* 296.
geniculatum All. 296.
marinum Huds. 296.
maritimum *With.* 296.
murinum *Linn.* 296.
nodosum Linn. 296.
pratense *Huds.* 296.
rigidum Roth. 296.
secalinum Willd. 296.
sylvaticum Huds. 297.
Hottonia *Linn.* 185.
palustris *Linn.* 185.
Humulus *Linn.* 219.
Lupulus *Linn.* 219.
Hutchinsia *R. Br.* 28.
petræa *R. Br.* 28.
Hyacinthus *Linn.* 270.
non scriptus *Linn.* 270.
racemosus Linn. 269.
HYDROCHARIDEÆ *Juss.* 254.
Hydrocharis *Linn.* 254.
Morsus ranæ *Linn.* 254.

Hydrochloa *Hartm.* 316.
aquatica *Hartm.* 316.
HYDROCOTYLINEÆ *Koch.* 127.
Hydrocotyle *Linn.* 127.
inundata Smith, 122.
vulgaris *Linn.* 128.
Hyoscyamus *Linn.* 181.
niger *Linn.* 181.
Hyoseris minima Linn. 157.
HYPERICINEÆ *Juss.* 41.
Hypericum *Linn.* 41.
Androsæmum Linn. 43.
barbatum *Jacq.* 42.
calycinum *Linn.* 41.
delphinense *Vill.* 42.
dubium *Leers,* 42.
elodes *Linn.* 42.
hirsutum *Linn.* 42.
humifusum *Linn.* 42.
maculatum Crantz, 42.
montanum *Linn.* 42.
perforatum *Linn.* 42.
pulchrum *Linn.* 42.
quadrangulum *Linn.* 42.
Hypochæris *Linn.* 161.
glabra *Linn.* 161.
maculata *Linn.* 161.
radicata *Linn.* 161.
Iberis *Linn.* 28.
amara *Linn.* 28.
nudicaulis Linn. 28.
Ilex *Linn.* 73.
Aquifolium *Linn.* 74.
ILICINEÆ *Brongn.* 73.
ILLECEBREÆ *R. Br.* 60.
Illecebrum *Linn.* 61.
verticillatum *Linn.* 61.
Impatiens *Linn.* 60.
Noli-me-tangere *Linn.* 60.
Imperatoria Ostruthium Linn. 116.
Inula *Linn.* 143.
crithmifolia Linn. 143.
crithmoides Linn. 143.
cylindrica With. 143.
dysenterica Linn. 143.
Helenium *Linn.* 143.
pulicaria Linn. 143.
uliginosa Sibth. 143.
IRIDEÆ *Juss.* 254.
Iris *Linn.* 255.
fœtidissima *Linn.* 255.
Pseud-acorus *Linn.* 255.
ISATIDEÆ *Dec.* 31.
Isatis *Linn.* 31.
tinctoria *Linn.* 32.
Isnardia *Linn.* 109.
palustris *Linn.* 109.
Isolepis *R. Br.* 283.
fluitans R. Br. 284.
Holoschœnus R. & S. 283.
setacea *R. Br.* 283.
Ixia Bulbocodium Linn. 255.
Jasione *Linn.* 137.
montana *Linn.* 137.
JUNCAGINEÆ *Rich.* 252.
JUNCEÆ *Dec.* 273.
Juncus *Linn.* 273.
acutiflorus *Ehr.* 275.
acutus *Linn.* 273.
arcticus *Willd.* 274.
articulatus E. Bot. 275.
biglumis *Linn.* 275.
bufonius *Linn.* 274.
bulbosus Linn. 274.
campestris Willd. 276.
capitatus *Weig.* 275.
castaneus *Smith,* 275.
cœnosus *Rich.* 274.
compressus *Jacq.* 274.
compressus Roth. 276.
conglomeratus *Linn.* 273.
effusus *Linn.* 273.
filiformis *Linn.* 274.
Forsteri Smith, 276.
Gesneri *Smith,* 274.
glaucus *Sibth.* 273.
gracilis Roth. 275.
gracilis Smith, 274.
inflexus Relh. 273.
lampocarpus *Ehr.* 275.
latifolius Jacq. 276.
liniger Purton, 277.
maritimus *Smith,* 273.
maximus Ehr. 276.
monanthus Jacq. 274.
nemorosus Sibth. 275.
obtusiflorus *Ehr.* 276.
pilosus *Linn.* 276.
polycephalus *D. Don,* 276.
setifolius Ehr. 275.

spicatus *Linn*. 277.
Sprengelii Willd. 274.
squarrosus *Linn*. 274.
subverticillatus *Wulf*. 275.
supinus Bich. 275.
supinus Don, 275.
sylvaticus Huds. 276.
sylvaticus Willd. 275.
tenuis Hook. 274.
trifidus *Linn*. 274.
triglumis *Linn*. 275.
uliginosus *Sibth*. 275.
vernalis Ehr. 276.
Juniperus *Linn*. 241.
communis *Linn*. 241.
nana *Linn*. 241.
Knappia agrostidea Sm. 301.
Knautia *Linn*. 140.
arvensis *Coulter*, 140.
Kobresia *Willd*. 284.
caricina *Willd*. 284.
Köleria cristata Pers. 307.
glauca Dec. 307.
Labiatæ *Juss*. 196.
Lactuca *Linn*. 156.
saligna *Linn*. 157.
Scariola *Linn*. 156.
sylvestris Lam. 156.
virosa *Linn*. 156.
Lagurus *Linn*. 299.
ovatus *Linn*. 299.
Lamium *Linn*. 202.
album *Linn*. 203.
amplexicaule *Linn*. 203.
dissectum With. 203.
incisum *Willd*. 203.
maculatum *Linn*. 203.
purpureum *Linn*. 203.
Lappa *Tourn*. 154.
glabra *Linn*. 154.
tomentosa *Allioni*, 154.
Lapsana *Linn*. 157.
chondrilloides Lam. 157.
communis *Linn*. 157.
minima Lam. 157.
pusilla *Willd*. 157.
Larbrea *St. Hil*. 51.
aquatica *St. Hil*. 52.
Lathræa *Linn*. 194.
squamaria *Linn*. 194.
Lathyrus *Linn*. 85.
Aphaca *Linn*. 86.
hirsutus *Linn*. 86.
latifolius *Linn*. 85.
Nissolia *Linn*. 86.
palustris *Linn*. 86.
pratensis *Linn*. 86.
sylvestris *Linn*. 85
Lavatera *Linn*. 41.
arborea *Linn*. 41.
Ledum *Linn*. 173.
palustre *Linn*. 173.
Leguminosæ *Juss*. 75.
Leiogyne *Don*, 67.
aizoides, 67.
cernua, 68.
granulata, 68.
nivalis, 68.
rivularis, 68.
Lemna *Linn*. 251.
gibba *Linn*. 252.
minor *Linn*. 252.
polyrhiza *Linn*. 252.
trisulca *Linn*. 251.
Lentibulariæ *Rich*. 186.
Leontodon *Linn*. 157.
autumnale Linn. 162.
hirtum Linn. 162.
hispidum Linn. 162.
montanum Lam. 162.
officinale With. 158.
palustre *Smith*, 158.
Taraxacum *Linn*. 158
Leonurus *Linn*. 198.
Cardiaca *Linn*. 199.
Galeobdolon Scop. 202.
Lepidineæ *Dec*. 30.
Lepidium *Linn*. 31.
campestre *R. Br*. 31.
didymum Linn. 31.
latifolium *Linn*. 31.
petræum Linn. 28.
ruderale *Linn*. 31.
Leucojum *Linn*. 265.
æstivum *Linn*. 265.
Ligusticum *Linn*. 118.
cornubien e Linn. 126.
Meum Crantz, 118.
scoticum *Linn*. 118.
vulgare *Linn*. 171.
Liliaceæ *Juss*. 266.

Limbarda *Cass.* 143.
tricuspis *Cass.* 143.
Limosella *Linn.* 192.
aquatica *Linn.* 192.
Linaria *Desf.* 191.
Cymbalaria *Mill.* 191.
Elatine *Desf.* 191.
ninor *Desf.* 192.
repens *H. Kew.* 191.
spuria *Mill.* 191.
vulgaris *Mœnch.* 191.
LINEÆ *Dec.* 53.
Linnæa *Gronov.* 132.
borealis *Gronov.* 132.
Linum *Linn.* 53.
angustifolium *Huds.* 53.
catharticum *Linn.* 54.
perenne *Linn.* 53.
Radiola Linn. 54.
tenuifolium With. 53.
usitatissimum *Linn.* 53.
ris *Linn.* 263.
Löselii *Rich.* 263.
Listera *R. Br.* 258.
cordata *R. Br.* 258.
ovata *R. Br.* 258.
Lithospermum *Linn.* 164.
arvense *Linn.* 164.
maritimum *Lehm.* 164
officinale *Linn.* 164.
purpuro-cæruleum *Linn.* 164.
Littorella *Linn.* 169.
lacustris *Linn.* 170.
LOBELIACEÆ *Juss.* 137.
Lobelia *Linn.* 137.
Dortmanna *Linn.* 137
urens *Linn.* 137.
Logfia brevifolia Cass. 145.
subulata Cass. 145.
Lolium *Linn.* 295.
arvense *With.* 295.
bromoides Huds. 314.
perenne *Linn.* 295.
temulentum *Linn.* 295.
tenue Linn. 295.
LONICEREÆ, 131.
Lonicera *Linn.* 132.
Caprifolium Linn. 131.
Periclymenum Linn. 131.
Xylosteum *Linn.* 132.
LORANTHEÆ *Rich.* 133.
LOTEÆ *Dec.* 76.
Lotus *Linn.* 81.
angustissimus *Linn.* 82.
corniculatus *Linn.* 81.
decumbens *Forster*, 82.
diffusus Smith, 82.
major *Scop.* 82.
Luzula *Dec.* 276.
arcuata *Smith*, 277.
campestris *Willd.* 276.
congesta *Dec.* 276.
Forsteri *Dec.* 276.
maxima Willd. 276.
pilosa *Willd.* 276.
spicata *Bich.* 277.
sylvatica *Bich.* 276.
Lychnis *Linn.* 46.
alpina *Linn.* 47.
Behen Scop. 45.
dioica *Linn.* 47.
dioica alba Smith, 47.
dioica rubra Smith, 47.
diurna *Sibth.* 47.
Flos Cuculi *Linn.* 47.
Githago Dec. 47.
sylvestris *Hoppe*, 47.
vespertina *Sibth.* 47.
viscaria *Linn.* 47.
Lycopsis *Linn.* 165.
arvensis *Linn.* 165.
Lycopus *Linn.* 197.
europæus *Linn.* 197.
Lysimachia *Linn.* 184.
nemorum *Linn.* 184.
Nummularia *Linn.* 184.
tenella Linn. 185.
thyrsiflora *Linn.* 184.
vulgaris *Linn.* 184.
Lythrum *Linn.* 72.
hyssopifolium *Sibth.* 72.
Salicaria *Linn.* 72.
MALAXIDEÆ *Lindley*, 263.
Malaxis *Linn.* 263.
Löselii Swartz, 263.
paludosa *Linn.* 263.
MALVACEÆ *Juss.* 40.
Malva *Linn.* 40.
moschata *Linn.* 40.
parviflora Hudson, 40.
pusilla Smith, 40.

rotundifolia *Linn.* 40.
sylvestris *Linn.* 40.
Marrubium *Linn.* 201.
vulgare *Linn.* 201.
Maruta *Cass.* 150.
fœtida *Cass.* 150.
Matricaria *Linn.* 148.
Chamomilla *Linn.* 149.
inodora Linn. 148.
Parthenium Linn. 148.
Matthiola *R. Br.* 22.
incana *R. Br.* 22.
sinuata *R. Br.* 22.
Meconopsis *Dec.* 17.
cambrica *Dec.* 17.
Medicago *Linn.* 82.
arabica With. 83.
falcata *Linn.* 82.
hispida Gærtn. 83.
lupulina *Linn.* 83.
maculata *Sibth.* 83.
minima *Willd.* 83.
muricata *Willd.* 83.
polymorpha Linn. 83.
sativa *Linn.* 82.
Megastachya rigida R. & S. 316.
MELAMPYRACEÆ *Rich.* 194.
Melampyrum *Linn.* 195.
arvense *Linn.* 195.
cristatum *Linn.* 195.
pratense *Linn.* 195.
sylvaticum *Linn.* 195.
MELANTHACEÆ *R. Br.* 264.
Melica *Linn.* 307.
cærulea Linn. 307.
decumbens Web. 311.
Lobelii Vill. 307.
montana Huds. 307.
nutans Huds. 307.
nutans *Linn.* 307.
uniflora *Retz*, 307.
Melilotus *Tourn.* 79.
officinalis *Willd.* 79.
Melissa Calamintha Linn. 205.
Nepeta Linn. 205.
Melittis *Linn.* 205.
grandiflora *Smith*, 205.
Melissophyllum *Linn.* 205.
Mentha *Linn.* 199.
acutifolia *Sm.* 200.
agrestis *Sole*, 201.
alopecuroides Hull, 199.
aquatica Huds. 200.
arvensis *Linn.* 201.
citrata *Ehr.* 200.
crispa Linn. 199.
gentilis *Linn.* 200.
gentilis E. Bot. 201.
gracilis *Smith*, 201.
gratissima Willd. 199.
hircina Hull, 200.
hirsuta *Linn.* 200.
longifolia Huds. 199.
nemorosa Willd. 199.
odorata Sole, 200.
officinalis Hull, 200.
paludosa Sole, 200.
palustris Sole, 200.
piperita *Smith*, 200.
præcox Sole, 201.
Pulegium *Linn.* 201.
rotundifolia Sole, 199.
rotundifolia *Linn.* 199.
rivalis Sole, 200.
rubra *Smith*, 200.
rubra Sole, 201.
rubra Huds. 201.
sativa Linn. 200.
sativa Sole, 200.
sylvestris *Smith*, 199.
sylvestris Sole, 199.
verticillata Linn. 200.
villosa Huds. 199.
viridis *Linn.* 199.
MENYANTHEÆ, 179.
Menyanthes *Linn.* 179.
trifoliata *Linn.* 179.
Menziesia *Smith*, 173.
cærulea *Swartz*, 173.
polifolia *Smith*, 173.
Mercurialis *Linn.* 223.
annua *Linn.* 223.
perennis *Linn.* 223.
Mespilus *Linn.* 104.
Cotoneaster Linn. 104.
germanica *Linn.* 104.
Oxyacantha Smith, 104.
Meum *Tourn.* 118.
athamanticum *Jacq.* 118.
Fœniculum Spreng. 119.

Meum inundatum Spreng. 122.
Mibora verna Beauv. 301.
Milium *Linn.* 301.
effusum *Linn.* 301.
lendigerum Linn. 302.
Mœnchia *Ehr.* 49.
erecta Smith, 49.
glauca *Pers.* 49.
olinia *Mœnch.* 306.
cærulea *Mœnch.* 307.
depauperata 307.
MONOCHLAMYDEÆ, *5.* 207.
MONOCOTYLEDONES *Juss.* 4. 245.
Monotropa *Linn.* 176.
Hypopithys *Linn.* 176.
Montia *Linn.* 62.
fontana *Linn.* 63.
Muscari *Dec.* 269.
racemosum *Dec.* 269.
Myagrum sativum Linn. 30.
Mycelis angulosa Cass. 157.
Mygalurus bromoides Link. 315.
caudatus Link. 314.
uniglumis Link. 314.
Myosotis *Linn.* 165.
alpestris *Lehm.* 166.
alpina Don, 166.
annua Mœnch. 166.
arvensis *Roth.* 166.
cæspitosa *Schulz.* 165.
intermedia *Link.* 166.
palustris *Roth.* 165.
rupicola Smith, 166.
scorpioides Willd. 165.
sylvatica *Lehm.* 166.
versicolor *Lehm.* 166.
Myosurus *Linn.* 10.
minimus *Linn.* 10.
MYRICEÆ *Linn.* 242.
Myrica *Linn.* 242.
Gale *Linn.* 242.
Myriophyllum *Linn.* 110.
spicatum *Linn.* 110.
verticillatum *Linn.* 110.
Myrrhis *Scop.* 125.
aromatica Spreng. 125.
aurea Spreng. 125.
odorata *Scop.* 125.
temula Spreng. 125.
Narcissus *Linn.* 265.
biflorus *Curt.* 265.
poeticus Huds. 265.
poeticus *Linn.* 265.
Pseudo-narcissus *Linn.* 265.
Nardus *Linn.* 296.
stricta *Linn.* 296.
Narthecium *Huds.* 277.
calyculatum Lam. 264.
ossifragum *Huds.* 277.
Nasmythia articulata Huds. 272.
Nasturtium *R. Br.* 23.
amphibium *R. Br.* 23.
officinale *R. Br.* 23.
palustre Dec. 23.
sylvestre *R. Br.* 23.
terrestre *R. Br.* 23.
Nemea Fries, 3.
NEOTTIEÆ *Lindley,* 257.
Neottia *Linn.* 257.
gemmipara Swartz, 257.
Nidus avis *Linn.* 258.
repens Swartz, 257.
spiralis Swartz, 257.
Nepeta *Linn.* 203.
cataria *Linn.* 203.
NOTORHIZEÆ *Dec.* 29.
Nuphar *Smith,* 15.
Kalmiana Hooker, 15.
lutea *Smith,* 15.
minima E. Bot. 15.
pumila *Hoffm.* 15.
NYMPHÆACEÆ *Dec.* 15.
Nymphæa *Linn.* 15.
alba *Linn.* 15.
Odontites lutea Spreng. 120.
tenuissima Spreng. 120.
Œnanthe *Linn.* 119.
aquatica Lam. 120.
crocata *Linn.* 120.
fistulosa *Linn.* 119.
peucedanifolia *Poll.* 120.
Phellandrium *Spreng.* 120.
pimpinelloides *Linn.* 119.
Œnothera *Linn.* 109.
biennis *Linn.* 109.
OLEINEÆ *H. L.* 171.
Oligosporus campestris Cass. 149.
ONAGRARIÆ *Juss.* 107.
Onobrychis *Tourn* 88.
sativa *Lam.* 88.
Ononis *Linn.* 78.
arvensis Linn. 78.

procurrens *Wallr.* 78.
repens Linn. 78.
spinosa *Linn.* 78.
Onopordum *Linn.* 152.
Acanthium *Linn.* 152.
Onotrophe palustris Cass. 152.
acaulis Cass. 153.
Ophiurus *Beauv.* 295.
incurvatus *Beauv.* 295.
OPHRYDEÆ *Lindley,* 259.
Ophrys *Linn.* 262.
anthropophora Linn. 262.
apifera *Huds.* 262.
arachnites *Willd.* 262.
aranifera *Huds.* 262.
corallorhiza Linn. 258.
cordata Linn. 258.
fucifera Curtis, 262.
fucifera *Smith,* 262.
liliifolia Huds. 263.
Löselii Linn. 263.
monorchis Linn. 263.
muscifera *Huds.* 262.
myodes Swartz, 262.
nidus avis Linn. 258.
ovata Linn. 258.
paludosa Linn. 263.
paludosa Fl. Dan. 263.
spiralis Linn. 257.
Oplismenus Crus-galli Dumort, 305.
ORCHIDEÆ *Juss.* 256.
Orchis *Linn.* 259.
albida Swartz, 261.
arachnites Scop. 262.
bifolia Linn. 261.
conopsea *Linn.* 261.
fusca *Jacq.* 260.
hircina *Scop.* 260.
latifolia *Linn.* 260.
maculata *Linn.* 260.
mascula *Linn.* 260.
militaris *Linn.* 260.
militaris E. Bot. 260.
Morio *Linn.* 260.
moravica Jacq. 260.
purpurea Huds. 260.
pyramidalis Rich. 261.
tephrosanthos *Vill.* 260.
ustulata *Linn.* 260.
viridis Swartz, 261.
Ornithogalum *Linn.* 268.
luteum Linn. 268.
nutans *Linn.* 269.
pyrenaicum *Linn.* 268.
umbellatum *Linn.* 269.
Ornithopus *Linn.* 87.
perpusillus *Linn.* 87.
ORTHOPLOCEÆ *Dec.* 32.
Origanum *Linn.* 206.
vulgare *Linn.* 206.
OROBANCHEÆ *Vent.* 193.
Orobanche *Linn.* 193.
cærulea *Vill.* 194.
elatior *Sutton,* 194.
major *Linn.* 193.
minor *Smith,* 194.
purpurea Jacq. 194.
ramosa *Linn.* 194.
rubra *Smith,* 194.
Orobus *Linn.* 86.
niger *Linn.* 87.
sylvaticus *Linn.* 87.
tenuifolius Roth. 87.
tuberosus *Linn.* 87.
Othonna palustris Linn. 147.
OXALIDEÆ *Dec.* 59.
Oxalis *Linn.* 59.
Acetosella *Linn.* 59.
corniculata *Linn.* 59.
Oxycoccus *Rich.* 134.
palustris *Rich.* 134.
Oxyria *R. Br.* 211.
reniformis *R. Br.* 211.
Oxytropis *Dec.* 79.
campestris *Dec.* 79.
uralensis *Dec.* 79.
Pæonia *Linn.* 14.
corallina *Retz,* 14.
Panicum Crus-galli Linn. 305.
Dactylon Linn. 298.
sanguinale Linn. 299.
verticillatum Linn. 309.
viride Linn. 309.
PAPAVERACEÆ *Juss.* 16.
Papaver *Linn.* 16.
Argemone *Linn.* 16.
cambricum Linn. 17.
dubium *Linn.* 17.
hybridum *Linn.* 16.
maritimum With. 16.
nudicaule *Linn.* 17.

Rhœas *Linn.* 17.
somniferum *Linn.* 17.
Parietaria *Linn.* 218.
officinalis *Linn.* 218.
Paris *Linn.* 271.
quadrifolia *Linn.* 271.
Parnassia *Linn.* 67.
palustris *Linn.* 67.
Pastinaca *Linn.* 116.
sativa *Linn.* 116.
Pedicularis *Linn.* 190.
palustris *Linn.* 190.
sylvatica *Linn.* 190.
Peplis *Linn.* 72.
Portula *Linn.* 72.
Petaloideæ, 245.
Petroselinum *Hoffm.* 123.
segetum *Koch.* 123.
Peucedanum *Linn.* 116.
officinale *Linn.* 117.
Ostruthium *Koch.* 116.
palustre *Mœnch.* 116.
Silaus Linn. 118.
Phaca campestris Wahl. 79.
uralensis Wahl. 79.
Phæcasium lampsanoides Cass. 157.
Phænogamia, 3.
Phalaris *Linn.* 300.
alpina Hænke, 301.
arenaria Willd. 300.
arundinacea Linn. 301.
aspera Retz, 301.
canariensis *Linn.* 300.
paniculata H. Kew. 301.
phleoides Linn. 301.
Phalona echinata Dumort, 306.
Phanerocotyledoneæ Ag. 4.
Phanerogamia, 3.
Phellandrium aquaticum Linn. 120.
Phleum *Linn.* 300.
alpinum *Linn.* 300.
arenarium, Linn. 300.
asperum *Jacq.* 301.
Böhmeri *Schrad.* 301.
crinitum Schreb. 302.
Michelii *All.* 301.
nodosum Linn. 300.
paniculatum Huds. 301.
pratense *Linn.* 300.
viride All. 301.
Phragmites communis Trin. 310.
Physospermum *Cusson.* 126.
commutatum *Spreng.* 126.
Phyteuma *Linn.* 135.
orbiculare *Linn.* 135.
spicatum *Linn.* 135.
Picris *Linn.* 159.
echioides Linn. 158.
hieracioides *Linn.* 159.
Taraxaci All. 162.
Pimpinella *Linn.* 121.
dioica Linn. 124.
dissecta Retz, 121.
magna *Linn.* 121.
major Huds. 121.
pumila Jacq. 124.
saxifraga *Linn.* 121.
Pinguicula *Linn.* 186.
grandiflora *Willd.* 186.
lusitanica *Linn.* 186.
villosa Huds. 186.
vulgaris *Linn.* 186.
Pinus *Linn.* 241.
sylvestris *Linn.* 241.
Pistiaceæ *Rich.* 251.
Pisum *Linn.* 83.
maritimum *Linn.* 84.
Plantagineæ *Juss.* 169.
Plantago *Linn.* 169.
Coronopus *Linn.* 169.
lanceolata *Linn.* 169.
major *Linn.* 169.
maritima *Linn.* 169.
media *Linn.* 169.
Platanthera *Rich.* 261.
albida 261.
bifolia 261.
viridis 261.
Pleurorhizeæ *Dec.* 22.
Plumbagineæ *Juss.* 170.
Poa *Linn.* 316.
alpina *Linn.* 316.
angustifolia Huds. 317.
angustifolia Linn. 317.
annua *Linn.* 317.
aquatica Linn. 316.
bulbosa *Linn.* 317.
cærulea Knapp, 317.
cæsia E. Bot. 317.
compressa *Linn.* 317.
cristata Willd. 307.
decumbens With. 311.

distans *Linn.* 318.
dubia Linn. 317.
flexuosa Sm. 317.
fluitans Scop. 315.
glabra Ehr. 317.
glauca *Fl. Dan.* 317.
humilis Ehr. 317.
laxa *Hænke,* 316.
loliacea Huds. 297.
maritima Huds. 315.
nemoralis *Linn.* 317.
pratensis *Linn.* 317.
procumbens Curt. 316.
retroflexa Curt. 318.
rigida Linn. 316.
rupestris With. 316.
salina Poll. 318.
scabra Ehr. 317.
setacea Huds. 317.
subcærulea Sm. 317.
sylvatica Poll. 313.
trinervata Ehr. 313.
trivialis *Linn.* 317.
POLEMONIACEÆ *Juss.* 168.
Polemonium *Linn.* 168.
cæruleum *Linn.* 168.
Polycarpon *Linn.* 61.
tetraphyllum *Linn.* 61.
POLYGALEÆ *Juss.* 39.
Polygala *Linn.* 39.
vulgaris *Linn.* 39.
POLYGONEÆ *Juss.* 209.
Polygonum *Linn.* 211.
amphibium *Linn.* 211.
aviculare *Linn.* 212.
Bistorta *Linn.* 212.
Convolvulus *Linn.* 212.
Fagopyrum *Linn.* 212.
Hydropiper *Linn.* 212.
incanum Willd. 211.
intermedium Ehr. 212.
lapathifolium *Linn.* 212.
minus *Huds.* 212.
pallidum With. 213.
pensylvanicum Curtis, 212.
Persicaria *Linn.* 211.
viviparum *Linn.* 212.
Polypogon *Desf.* 302.
Lagascæ Trin. 302.
littoralis *Smith,* 302.
monspeliensis *Desf.* 302.
POMACEÆ *Juss.* 103.
Populus *Linn.* 238.
alba *Linn.* 238.
canescens *Linn.* 238.
nigra *Linn.* 238.
nivea Willd. 238.
tremula *Linn.* 238.
PORTULACEÆ *Juss.* 62.
Potameæ Juss. 248.
Potamogeton *Linn.* 248.
acutifolium Presl. 249.
acutifolius Link. 249.
affine Bonnigh. 250.
alpinum Balb. 250.
annulatum Bell. 250.
augustanum Balb. 25
coloratus Hornem. 250.
complanatus Willd. 249.
compressum Teesd. 249.
compressus Linn. 249.
crispus *Linn.* 249.
cuspidatum Schrad. 249.
densus *Linn.* 248.
denticulatum Link. 249.
distachyum Bell. 250.
filiformis Pers. 248.
fluitans Smith, 250.
gramineum Merat. 249.
gramineum Sm. 249.
heterophyllum Schreb. 250.
hybridum Petagn. 250.
interruptus Kitaib. 249.
lanceolatus *Sm.* 250.
Löselii R. & S. 249.
lucens Linn. 250.
marinum Linn. 249.
natans *Linn.* 250.
oblongus *Viv.* 250.
obscurum Dec. 250.
obtusifolius *M. & K.* 249.
oppositifolius Dec. 248.
parnassifolius Schrad. 250.
pauciflorus Lam. 248.
pectinatus *Linn.* 248.
perfoliatus *Linn.* 249.
Plantago Bat. 250.
polygonifolius Pour. 250.
prælongas Wulff. 249.
Proteus *Schlecht,* 250.
pusillus *Linn.* 249.
rufescens *Schrad.* 250.

serratus Lam. 249.
serratus Linn. 248.
setaceus Lam. 248.
tenuifolium H. & B. 249.
trichoides Schlecht, 249.
uliginosum Bonnigh. 250.
Vaillantii R. & S. 249.
zosterifolius *Schum.* 249.
Potentilla *Linn.* 96.
alba *Linn.* 97.
alpestris *Hall.* 96.
anserina *Linn.* 96.
argentea *Linn.* 96.
aurea Smith, 96.
Comarum *Scop.* 97.
Fragaria *Dec.* 97.
Fragariastrum Ehr. 97.
Fragarioides Vill. 97.
fruticosa *Linn.* 96.
nemoralis Nestl. 98.
opaca *Linn.* 97.
palustris Lehm. 97.
procumbens Sibth. 98.
reptans *Linn.* 97.
reptans Dec. 98.
rubra Hall. 97.
rupestris *Linn.* 96.
salisburgensis Hænke, 96.
tetrapetala Hall. 98.
Tormentilla *Sibth.* 97.
tridentata *Sol.* 97.
verna *Smith,* 97.
Poterium *Linn.* 103.
Sanguisorba *Linn.* 103.
Prenanthes *Linn.* 157.
hieracifolia *Willd.* 157.
muralis Linn. 157.
pulchra Dec. 157.
Primulaceæ *Vent.* 182.
Primula *Linn.* 183.
acaulis Jacq. 184.
elatior *With.* 184.
farinosa *Linn.* 184.
officinalis Jacq. 184.
scotica *Hook.* 184.
veris *Linn.* 184.
vulgaris *Huds.* 184.
Prismatocarpus *L'Herit.* 135.
hybridus *L'Herit.* 135.
Prunella *Linn.* 205.
vulgaris *Linn.* 205.
Prunus *Tournef.* 90.
avium Linn. 90.
domestica *Linn.* 90.
insititia Linn. 90.
Padus Linn. 90.
spinosa *Linn.* 90.
Pseudocotyledoneæ Ag. 3.
Pulicaria *Cass.* 143.
dysenterica *Cass.* 143.
vulgaris *Cass.* 143.
Pulmonaria *Linn.* 163.
angustifolia *Linn.* 164.
maritima Linn. 164.
officinalis *Linn.* 164.
Pyrethrum inodorum Smith, 148.
maritimum Smith, 148.
Parthenium Smith, 148.
Pyroleæ, 175.
Pyrola *Linn.* 175.
media *Swartz,* 175.
minor *Linn.* 175.
rosea E. Bot. 175.
rotundifolia *Linn.* 175.
secunda *Linn.* 175.
uniflora *Linn.* 175.
Pyrus *Linn.* 105.
Aria *Smith,* 105.
aucuparia *Gærtn.* 106
communis *Linn.* 105.
domestica *Smith,* 105.
intermedia *Ehr.* 105.
Malus *Linn.* 105.
pinnatifida *Ehr.* 105.
torminalis *Sm.* 105.
Quercineæ Juss. 239.
Quercus *Linn.* 239.
femina With. 240.
pedunculata Willd. 240.
Robur *Linn.* 240.
Robur Willd. 240.
sessiliflora *Salisb.* 240.
sessilis Ehr. 240.
Radiola *Gmel.* 54.
linoides *Gmel.* 54.
Millegrana Smith, 54.
Ranunculaceæ *Juss.* 7.
Ranunculus *Linn.* 10.
acris *Linn.* 11.
alpestris *Linn.* 11.
aquatilis *Linn.* 12.
arvensis *Linn.* 11.

auricomus *Linn.* 10.
bulbosus *Linn.* 11.
circinatus Sibth. 12.
Ficaria *Linn.* 10.
Flammula *Linn.* 10.
fluviatilis Wigg. 12.
gramineus *Linn.* 10.
hederaceus *Linn.* 11.
hirsutus Curtis, 11.
Lingua *Linn.* 10.
pantothrix *Dec.* 12.
parviflorus *Linn.* 11.
parvulus Linn. 11.
Philonotis *Ehr.* 11.
repens *Linn.* 11.
sceleratus *Linn.* 11.
RAPHANEÆ *Dec.* 34.
Raphanus *Linn.* 34.
maritimus *Smith,* 34.
Raphanistrum *Linn.* 34.
RESEDACEÆ *Lindley,* 219.
Reseda *Linn.* 219.
lutea *Linn.* 219.
Luteola *Linn.* 219.
RESTIACEÆ *R. Br.* 272.
RHAMNEÆ *Juss.* 72.
Rhamnus *Linn.* 73.
catharticus *Linn.* 73.
Frangula *Linn.* 73.
Rheum digynum Wahl. 211.
Rhinanthus *Linn.* 190.
Crista galli *Linn.* 190.
major *Ehr.* 190.
Rhodiola odorata Lam. 64.
rosea Linn. 64.
Rhynchospora *Vahl.* 279.
alba *Vahl.* 279.
fusca *Smith,* 279.
Ribes *Linn.* 106.
alpinum *Linn.* 107.
Grossularia *Linn.* 107.
nigrum *Linn.* 107.
petræum *Wulf.* 106.
reclinatum Linn. 107.
rubrum *Linn.* 106.
spicatum *Robson,* 106.
Uva crispa Linn. 107.
Robertsonia *Haw.* 70.
Geum, 70.
hirsuta, 71.
stellaris, 71.
umbrosa, 71.
Römeria *Dec.* 18.
hybrida *Dec.* 18.
ROSACEÆ *Juss.* 88.
ROSEÆ *Dec.* 99.
Rosa *Linn.* 99.
arvensis *Huds.* 102.
Borreri Woods, 101.
bractescens Woods, 102.
cæsia *Smith,* 102.
canina *Linn.* 101.
collina *Jacq.* 102.
collina Smith, 102.
Dicksoniana *Lindley,* 99
Doniana *Woods,* 100.
dumetorum E. Bot. 101
dumetorum *Thuill.* 102.
eglanteria Huds. 101.
fœtida Bat. 100.
Forsteri Smith, 102.
glaucophylla Winch, 102.
gracilis Woods, 100.
heterophylla Woods, 100.
hibernica *Smith,* 100.
inodora Agardh. 101.
involuta Winch. 100.
involuta *Smith,* 100.
micrantha Smith, 101.
mollis *Smith,* 100.
nivalis Donn, 100.
pulchella Woods, 100.
repens Ehr. 102.
rubella *Smith,* 99.
rubiginosa *Linn.* 101.
Sabini *Woods,* 100.
sarmentacea Woods, 102.
scabriuscula E. Bot. 100.
sepium *Thuill.* 101.
spinosissima *Linn.* 100.
stylosa Desv. 102.
subglobosa Smith, 100.
surculosa Woods, 101.
sylvestris, 101.
systyla *Batard,* 102.
tomentosa *Smith,* 100.
umbellata Leers, 101.
villosa Smith, 100.
Rottböllia incurvata Linn. 295.
Rubia *Linn.* 131.
peregrina *Linn.* 131.

Rubus *Linn.* 90.
abruptus, 92.
affinis *W. & N.* 91.
arcticus *Linn.* 95.
cæsius *Linn.* 95.
Chamæmorus *Linn.* 95.
collinus Dec. 92.
cordifolius *W. & N.* 92.
discolor *W. & N.* 93.
diversifolius, 93.
dumetorum *W. & N.* 95.
echinatus, 94.
fastigiatus *W. & N.* 91.
fruticosus *Linn.* 92.
fusco-ater *W. & N.* 94.
hirtus *W. & K.* 94.
idæus *Linn.* 95.
Köhleri *W. & N.* 94.
leucostachys *Sm.* 93.
macrophyllus *W. & N.* 93.
nemorosus *Willd.* 95.
nessensis Hall. 91.
nitidus *W. & N.* 92.
pallidus *W. & N.* 94.
plicatus *W. & N.* 91.
rhamnifolius *W. & N.* 92.
rudis *W. & N.* 94.
saxatilis *Linn.* 93.
suberectus *Anders.* 91.
vulgaris *W. & N.* 93.

Rumex *Linn.* 209.
Acetosa *Linn.* 211.
Acetosella *Linn.* 211.
acutus Linn. 210.
aquaticus Smith, 211.
Britannica Huds. 211.
crispus *Linn.* 211.
digynus Linn. 211.
glomeratus *Spreng.* 210.
Hydrolapathum *Huds.* 211.
maritimus *Linn.* 209.
maritimus Huds. 210.
Nemolapathum *Linn.* 210.
obtusifolius *Linn.* 210.
paludosus With. 210.
palustris *Smith,* 210.
pulcher *Linn.* 210.
sanguineus *Linn.* 210.

Ruppia *Linn.* 251.
maritima *Linn.* 251.

Ruscus *Linn.* 271.
aculeatus *Linn.* 271.
laxus Smith, 271.
Sagina *Linn.* 49.
apetala *Linn.* 49.
cerastoides Smith, 51.
erecta Linn. 49.
maritima *Don,* 49.
procumbens *Linn.* 49.
Sagittaria *Linn.* 253.
sagittifolia *Linn.* 253.
SALICARIÆ *Juss.* 71.
Salicornia *Linn.* 214.
annua Smith, 214.
fruticosa *Linn.* 214.
herbacea *Linn.* 214.
herbacea Ehr. 214.
procumbens *Smith,* 214.
radicans *Smith,* 214.
SALICINEÆ *Rich.* 229.
Salix *Linn.* 229.
acuminata Hoffm. 236.
acuminata *Sm.* 237.
alba *Linn.* 238.
amygdalina *Linn.* 230.
Andersoniana *Sm.* 237.
angustifolia Wulf. 235.
appendiculata Fl. Dan. 234.
aquatica *Sm.* 236.
Arbuscula *Linn.* 233.
arbutifolia Willd. 233.
arenaria *Linn.* 234.
arenaria Lightf. 235.
argentea *Sm.* 234.
ascendens Sm. 235.
aurita *Linn.* 236.
aurita Hoffm. 236.
bicolor *Ehr.* 230.
Borreriana *Sm.* 230.
cærulea Sm. 238.
caprea *Linn.* 237.
carinata *Sm.* 233.
chrysanthos Fl. Dan. 234.
cinerea *Linn.* 236.
cinerea With. 236.
cotinifolia *Sm.* 236.
Croweana *Sm.* 233.
daphnoides Vill. 236.
Davalliana *Sm.* 230.
decipiens *Hoffm.* 232.
depressa Hoffm. 235.

Dicksoniana *Sm.* 233.
Doniana *Sm.* 235.
fissa Relh. 232.
fissa Hoffm. 233.
fœtida *Sm.* 235.
Forbiana *Sm.* 232.
Forsteriana *Sm.* 237.
fragilis *Linn.* 232.
fusca *Linn.* 235.
glauca *Linn.* 234.
Helix *Linn.* 232.
Helvetica Vill. 234.
herbacea *Linn.* 234.
hirta *Sm.* 236.
Hoffmanniana *Sm.* 229.
incubacea *Linn.* 235.
Lambertiana *Sm.* 232.
lanata Roth. 235.
lanata Lightf. 237.
lanata *Linn.* 234.
lanceolata *Sm.* 230.
Lapponum Lightf. 234.
laurina Sm. 231.
limosa Wahl. 234.
livida *Wahl.* 234.
malifolia *Sm.* 231.
mollissima *Sm.* 238.
monandra Ehr. 232.
monandra Hoffm. 232.
myrsinites *Linn.* 223.
myrsinites Lightf. 233.
myrtilloides Fl. Brit. 233.
nigricans *Smith,* 230.
nitens *Anders.* 230.
oleifolia *Sm.* 236.
parvifolia Sm. 235.
pentandra *Linn.* 230.
petiolaris *Sm.* 231.
phylicifolia *Linn.* 230.
phylicifolia Host. 231.
phylicifolia Willd. 231.
polymorpha Ehr. 235.
prostrata *Sm.* 235.
prunifolia *Sm.* 233.
purpurea *Linn.* 232.
radicans Sm. 230.
repens *Linn.* 235.
reticulata *Linn.* 234.
retusa Dicks. 233.
rosmarinifolia *Linn.* 236.
rubra *Huds.* 232.
rupestris *Don,* 237.
Russelliana *Linn.* 232
Smithiana *Willd.* 237.
spadicea Vill. 236.
sphacelata *Sm.* 237.
stipularis *Sm.* 238.
Stuartiana *Sm.* 234.
tenuifolia *Sm.* 231.
tetrapla *Walk.* 231
Timmii Schk. 236.
triandra *Linn.* 229.
triandra Hoffm. 230.
uliginosa Willd. 236.
ulmifolia Vill. 236.
vacciniifolia *Sm.* 233.
venulosa *Sm.* 233.
viminalis *Linn.* 237.
virescens Vill. 233.
vitellina *Linn.* 231.
Wulfeniana *Willd.* 231.
Salsola *Linn.* 214.
fruticosa Linn. 216.
Kali *Linn.* 214.
Salvia *Linn.* 197.
pratensis *Linn.* 197.
verbenaca *Linn.* 197.
Sambucus *Linn.* 132.
Ebulus *Linn.* 132.
nigra *Linn.* 132.
Samolus *Linn.* 185.
Valerandi *Linn.* 185.
SANGUISORBEÆ *Juss.* 102.
Sanguisorba *Linn.* 103.
media *Linn.* 103.
officinalis *Linn.* 103.
SANICULEÆ *Koch.* 127.
Sanicula *Linn.* 127.
europæa *Linn.* 127.
SANTALACEÆ *R. Br.* 207.
Santolina maritima Linn. 150.
Satyrium albidum Linn. 261.
hircinum Linn. 260.
repens Linn. 257.
viride Linn. 262.
SAXIFRAGEÆ *Linn.* 68.
Saxifraga *Linn.* 68.
affinis *Don,* 69.
aizoides Linn. 67.
autumnalis Willd. 67
cæspitosa Huds. 68.
cæspitosa *Linn* 69.

cernua Linn. 68.
condensata Gmel. 70
decipiens Ehr. 69.
denudata *Don*, 69.
elongella Sm. 70.
Geum Linn. 70.
granulata Linn. 68.
grönlandica Linn. 69.
hirculus Linn. 67.
hirsuta Linn. 71.
hirta *Don*, 69.
hypnoides *Linn*. 69.
incurvifolia *Don*, 69.
lævis Don, 69.
lætevirens *Don*, 70.
leptophylla *Pers*. 70.
moschata E. Bot. 69
moschata With. 68.
muscoides Hook. 69.
muscoides *Wulff*. 68.
nivalis Linn. 68.
oppositifolia *Linn*. 68.
palmata Smith, 69.
pedatifida *Ehr*. 70.
petræa With. 69.
platypetala *Sm*. 69.
pygmæa *Haw*. 68.
quinquefida Don, 70.
rivularis Linn. 68.
stellaris Linn. 71.
tridactylites *Linn*. 68.
umbrosa Linn. 71.
Scabiosa *Linn*. 139.
arvensis Linn. 140.
columbaria *Linn*. 140.
succisa *Linn*. 139.
SCANDICINEÆ *Koch*. 124.
Scandix *Linn*. 124.
Anthriscus Linn. 124.
Cerefolium Linn. 124.
infesta Linn. 114.
odorata Linn. 125.
Pecten-Veneris *Linn*. 125.
Schedonorus *Beauv*. 312.
elatior, 313.
loliaceus *Dumort*, 313.
pratensis *Beauv*. 312.
radicans Dumort, 313.
sylvaticus *Beauv*. 313.
Scheuchzeria *Linn*. 252.
palustris *Linn*. 252.
Schœnus *Linn*. 280.
albus Linn. 279.
compressus Linn. 280
fuscus Linn. 279.
Mariscus Linn. 283.
minimus Forst. 275.
monoicus Smith, 284.
nigricans *Linn*. 280.
rufus *Huds*. 280.
Scilla *Linn*. 269.
autumnalis *Linn*. 269.
bifolia *Linn*. 269.
non scripta Redouté, 270.
nutans Smith, 270.
verna *Huds*. 269.
Scirpus *Linn*. 281.
acicularis Linn. 280.
australis Linn. 283.
Bæothryon Ehr. 281.
cæspitosus Linn. 281.
campestris Roth. 281.
caricinus Schrad. 280.
carinatus *Sm*. 281.
compressus Pers. 280.
fluitans Linn. 283.
glaucus *Smith*, 281.
Holoschœnus Linn. 283.
lacustris *Linn*. 281.
maritimus *Linn*. 281.
multicaulis Sm. 280.
palustris Linn. 280.
pauciflorus Lightf. 281.
pungens Vahl. 281.
romanus Linn. 283.
rufus Schrad. 280.
setaceus Linn. 283.
sylvaticus *Linn*. 281.
triqueter *Linn*. 281.
tuberosus Desf. 281.
SCLERANTHEÆ *Link*. 217.
Scleranthus *Linn*. 218.
annuus *Linn*. 218.
perennis *Linn*. 218.
Sclerochloa *Beauv*. 315.
? *dichotoma* Link. 315.
maritima, 315.
procumbens *Beauv*. 316.
rigida *Panz*. 316.
SCROPHULARINEÆ *Juss*. 187.
Scrophularia *Linn*. 192.
aquatica *Linn*. 193.

nodosa *Linn.* 193.
Scorodonia *Linn.* 193.
vernalis *Linn.* 193.
Scutellaria *Linn.* 204.
galericulata *Linn.* 204.
minor *Linn.* 204.
Sedum *Linn.* 64.
albescens *Haw.* 65.
album *Linn.* 65.
acre *Linn.* 64.
anglicum *Huds.* 64.
annuum Huds. 64.
dasyphyllum *Linn.* 64.
Forsterianum *Smith,* 65.
reflexum *Linn.* 65.
Rhodiola *Dec.* 64.
rupestre *Linn.* 65.
sexangulare *Linn.* 64.
Telephium *Linn.* 64.
villosum *Linn.* 65.
Selineæ *Koch.* 116.
Selinum palustre Linn. 117.
Seminiferæ Ag. 4.
Sempervivum *Linn.* 65.
tectorum *Linn.* 65.
Senebiera coronopus Dec. 30.
didyma Smith, 31.
pinnatifida Dec. 31.
Senecio *Linn.* 146.
aquaticus *Huds.* 146.
chrysanthemifolius Bivona, 146.
erucifolius Huds. 146.
Jacobæa *Linn.* 146.
lividus *Linn.* 146.
paludosus *Linn.* 146.
saracenicus *Linn.* 147.
squalidus *Linn.* 146.
sylvaticus *Linn.* 146.
tenuifolius *Jacq.* 146.
viscosus *Linn.* 146.
vulgaris *Linn.* 146.
Serapias ensifolia Murr. 259.
grandiflora Linn. 259.
grandiflora Fl. Dan. 259.
lancifolia Murr. 259.
latifolia Linn. 259.
longifolia Huds. 259.
palustris Scop. 259.
rubra Linn. 259.
Serratula *Linn.* 154.
alpina Linn. 152.
tinctoria *Linn.* 154.
Seselineæ *Koch.* 117.
Seseli *Linn.* 119.
Libanotis *Koch.* 119.
Sesleria *Arduin,* 309.
cærulea *Scop.* 309.
Setaria *Beauv.* 309.
verticillata *Beauv.* 309
viridis *Beauv.* 309.
Sherardia *Linn.* 130.
arvensis *Linn.* 130.
Sibbaldia *Linn.* 98.
procumbens *Linn.* 98.
Sibthorpia *Linn.* 192.
europæa *Linn.* 192.
Silaus *Bess.* 118.
pratensis *Bess.* 118.
Sileneæ *Dec.* 44.
Silene *Linn.* 45.
acaulis *Linn.* 46.
amæna Huds. 45.
anglica *Linn.* 45.
Armeria *Linn.* 46.
conica *Linn.* 45.
conoidea *Linn.* 46.
inflata *Smith,* 45.
inflata β. Hook. 45.
inflata uniflora Otth. 45.
maritima *With.* 45.
maritima Dec. 45.
noctiflora *Linn.* 46.
nutans *Linn.* 46.
Otites *Smith,* 46.
paradoxa Smith, 46.
quinquevulnera *Linn.* 45.
uniflora Roth. 45.
Silybum *Vaill.* 154.
marianum *Gærtn.* 154.
Sinapis *Linn.* 33.
alba *Linn.* 33.
arvensis *Linn.* 33.
muralis R. Br. 33.
nigra *Linn.* 33.
tenuifolia R. Br. 33.
Sison *Linn.* 122.
Amomum *Linn.* 122.
inundatum Wigg. 122.
Podagraria Spreng. 123.

segetum Linn. 123.
verticillatum Linn. 122.
Sisymbrieæ *Dec.* 29.
Sisymbrium *Linn.* 29.
amphibium Linn. 23.
Irio *Linn.* 29.
murale Linn. 33.
Nasturtium Linn. 23
officinale *Scop.* 29.
Sophia *Linn.* 29.
sylvestre Linn. 23.
tenuifolium Linn. 33.
terrestre Linn. 23.
Sium *Linn.* 121.
angustifolium *Linn.* 121.
erectum *Huds.* 121.
inundatum Wigg. 122
latifolium *Linn.* 121.
nodiflorum Linn. 122.
repens Linn. 122.
verticillatum Lam. 122.
Smilaceæ *R. Br.* 270.
Smyrnieæ *Koch.* 125.
Smyrnium *Linn.* 126.
Olusatrum *Linn.* 126
Solaneæ *Juss.* 180.
Solanum *Linn.* 182.
Dulcamara *Linn.* 182.
nigrum *Linn.* 182.
Solidago *Linn.* 144.
cambrica *Huds.* 144.
Virgaurea *Linn.* 144.
Sonchus *Linn.* 156.
alpinus *Willd.* 156.
arvensis *Linn.* 156.
cæruleus Smith, 156.
canadensis Linn. 156.
oleraceus *Linn.* 156.
palustris *Linn.* 156.
Sorbus aucuparia Linn. 106.
domestica Linn. 105.
hybrida Linn. 105.
hybrida Huds. 105.
scandica Fries, 105.
Sparganium *Linn.* 247.
erectum Linn. 247.
natans *Linn.* 248.
ramosum *Huds.* 247.
simplex *Huds.* 247.
superaxillare Ehr. 248.
Spartina *Willd.* 298.
stricta *Smith*, 298.
Spartium scoparium Linn. 77.
Spergula *Linn.* 48.
arvensis *Linn.* 48.
laricina Huds. 48.
nodosa *Linn.* 48.
pentandra Linn. 48.
saginoides *Linn.* 48.
saginoides Curtis, 48.
subulata *Swartz*, 48.
Spiræaceæ *Dec.* 89.
Spiræa *Linn.* 89.
Filipendula *Linn.* 89.
salicifolia *Linn.* 89.
Ulmaria *Linn.* 89.
Spiranthes *Rich.* 257.
autumnalis *Rich.* 257.
gemmipara 257.
Stachys *Linn.* 201.
ambigua *Smith*, 202.
arvensis *Linn.* 202.
germanica *Linn.* 202.
palustris *Linn.* 202.
sylvatica *Linn.* 202.
Staphyleaceæ, 75.
Staphylea *Linn.* 75.
pinnata *Linn.* 75.
Statice *Linn.* 170.
Armeria Linn. 170.
Limonium *Linn.* 170.
linearifolia Laterr. 170.
reticulata *Linn.* 170.
Stellatæ, 128.
Stellaria *Linn.* 52.
Alsine Willd. 52.
aquatica Poll. 52
cerastoides *Linn.* 53.
Dilleniana Leers, 52.
fontana Jacq. 52.
glauca *With.* 52.
graminea *Linn.* 52.
Holostea *Linn.* 52.
hypericifolia Wigg. 52.
lateriflora Krock. 52.
media *With.* 52.
media Sibth. 52.
nemorum *Linn.* 52.
palustris Retz, 52.
scapigera *Willd.* 52.
uliginosa Smith, 52.

Stipa *Linn.* 302.
membranacea Linn. 314.
pennata *Linn.* 302.
Stratiotes *Linn.* 254.
aloides *Linn.* 254.
Sturmia minima Hoppe, 301.
SUBULARIEÆ *Dec.* 34.
Subularia *Linn.* 34.
aquatica *Linn.* 34.
Swertia *Linn.* 179.
perennis *Linn.* 179.
Symphytum *Linn.* 164.
officinale *Linn.* 164.
patens Sibth. 164.
tuberosum *Linn.* 164.
Synorhizeæ Rich. 4.
Syntherisma vulgare Schreb. 299.
TAMARISCINEÆ *Desv.* 61.
Tamarix *Linn.* 62.
gallica *Linn.* 62.
Tamus *Linn.* 271.
communis *Linn.* 271.
Tanacetum *Linn.* 149.
vulgare *Linn.* 149.
Taraxacum Dens Leonis Desf. 158.
officinale Sibth. 158.
palustre Dec. 158.
Taxus *Linn.* 241.
baccata *Linn.* 241.
fastigiata, 241.
Teesdalia *R. Br.* 28.
Iberis *Dec.* 28.
nudicaulis R. Br. 28.
Teucrium *Linn.* 198.
Chamædrys *Linn.* 198.
Chamæpitys Linn. 198.
Scorodonia *Linn.* 198.
Scordium *Linn.* 198.
Thæmatia § 133.
Thalictrum *Linn.* 8.
alpinum *Linn.* 8.
flavum *Linn.* 9.
majus *Crantz,* 9.
minus *Linn.* 9.
Thesium *Linn.* 208.
linophyllum *Linn.* 208.
pratense Ehr. 208.
THLASPIDEÆ *Dec.* 27.
Thlaspi *Linn.* 27.
alpestre *Linn.* 28.
arvense *Linn.* 27.
Bursa Pastoris Linn. 31.
campestre Linn. 31.
hirtum Linn. 31.
perfoliatum *Linn.* 27.
Thrincia *Roth.* 162.
hirta *Roth.* 162.
THYMELEÆ *Juss.* 208.
Thymus *Linn.* 204.
Acinos *Linn.* 205.
Calamintha *Scop.* 205.
nepeta *Smith,* 205.
Serpyllum *Linn.* 204.
Thysselinum palustre. Tourn. 17
Plinii Spreng. 117.
TILIACEÆ *Juss.* 54.
Tilia *Linn.* 54.
corallina Smith, 55.
cordifolia Bess. 55.
europæa Linn. 54.
grandifolia *Ehr.* 54.
intermedia *Dec.* 54.
microphylla Vent. 55.
parvifolia *Ehr.* 55.
platyphylla Dec. 55.
rubra *Dec.* 55.
Tillæa *Linn.* 63.
muscosa *Linn.* 63.
Tofieldia *Huds.* 264.
borealis Wahl. 264.
palustris *Huds.* 264.
TORDYLINEÆ *Koch.* 115.
Tordylium *Linn.* 115.
latifolium Linn. 114.
maximum *Linn.* 115.
officinale Linn. 115.
Torilis *Adans.* 114.
Anthriscus *Gærtn.* 114.
infesta *Spreng.* 114.
nodosa *Gærtn.* 114.
Tormentilla erecta Linn. 97.
officinalis Smith, 98.
reptans Linn. 98.
Tragopogon *Linn.* 161.
parvifolius *Linn.* 161.
pratensis *Linn.* 161.
Trichodium *Schrad.* 303.
caninum *Schrad.* 303.
setaceum *R. & S.* 303.
Trichonema *Ker.* 255.
Bulbocodium *Ker.* 255.
Trientalis *Linn.* 185.
europæa *Linn.* 185.

Trifolium *Linn.* 79.
agrarium Huds. 81.
alpestre Huds. 80.
arvense *Linn.* 80.
dubium Sibth. 81.
filiforme *Linn.* 81.
flexuosum Jacq. 80.
fragiferum *Linn.* 81.
glomeratum *Linn.* 81.
maritimum *Huds.* 80.
medium *Linn.* 80.
minus *Relh.* 81.
ochroleucum *Linn.* 80.
officinale Smith, 79.
ornithopodioides Linn. 82.
pratense *Linn.* 80.
procumbens *Linn.* 81.
procumbens Huds. 81.
repens *Linn.* 79.
scabrum *Linn.* 80.
squarrosum *Linn.* 80.
stellatum *Linn.* 80.
stellatum Huds. 80.
striatum *Linn.* 81.
subterraneum *Linn.* 80.
suffocatum *Linn.* 79.
Triglochin *Linn.* 252.
maritimum *Linn.* 252.
palustre *Linn.* 252.
Trigonella *Linn.* 82.
ornithopodioides *Dec.* 82.
Trinia *Hoffm.* 123.
glaberrima *Hoffm.* 124.
Triodia *R. Br.* 311.
decumbens *Beauv.* 311.
Trisetum *Pers.* 308.
flavescens *Beauv.* 309.
pubescens *Pers.* 308.
Triticum caninum Huds. 298.
cristatum Linn. 298.
junceum Linn. 298.
loliaceum Sm. 297.
repens Linn. 298.
unilaterale H. Kew. 297.
Trollius *Linn.* 12.
europæus *Linn.* 12.
Tulipaceæ Dec. 266.
Tulipa *Linn.* 266.
sylvestris *Linn.* 266.
Turritis *Linn.* 23.
alpina Linn. 24.
ciliata Willd. 24.
glabra *Linn.* 24.
hirsuta Linn. 24.
Tussilago *Linn.* 147.
Farfara *Linn.* 147.
hybrida Linn. 147.
Petasites *Linn.* 147
TYPHACEÆ. *Juss.* 247.
Typha *Linn.* 247.
angustifolia *Linn.* 247.
latifolia *Linn.* 247.
major Curtis, 247.
minor Curtis, 247.
minor *Smith,* 247.
Ulex *Linn.* 76.
europæus *Linn.* 77.
nanus *Forst.* 77.
ULMACEÆ *Mirb.* 225.
Ulmus *Linn.* 226.
campestris *Linn.* 226.
campestris Willd. 227.
carpinifolia 226.
effusa Sibth. 227.
glabra *Miller,* 226.
glabra Huds. 227.
hollandica Miller, 226.
major *Smith,* 226.
montana *Bauh.* 227
nuda Ehr. 227.
stricta, 227.
suberosa *Ehr.* 226.
UMBELLIFERÆ *Juss.* 111.
Umbilicus *Dec.* 63.
erectus *Dec.* 64.
pendulinus *Dec.* 64.
URTICEÆ *Juss.* 218.
Urtica *Linn.* 218.
dioica *Linn.* 219.
pilulifera *Linn.* 219.
repens *Linn.* 219.
Utricularia *Linn.* 186.
intermedia *Hayne,* 186.
minor *Linn.* 186.
vulgaris *Linn.* 186.
VACCINIEÆ *Dec.* 134.
Vaccinium *Linn.* 134.
Myrtillus *Linn.* 134.
oxycoccus Linn. 134.
uliginosum *Linn.* 134.
Vitis Idæa *Linn.* 134.
Vaccinium cantabricum Huds. 175.
Valantia Aparine Linn. 129.

VALERIANEÆ *Dec.* 137.
Valeriana *Linn.* 138.
dentata Willd. 138.
dioica *Linn.* 138.
officinalis *Linn.* 138.
olitoria Linn. 138.
pyrenaica *Linn.* 138
rubra Linn. 139.
Valerianella *Tourn.* 138.
dentata *Dec.* 138.
olitoria *Mönch.* 138.
VASCULARES *Dec.* 3.
Vella annua Linn. 33.
VELLEÆ *Dec.* 33.
Verbascum *Linn.* 181.
Blattaria *Linn.* 181.
Lychnitis *Linn.* 181.
nigrum *Linn.* 181.
pulverulentum *Vill.* 181.
thapsiforme *Schrad.* 181.
thapsoides Willd. 181.
Thapsus *Linn.* 181.
virgatum *With.* 181.
VERBENACEÆ *Juss.* 195.
Verbena *Linn.* 196.
officinalis *Linn.* 196
Veronica *Linn.* 187.
agrestis *Linn.* 189.
Allionii Hook. 189.
alpina *Linn.* 188.
Anagallis *Linn.* 188.
arvensis *Linn.* 189.
Beccabunga *Linn.* 188
Bellardi Willd. 190.
Chamædrys *Linn.* 189.
fruticulosa *Linn.* 188.
hederifolia *Linn.* 189
hirsuta *Hopk.* 189.
humifusa Dicks. 188.
hybrida *Linn.* 188.
montana *Linn.* 189.
officinalis *Linn.* 189.
parmularia Poit. 189.
pumila All. 188.
saxatilis *Linn.* 188.
scutellata *Linn.* 188.
serpyllifolia *Linn.* 188.
setigera D. Don. 189.
spicata *Linn.* 188.
succulenta Willd. 190.
triphyllos *Linn.* 189.
verna *Linn.* 189.
Viburnum *Linn.* 132.
Lantana *Linn.* 132.
Opulus *Linn.* 132.
VICIEÆ *Dec.* 83.
Vicia *Linn.* 84.
angustifolia *Sibth.* 84.
angustifolia Willd. 84.
bithynica *Linn.* 85.
Cracca *Linn.* 84.
hybrida *Linn.* 85.
hybrida Huds. 85.
lævigata *Smith,* 85.
lathyroides *Linn.* 84.
lathyroides Huds. 84.
lutea *Linn.* 85.
sativa *Linn.* 84.
sepium *Linn.* 85.
sylvatica *Linn.* 84.
Villarsia *Vent.* 179.
nymphæoides *Vent.* 180.
Vinca *Linn.* 176.
major *Linn.* 176.
minor *Linn.* 176.
VIOLACEÆ *Juss.* 35.
Viola *Linn.* 35.
arvensis Sibth. 36.
canina *Linn.* 35.
flavicornis *Smith,* 36.
grandiflora Huds. 36.
hirta *Linn.* 35.
lactea *Smith,* 35.
lutea Hudson, 36.
odorata *Linn.* 35.
palustris *Linn.* 35.
tricolor *Linn.* 36.
Viscum *Linn.* 133.
album *Linn.* 133.
Vulpia *Gmel.* 314.
bromoides *Dumort,* 315.
membranacea Link. 314.
Myurus *Gmel.* 314.
uniglumis *Dumort,* 314.
Xanthium *Linn.* 151.
strumarium *Linn.* 151
Zannichellia *Linn.* 251.
palustris *Linn.* 251.
Zostera *Linn.* 251.
marina *Linn.* 251.

INDEX

OF

THE ENGLISH NAMES

IN

THE FIRST VOLUME.

ABELE-TREE, 238.
Agrimony, 99.
Alder, 229.
Alehoof, 199.
Alexanders, 126.
Alkanet, 165.
All-seed, 61.
Arrow-grass, 252.
Arrow-head, 253.
Asarabacca, 224.
Ash-tree, 171.
Asparagus, 267.
Aspen, 238.
Avens, 98.
Awl-wort, 34.
Bald-money, 118.
Bane-berries, 13.
Barley, Mouse, 296.
——, Sea, 296.
——, Wall, 296.
Barren-wort, 14.
Base Rocket, 219.
Bastard Alkanet, 164.
—— Balm, 205.
—— Pimpernel, 183.
—— Toad-flax, 208.
Beam-tree, 105.
Bear's-foot, 13.
Bee Nettle, 204.
Beech, 239.
Belleisle Cress, 23.
Bent-grass, 303.
Berberry, 14.
Berry-bearing Alder, 73.
Betony, 202.
Bilberry, 134.
Bindweed, 167.
Birch, Common, 229.
——, Weeping, 229.
Bird's-eye Primrose, 184.
Bird's-foot, 87.
Bird's-foot Trefoil, 81.
Birthwort, 225.
Bistort, 212.
Black Bryony, 271.
—— Bindweed, 212.
—— Nonsuch, 83.
—— Saltwort, 183.
Blackthorn, 90.
Bladder Nut, 75.
Bladderwort, 186.
Blue-bottle, 155.
Bog-bean, 179.
Borage, 164.
Box-tree, 223.
Brank, 212.
Brome-grass, 311.
—— ——, rye, 311.
Brooklime, 188.
Brook-weed, 185.
Broom-rape, 193.
Broom, 77.
Bryony, 319.
Buckbean, 179.

Buckthorn, 73.
Buck-wheat, 212.
Bugle, 198.
Bugloss, 165.
Bullace, 90.
Bull-rush, 281. 247.
Bur-marigold, 151.
Burnet, 103.
Burnet-saxifrage, 121.
Bur-reed, 247.
Bur-weed, 151.
Butcher's Broom, 271.
Butter-bur, 147.
Buttercup, 10.
Butterwort, 186.
Cabbage, 32.
Calamint, 205.
Calathian Violet, 178.
Canary-grass, 300.
Candy-tuft, 28.
Caraway, 122.
Carnation, 44.
Carrot, Wild, 113.
Catchfly, Nottingham, 46.
———, Red German, 47.
———, Spanish, 46.
Cat-mint, 203.
Cat's-ear, 161.
Cat's-tail, 247.
Cat's-tail grass, 300.
Celandine, 18.
Celery, 123.
Chaff-weed, 183.
Charlock, 33.
Cherry, Wild, 90.
———, Bird, 90.
Chervil, 124.
Chickweed, Common, 52.
————, Mouse-ear, 49.
————, Sea, 49.
————, Winter-green, 185.
Chives, 268.
Cinquefoil, 96.
Cleavers, 130.
Clover, Common Purple, 80.
———, Dutch, 79.
Cloudberry, 95.
Cock's-foot Finger-grass, 299.
Cock's-foot-grass, 310.
Codlings and Cream, 108.
Cole-seed, 32.
Colt's-foot, 147.
Columbine, 13.
Comfrey, 164.
Coral-root, 258.
Coralwort, 25.
Coriander, 115.
Corn Cockle, 47.
—— Marigold, 148.
—— Salad, 138.
Cornish Moneywort, 192.
Cotton-grass, 282.
Cotton-thistle, 152.
Cotton-weed, 150.
Couch-grass, 298.
Cow-berry, 134.
Cow Parsnep, 116.
Cowslip, 184.
Cow-wheat, 195.
Crab-tree, 105.
Crake-berry, 224.
Cran-berry, 134.
Crane's-bill, 56.
Crow-berry, 224.
Crowfoot, 10.
Cuckow-pint, 246.
Currants, 106.
Daffodil, 265.
———, Chequered, 266.
Daisy, 148.
Dame's Violet, 29.
Dandelion, 158.
Danewort, 132.
Darnel, 295.
Deadly Nightshade, 182.
Dead-nettle, Red, 203.
————, White, 203.
Devil's-bit, 139.
Dock, 209.
Dodder, 168.
Dog Violet, 35.
Dog-wood, 133.
Dog's-tail-grass, 306.
Dog's-tooth-grass, 298.
Dropwort, 89.
Duck-weed, 251.
Dutch Myrtle, 242.
Dwale, 182.
Dyer's Rocket, 219.
——- Woad, 32.
Earth-nut, 121
Elder, 132.

Elecampane, 143.
Elm, Chichester, 227.
——, Cornish, 227.
——, Downton, 227.
——, Giant, 227.
——, Hertfordshire, 226.
——,Narrow-leaved Engli6.
——, Scampston, 227.
——, Witch, 227.
Enchanter's Nightshade, 109.
English Galingale, 279.
Evening Primrose, 109.
Eye-bright, 191.
Feather-grass, 302.
Fescue-grass, Creeping, 314.
————, Hard, 314.
————, Sheep's, 313.
————, Tall, 313.
Feverfew, 148.
Figwort, 193.
Fiorin-grass, 303.
Flax, 53.
Flax-seed, 29.
Flowering-rush, 272.
Fly Honeysuckle, 132.
Fool's-parsley, 119.
French Willow, 108.
Fritillary, 266.
Frog-bit, 254.
Fuller's Teasel, 139.
Fumitory, 19.
Furze, 77.
Garden Angelica, 117.
Germander, 189.
German Madwort, 165.
Gill, 199.
Gipsy-wort, 197.
Gladwyn, 255.
Globe-flower, 12.
Goat's-beard, 161.
Gold of Pleasure, 30.
Golden Rod, 144.
——— Samphire, 143.
——— Saxifrage, 66.
Goldilocks, 10. 143.
Gooseberries, Common, 107.
————, Lancashire, 107.
Goosefoot, 215.
Goose-grass, 130.
Goose-tongue, 151.
Gorse, 77.
Gout-weed, 123.
Grass-wrack, 251.
Great Henbit, 203.
——— Valerian, 168.
Grey Mill, 164.
Gromwell, 164.
Ground-ivy, 199.
Ground Pine, 198.
Groundsel, 146.
Guelder Rose, 132.
Hare's-ear, 120.
Hare's-tail-grass, 299.
Haver, 310.
Hawthorn, 104.
Hazel-nut, 240.
Heart's-ease, 36.
Heath, 173.
Hedge Bed-straw, 130.
Hedge-mustard, 29.
Hellebore, 12.
Helleborine, 258.
Hemlock, Common, 126.
————, Water, 123.
Hemp-agrimony, 142.
Hemp-nettle, 204.
Henbane, 181.
Herb Bennet, 98.
—— Christopher, 13.
—— Gerarde, 123.
—— Paris, 271.
—— Robert, 57.
—— Twopence, 184.
High taper, 181.
Hogweed, 116.
Holly, 74.
Holy-grass, 306.
Honewort, 122.
Honeysuckle, 131.
Hooded Milfoil, 18
Hop, 219.
Horehound, Black, 201.
————, White, 201.
Hornbeam, 240.
Horned Pondweed, 251.
Horned Poppy, 17.
Hornwort, 225.
Horse-radish, 27.
Hound's-tongue, 166.
Houseleek, 65.
Ivy, 133.
Jack by the Hedge, 29.

Jacob's Ladder, 168.
Jointed Charlock, 34.
Jointed Glasswort, 214.
Juniper, 241.
Kidney Vetch, 78.
Kipper, 121.
Knapweed, 155.
Knawel, 218.
Knot-grass, 61. 212.
Ladies'-finger, 78.
——— Mantle, 103.
——— Slipper, 263.
Ladies'-smock, 25.
Ladies' Traces, 257.
Lamb's Lettuce, 138.
Lancashire Bog-asphodel, 277.
Larkspur, 13.
Leopard's-bane, 147.
Lily of the Valley, 270.
Lime-tree, 54.
Ling, 173.
London Rocket, 29.
Louse-wort, 190.
Lucerne, 82.
Lungwort, 164.
Madder, Wild, 131.
Maiden's Hair, 315.
Mallow, 40.
Manna Grass, 315.
Maple, 55.
Mare's-tail, 110.
Marjoram, 206.
Marram, 303.
Marsh Felwort, 179.
——— Flea-wort, 147.
——— Mallow, 41.
Marsh-marigold, 12.
Marsh Pennywort, 128
——— Samphire, 214.
——— Trefoil, 179.
Masterwort, Great, 116.
Mat-grass, 296.
Mayweed, 148.
Meadow Clary, 197.
——— Fox-tail-grass, 299.
Meadow-grass, 317.
Meadow-rue, 8.
Meadow-saffron, 264.
Meadow-sweet, 89.
Medlar, 104.
Melic-grass, 307.
Melilot, 79.
Mercury, 223.
Meu, 118.
Mezereon, 209.
Mignonette, Wild, 21[illegible]
Milfoil, 151.
Milk Thistle, 154.
Milkwort, 39.
Mill-mountain, 54.
Mint, Bergamot, 200.
——, Horse, 199.
——, Pepper, 200.
——, Spear, 199.
Misseltoe, 133.
Mithridate Mustard, 27.
Moneywort, 184.
Monks'-hood, 13.
Moon Daisy, 148.
Moor-grass, 309.
Moss Campion, 46.
Motherwort, 199.
Mountain-ash, 106.
Mountain-sorrel, 2
Mountain Spiderwort, 269.
Mouse-ear Hawkweed, 159
Mouse-tail, 10.
Mudwort, 192.
Mug-weed, 128.
Mugwort, 149.
Musk Thistle, 155.
Mustard, Brown, 33.
———, White, 33.
Navelwort, 63.
Nep, 203.
Nettle, Great, 219.
——, Roman, 219.
——, Small, 219.
Nightshade, 182
Nipplewort, 157
Oak, 240.
Oat, Wild, 310.
Orchis, Bee, 262.
——, Bog, 263.
——, Butterfly, 261.
——, Drone, 262.
——, Fly, 262.
——, Frog, 261.
——, Green Man, 262.
——, Green Musk, 263.
——, Late Spider, 262.
——, Lizard, 260.

Orchis, Military, 260.
——, Monkey, 260.
——, Spider, 262.
Orpine, 64.
Osier, Common, 237.
——, Golden, 230.
Oxlip, 184.
Pæony, 14.
Paigle, 184.
Pansy, 36.
Parsley Piert, 103.
Parsnep, 116.
Pasque-flower, 9.
Paul's Betony, 188.
Pea, 83.
Pearl-wort, 49.
Pear-tree, 105.
Penny Cress, 27.
Penny-royal, 201.
Periwinkle, 176.
Persian Willow, 108.
Persicaria, Spotted, 211.
Petty Whin, 77.
Pheasant's-eye, 9.
Pig-nut, 121.
Pilewort, 10.
Pimpernel, 185.
Pink, Clove, 44.
——, Deptford, 44.
Pipewort, 272.
Plantain, 169.
—— Shore-weed, 170.
Plowman's Spikenard, 142.
Plum, 90.
Poplar, Black, 238.
——, White, 238.
Poppy, 16.
Prickly Samphire, 126.
Prickwood, 74.
Primrose, 184.
—— peerless, 265.
Privet, 171.
Quaking-grass, 315.
Queen of the Meadows, 89.
Quicken-tree, 106.
Ragged Robin, 47.
Ragwort, 146.
Rampion, 136.
Ramsons, 268.
Rape, 32.
Raspberry, 95.
Red Rattle, 190.
Reed-mace, 247.
Reed, 310.
Rest Harrow, 78.
Roan-tree, 106.
Roast-beef plant, 255.
Rock Rose, 36.
Rose, 99.
Rupture-wort, 61.
Rush, 272.
Rye-grass, 295.
Saintfoin, 88.
St. Barnaby's Thistle, 155.
St. John's-wort, 41.
St. Peter's-wort, 42.
Saffron Crocus, 255.
Sallow, 236.
Saltwort, 214.
Samphire, 118.
Sanicle, Wood, 127.
Sauce Alone, 29.
Saw-wort, 154.
Saxifrage, 68.
Scotch Fir, 241.
Scottish Asphodel, 264.
Scurvy-grass, 27.
Sea Beet, 216.
— Buck-thorn, 208.
— Gilliflower, 170.
— Hard-grass, 295.
— Heath, 39.
— Holly, 127.
— Kale, 34.
— Lavender, 170.
— Lyme-grass, 296.
— Mat-grass, 303.
— Milkwort, 183.
— Purslane, 216.
— Radish, 34.
— Reed, 303.
— Rocket, 28.
— Star-wort, 143.
— Sulphur-wort, 117.
Sedge, 284.
Self-heal, 205.
Service-tree, 105.
Setter-wort, 13.
Sheep's-bit, 137.
Sheep's Scabious, 137.
Shepherd's Needle, 125.
—— Purse, 31.

Shepherd's Staff, 139.
Simpson, 146.
Skull-cap, 204.
Sloe, 90.
Slough-heal, 205.
Snake-weed, 212.
Sneeze-wort, 151.
Snowdrop, 265.
Snowflake, 265.
Soapwort, 45.
Soft-grass, 305.
Solomon's Seal, 271.
Sorrel, 211.
Sow-thistle, 156.
Speedwell, 189.
Spignel, 118.
Spindle-tree, 74.
Spurge-olive, 209.
Spurge-laurel, 209.
Spurrey, 48.
Squinancy-wort, 130.
Squirrel-tail-grass, 296.
Star of the Earth, 169.
—— of Bethlehem, 269.
Star-thistle, 155.
Starch Hyacinth, 269.
Stitchwort, 52.
Stock, 22.
Stork's-bill, 58.
Strapwort, 60.
Strawberry, Hautboy, 96.
———— Tree, 174.
————, Wood, 95.
Sun-dew, 38.
Sweet Alyssum, 26.
—— Cicely, 125.
—— Flag, 246.
—— Gale, 242.
Sweet-scented Vernal grass, 306.
Swine's Succory, 157.
Sycamore, 55
Tamarisk, 62.
Tansy, 149.
Tare, 83.
Thorn-apple, 181.
Thorow-wax, 120.
Thrift, 170.
Thyme, 204.
Timothy-grass, 300.
Toothwort, 194.
Touch me not, 60.
Tower-mustard, 23.
Traveller's Joy, 8.
Tulip, 266.
Turnip, 32.
Tutsan, 43.
Twayblade, 258.
Valerian, Red, 139.
Venus's Comb, 125.
Vervain, 196.
Vetch, 84.
Violet, 35.
Viper's Bugloss, 163.
Wake Robin, 246.
Wallflower, 22.
Wall-pellitory, 218.
Water Aloe, 254.
—— Betony, 193.
—— Blinks, 63.
Water-cress, 23.
Water Dropwort, 119.
—— Featherfoil, 185.
—— Hair-grass, 306.
—— Horehound, 197.
Water-lily, White, 15.
————, Yellow, 15.
Water Milfoil, 110.
Water-parsnep, 121.
Water-plaintain, 253.
Water-soldier, 254.
Water-starwort, 243.
Water-violet, 185.
Waterwort, 48.
Way Bennet, 296.
Wayfaring-tree, 132.
Weld, 219.
Whin, 77.
Whip-tongue, 130.
White Mullein, 181.
—— Ox-eye, 148.
White-rot, 128.
Whitlow-grass, 26.
Wild Basil, 206.
—— Navew, 32.
—— Rosemary, 173.
—— Succory, 162.
Willow, Bedford, 232.
—— Rose, 232.
——, Sweet, 230.
——, Varnished, 232.
Willow, White Welsh, 232.
Winter Green, 175.

Wolf's-bane, 13.
Woodbine, 131.
Woodruff, Sweet, 130.
Wood Sage, 198.
—— Sorrel, 59.
Wormwood, 149.
Yarrow, 151.
Yellow Archangel, 202.
Yellow Bird's Nest, 176.
—— Bugle, 198.
—— Loosestrife, 184.
—— Ox-eye, 148.
—— Pimpernel, 184.
Yellow-weed, 219.
Yew, Common, 241.
——, Irish, 241.

END OF THE FIRST VOLUME.

LONDON:
Printed by A. & R. Spottiswoode,
New-Street-Square.

Printed in the United States
By Bookmasters